Recent Progress in Urban Forest Planning and Monitoring

Guest Editors

Giulia Capotorti
Simone Valeri

Basel • Beijing • Wuhan • Barcelona • Belgrade • Novi Sad • Cluj • Manchester

Guest Editors

Giulia Capotorti
Department of
Environmental Biology
Sapienza University of Rome
Rome
Italy

Simone Valeri
Department of
Environmental Biology
Sapienza University of Rome
Rome
Italy

Editorial Office
MDPI AG
Grosspeteranlage 5
4052 Basel, Switzerland

This is a reprint of the Special Issue, published open access by the journal *Land* (ISSN 2073-445X), freely accessible at: https://www.mdpi.com/journal/land/special_issues/urban_forest_plan_monitor.

For citation purposes, cite each article independently as indicated on the article page online and as indicated below:

Lastname, A.A.; Lastname, B.B. Article Title. *Journal Name* **Year**, *Volume Number*, Page Range.

ISBN 978-3-7258-4249-0 (Hbk)
ISBN 978-3-7258-4250-6 (PDF)
https://doi.org/10.3390/books978-3-7258-4250-6

Contents

land

MDPI

Article

Ecological Connectivity in Agricultural Green Infrastructure: Suggested Criteria for Fine Scale Assessment and Planning

Simone Valeri *, Laura Zavattero and Giulia Capotorti

Department of Environmental Biology, Sapienza University of Rome, Piazzale Aldo Moro 5, 00185 Rome, Italy; laura.zavattero@uniroma1.it (L.Z.); giulia.capotorti@uniroma1.it (G.C.)
* Correspondence: simone.valeri@uniroma1.it

Abstract: In promoting biodiversity conservation and ecosystem service capacity, landscape connectivity is considered a critical feature to counteract the negative effects of fragmentation. Under a Green Infrastructure (GI) perspective, this is especially true in rural and peri-urban areas where a high degree of connectivity may be associated with the enhancement of agriculture multifunctionality and sustainability. With respect to GI planning and connectivity assessment, the role of dispersal traits of tree species is gaining increasing attention. However, little evidence is available on how to select plant species to be primarily favored, as well as on the role of landscape heterogeneity and habitat quality in driving the dispersal success. The present work is aimed at suggesting a methodological approach for addressing these knowledge gaps, at fine scales and for peri-urban agricultural landscapes, by means of a case study in the Metropolitan City of Rome. The study area was stratified into Environmental Units, each supporting a unique type of Potential Natural Vegetation (PNV), and a multi-step procedure was designed for setting priorities aimed at enhancing connectivity. First, GI components were defined based on the selection of the target species to be supported, on a fine scale land cover mapping and on the assessment of land cover type naturalness. Second, the study area was characterized by a Morphological Spatial Pattern Analysis (MSPA) and connectivity was assessed by Number of Components (NC) and functional connectivity metrics. Third, conservation and restoration measures have been prioritized and statistically validated. Notwithstanding the recognized limits, the approach proved to be functional in the considered context and at the adopted level of detail. Therefore, it could give useful methodological hints for the requalification of transitional urban–rural areas and for the achievement of related sustainable development goals in metropolitan regions.

Keywords: peri-urban landscapes; metropolitan areas; MSPA; fragmentation; native woody species; environmental units; naturalness; ecological corridors; conservation and restoration priorities

1. Introduction

Connectivity represents an emergent property of landscapes with respect to species dispersal and ecological processes [1,2]. As such, it is increasingly recognized as a fundamental feature for enhancing biodiversity conservation and ecosystem service capacity against fragmentation, in both ecological networks and GI planning [3,4]. These roles of connectivity have been quite thoroughly disentangled in urban areas as well as in rural landscapes [5–7], while additional values are emerging for peri-urban transitional contexts, spanning from the reconnection between cities and their countryside to the enhancement of agriculture multifunctionality and sustainable development of metropolitan regions [8,9]. Pragmatically, ecological connectivity analyses focus on structural, functional, and dynamic individual characteristics and mutual relationships between patches, matrix, and corridors in order to assess landscape permeability to species movement [5,10]. As regards agricultural landscapes, current research is increasingly addressing the vegetation component of biodiversity in addition to the faunistic one, which represents a more traditional target of investigation [11]. Both the impact of plant community composition on connectivity [12] and, vice versa, the impact of connectivity features on taxonomic and functional structure of plant communities [13] have been explored. Native status and dispersal traits of plants, corridor suitability, and patch/matrix resistance to dispersal represent the more frequently investigated attributes at the species, community, and landscape level [14–16]. Nevertheless, especially in the context of the European GI Strategy implementation [17], little evidence is available on how to select plant species to be primarily favored in dispersal and on the role of environmental heterogeneity and quality of habitat patches and corridors in facilitating/impairing such a dispersal.

The present work is aimed at suggesting a methodological approach for addressing these knowledge gaps at fine scales and for peri-urban agricultural landscapes. The approach was tested in a Natural Reserve in the Metropolitan City of Rome (Italy), within which urbanization pressure and rural landscape homogenization may impair the resilience of the rural system and its capacity to provide valuable ES despite the legally protected status [18,19]. Our findings suggest that, in such a context, the prioritization of GI actions for enhancing biodiversity and connectivity may be suitably driven by (i) the selection of target plant species according to the vegetation potential, (ii) the stratification of land into homogeneous environmental units, and (iii) the assessment of naturalness of the landscape mosaic components.

2. Materials and Methods

2.1. Study Area

The Marcigliana Nature Reserve is located in the northeastern peri-urban sector of the Metropolitan City of Rome (42°00′18.72″ N 12°35′13.92″ E/42.0052° N 12.5872° E), Italy, and covers an area of 4696 hectares (Figure 1). It belongs to a system of protected areas in the Municipality of Rome, managed by the RomaNatura regional body, that hosts biodiversity of conservation interest at the species, ecosystem and/or genetic level (L.R. n. 29/97). The Reserve, as the whole municipality, is embedded within the ecoregional subsection of the "Roman Area", characterized by coastal Mediterranean and hilly transitional bioclimate, composite sedimentary and volcanic litho-morphology, and prevailing PNV for deciduous oak forests [20].

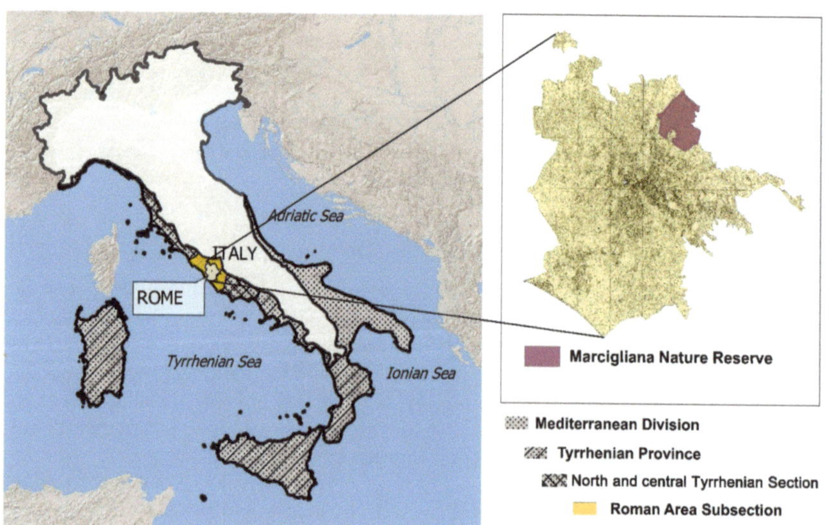

Figure 1. Study area. Ecoregional setting of the Municipality of Rome, from the division to the subsection level (on the **left**), and location of the Marcigliana Nature Reserve within the Municipality of Rome (on the **right**).

More in detail, the Reserve shows a varied pattern of different Environmental Unit types (EUN), i.e., homogeneous portions of land, with respect to climatic, lithologic, morphologic and PNV features, hosting a unique type of mature vegetation together with semi-natural and anthropogenic seral stages. The occurring EUNs include: (i) Volcanic Plateaux (VPL), supporting Turkey oak and eastern hornbeam forest potential (*Carpino orientalis-Quercetum cerris* vegetation series) (66% of the site); (ii) Alluvial Valleys (AV), supporting hygrophilous and meso-hygrophilous forest potential (*Querco roboris-Ulmetum minoris/Salicetum albae* vegetation complex) (17%); and (iii) Sandy-Clayey Slopes (SCS), supporting Virgilian oak and Turkey oak forest potential (*Carpino orientalis-Quercetum cerris varietas quercetosum virgilianae* vegetation series) (17%) [21]. With respect to this potential arrangement, the present land use and land cover is starkly dominated by agricultural areas, without clear trends upon abandonment [22]. On the contrary, urban sprawl and soil consumption are threatening the rurality of the Reserve especially at its borders [23,24], with artificial areas representing about 4% of the site. Natural and semi-natural vegetation is therefore reduced to minor remnants, with the mature stages of the most widespread vegetation series types, i.e., *Quercus cerris* woods, accounting for about 10% of the site. Owing to the agricultural vocation, environmental protection rules, recognized role as a metropolitan ecological network buffer zone, and geographic position between the consolidating city and traditional rural landscapes of the countryside [25–27], the Reserve has been selected as a suitable case study for addressing the connectivity issue in support of peri-urban GI planning.

2.2. Research Design

In keeping with the principles proposed for local scale GI planning [28], a multi-step procedure was designed for setting priority measures aimed at enhancing the ecological connectivity in a peri-urban agricultural landscape (Figure 2).

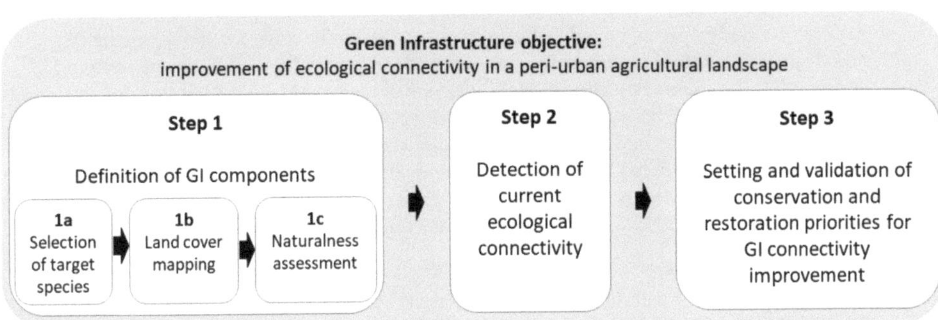

Figure 2. Multi-step procedure aimed at setting conservation and restoration priorities for GI connectivity improvement in the study area.

First, the current GI components were defined based on the selection of the target species to be supported (step 1a), on a fine scale land cover mapping (including natural and semi-natural ecosystem patches as well as linear vegetation elements) (step 1b), and on the assessment of land cover type naturalness (step 1c). Second, current ecological connectivity was assessed in both structural and functional terms (step 2) and, third, conservation and restoration measures have been prioritized and validated by means of statistical correlation with the observed occurrence of target species (step 3).

More in-depth information on the definitions of ecosystem naturalness and ecological connectivity adopted for the research [29–33] is provided in Supplementary Material Table S1.

2.3. First Step: Definition of GI Components According to Target Species, Ecosystem Occurrence, and Naturalness

Assuming that the dispersal of trees representative of the mature vegetation communities may facilitate the resistance and resilience of natural forest ecosystems in a rural landscape [34], the woody plants with a limited dispersal capacity and that are characteristic of the PNV types occurring in the Marcigliana Natural Reserve have been selected as target species (step 1a, Figure 2). These include three oak species, namely *Quercus cerris*, *Q. robur*, and *Q. virgiliana*, that are barochore and zoochore and may be effectively dispersed by the jay (*Garrulus glandarius* L.) or by hoarding rodents [35]. Since the presence of the jay in the Reserve is not ascertained [36], it was assumed that occurring small rodents, such as *Apodemus sylvaticus* L., may act as main dispersers [37,38]. The dispersal distance mediated by the wild mouse increases, up to a little more than 100 m, as the number of successive movements increases (re-dispersal) and is more affected by the distance from shelter habitats rather than by the weight of the acorn [39].

Current GI components have been then recognized according to the capacity of different land cover types to sustain the persistence, dispersal or spontaneous colonization of target species. Therefore, all the ecosystems occurring in the study area have been mapped in a GIS environment (Quantum GIS) (step 1b, Figure 2) and typified. For a finer scale definition of the GI components, ecosystem and other land cover typology was defined by detailing the legend classes of the Actual Vegetation Map of the Province of Rome (1:25,000 scale) [40]. Based on these detailed classes, an original map was drawn at 1:2000 scale by means of Google Satellite Imagery visual interpretation, with a minimum mapping unit of 0.15 ha. The woody hedgerows occurring in the agricultural matrix, important for target species as natural and semi-natural ecosystem patches, were first drawn as polylines, then converted into polygons by a 5 m buffer either side and finally integrated in the main map. Photointerpretation was validated with field checks for all the accessible sites, and with open-source geo-visualization tools (Google Street View and Bing Maps) and comparison with the Forest Copernicus High Resolution Layer [41] for inaccessible sites.

Both the areal and linear elements occurring in the landscape mosaic and dominated by woody species have been assumed as suitable habitats for oak persistence and dispersal, but their performance was supposed to be conditioned by the respective degree of naturalness. Specifically, naturalness has been assessed accounting for the physiognomic and structural features of the mapped woody elements with respect to those of the PNV [42] (step 1c, Figure 2; Supplementary Material Table S1): areal and linear elements dominated by non-native species and/or with a regular structure due to plantation activities were considered less natural than those dominated by the native species typical of the PNV and showing a spontaneous cover pattern.

2.4. Second Step: Detection of Current Connectivity

Current structural and functional connectivity was investigated at different levels of detail by considering as suitable habitats either just areal or both areal and linear components, and whether or not their degree of naturalness is accounted for:

- Level 1—Areal components, with both high and low degree of naturalness;
- Level 2—Both areal and linear components, with all degrees of naturalness;
- Level 3—Areal components with just a high degree of naturalness;
- Level 4—Both areal and linear components with just a high degree of naturalness.

Moreover, the three EUNs occurring in the Reserve (i.e.,: VPL, with *Quercus cerris* and *Carpinus orientalis* forest potential; AV, with meso- and hygrophilous forest potential; and SCS, with *Quercus virgiliana* and *Q. cerris* forest potential) were individually investigated at the level assumed as most suitable among these four (Level 4). Thus, the 7 maps (one for each level of investigation, and three for the Level 4 stratified per EUN) were converted into binary rasters (1 = habitat; 0 = non habitat) with a spatial resolution of 5 m.

For structural connectivity detection, a MSPA along with a Network Analysis were performed. MSPA, a useful tool for describing pattern structures and automatically detecting connectivity pathways, was carried out by means of the GUIDOS Toolbox [43] with the following settings: 8-connectivity, so that foreground connectivity was based on both border and corner sharing between pixels of habitats (that sometimes have a very small extent in the source map); Transition turned on, so that more importance was posed on the role of linear elements as connectors rather than on the continuity of patch edges; Intext = 1, so that the perforations of habitat patches due to enclosed features, very rare in the study area, were neglected; Edge width = 2 pixels (10 m), so that linear elements were prevented to be recruited as areal habitats. The MSPA returned a categorization of the habitats into cores, islets, perforations, edges, loops, bridges and branches. With the same GUIDOS Toolbox, a Network Analysis was performed in order to estimate the NC in the landscape mosaic. An individual component represents a region of interconnected nodes and links, respectively generated by core and bridge MSPA categories, so that a landscape can be considered as more connected as the NC is fewer [44,45].

For functional connectivity assessment, the Integral Index of Connectivity (IIC) was estimated (Conefor 2.6 software). The index, widely recommended for habitat and link prioritization [46], provides a measure of connectivity between nodes according to a threshold distance. Given the wild mouse-mediated dispersal capacity of the target species, such a distance was approximated at 100 m. The IIC varies between 0 and 1 and positively increases with connectivity:

$$IIC = \frac{\sum_{i=1}^{n} \cdot \sum_{j=1}^{n} \cdot \frac{a_i a_j}{1+nl_{ij}}}{A_L^2} \quad (1)$$

where n is the total number of nodes in the landscape, a_i and a_j are the attributes (i.e., the extent) of nodes i and j, nl_{ij} is the number of links in the shortest path (topological distance) between patches i and j, and A_L is the maximum landscape attribute (i.e., the extent of a habitat patch covering all the landscape).

2.5. Third Step: Prioritization of Conservation and Restoration Measures

By combining multiple indicators, alternatively fitting with areal or linear components, conservation priorities for the maintenance of landscape connectivity were assigned to habitat patches and corridors at the Level 4 stratified per EUN. The values for each indicator were then scored and added together for the assignment of a comprehensive priority to each component.

Specifically, habitat patches were prioritized according to:

(a) Node Importance [47], calculated as

$$IIC(\%) = 100 \cdot \frac{IIC - IICremove}{IIC} \quad (2)$$

where *IIC* is the index value when the overall existing nodes are considered, and IIC_{remove} is the index value after the removal of that single node from the landscape. Priority scores for Node Importance were assigned following the distribution of the indicator values into quartiles;

(b) Condition of the EUN of occurrence, derived from the previous methodological step and qualitatively scored, with a null value assigned to the less critical EUNs and a unit value assigned to the most critical one.

Corridors, prevalently links (*bridges*), but also the other linear MSPA categories, were evaluated by means of:

Condition of the EUN of occurrence, as for nodes;

(c) Link Removal indicator, so that the removal of each *bridge* was simulated, the respective impact (dIIC) calculated as for nodes, and the priority quantitatively scored according to distribution of the indicator values into quartiles;

(d) Conservation priority of the nodes connected by the link, derived from Node Importance (criterion a) and qualitatively scored in compliance with every emerging combination (i.e., the higher the importance of nodes, the higher the score assigned to the connector);

(e) Connection importance, assigned to links that, if removed, originate a new component;
(f) Structural contiguity and singularity of connections, so that a higher priority was assigned to bridges and branches with respect to isolated islets (due to less contiguity) and to loops (due to connection redundancy).

In order to validate conservation priorities for links, the presence and abundance of *Quercus* specimens were estimated by means of physiognomic structural surveys of the linear woody elements at accessible sites. Subsequently, the correlation between abundance and conservation priority was assessed with the Kendall Tau-b statistic [48], whose values range from −1 (100% negative association) to +1 (100% positive association) with 0 indicating absence of association. The Kendall Tau-b coefficient is defined as:

$$\tau_B = \frac{f_c - f_d}{\sqrt{(f_c + f_1 + E_x)(f_c + f_1 + E_y)}} \quad (3)$$

where f_c are the concordance frequencies; f_d is the frequency of discrepancies; E_x (y) are the bonds of the independent (and dependent) variable. Owing to the difficulties encountered in making many surveys, it was reasonable to set a level of significance $p \leq 0.10$.

With respect to restoration, the criteria for setting priorities were defined in order to minimize conflicts with primary production [49,50], so that the boosting of links, especially the conversion of branches into bridges, was preferred to the creation of new forest patches. Moreover, such a conversion was simulated by favoring restoration of tree cover in pre-existing paths or along linear element residuals between cultivated fields (such as unpaved road edges or grass verges) (Figure 3) and by limiting the development of redundant links between nodes (i.e., loops).

(a) **(b)**

Figure 3. Simulated conversion of two *branches* (**a**) into a single *bridge* according to a pre-existing path (**b**).

The improvement in connectivity, potentially determined by the simulated restoration, was then assessed by means of Conefor connector-based (not distance-based) measures. Namely, the IIC and the NC were measured ex ante and ex post the conversion of branches into bridges.

3. Results

3.1. Current GI Components

The landscape matrix of the Reserve is represented by agricultural surfaces (80%, mainly arable lands), with interspersed natural patches (16%, mainly *Quercus cerris* woods), artificial surfaces (4%, prevalently with constructions related to agricultural activities) and woody linear elements (206, 163 of which are natural and 43 artificial, with a density of 14.78 m/ha) (Figure 4).

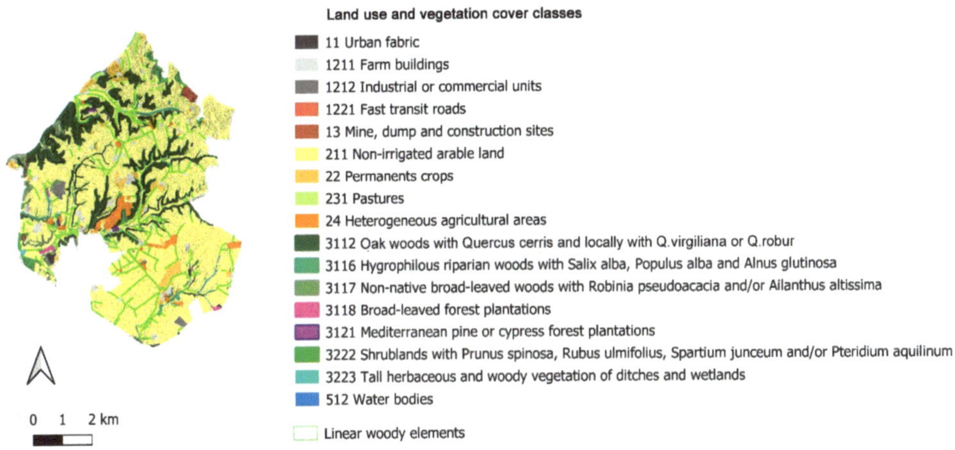

Figure 4. Land use and vegetation cover in the Marcigliana Nature Reserve.

Such an arrangement is coarsely confirmed at the EUN level, but with a varying prevalence of agricultural surfaces over natural vegetation and a different density of linear elements (Table 1).

Table 1. Landscape features (percent coverage of main land cover types and density of linear woody elements) of the EUNs occurring in the Marcigliana Nature Reserve.

Environmental Unit	Total Surface (ha)	Agricultural Surfaces (%)	Natural Surfaces (%)	Artificial Surfaces (%)	Density of Linear Elements (m/ha)
Volcanic Plateaux (VPL)	2719	86%	10%	4%	11.7
Alluvial Valleys (AV)	602	75%	22%	3%	44.5
Sandy-Clayey Slopes (SCS)	1138	65%	32%	3%	6.5

In all the EUNs, the woody vegetation types, together with the linear woody elements occurring in the agricultural matrix, have been a priori selected as suitable GI components for supporting native oak species. According to their naturalness, these components were arranged into the following classes (Figure 5):

- areal "Natural-high", including Oak woods with *Quercus cerris* and locally with *Q. virgiliana* or *Q robur* (map code 3112); Hygrophilous woods with *Populus* sp.pl., *Salix* sp.pl., *Alnus glutinosa* and *Fraxinus oxycarpa* (3116); Shrublands with *Prunus spinosa, Rubus ulmifolius, Spartium junceum* and/or *Pteridium aquilinum* (3222); and Tall herbaceous and woody vegetation of ditches and wetlands (3223);
- areal "Natural-low", including Non-native broad-leaved woods with *Robinia pseudoacacia* and/or *Ailanthus altissima* (3117); Broad-leaved forest plantations 3118); and Mediterranean pine or cypress forest plantations (3121);

- linear "Natural", when dominated by spontaneous woody species;
- linear "Artificial", when dominated by planted woody species.

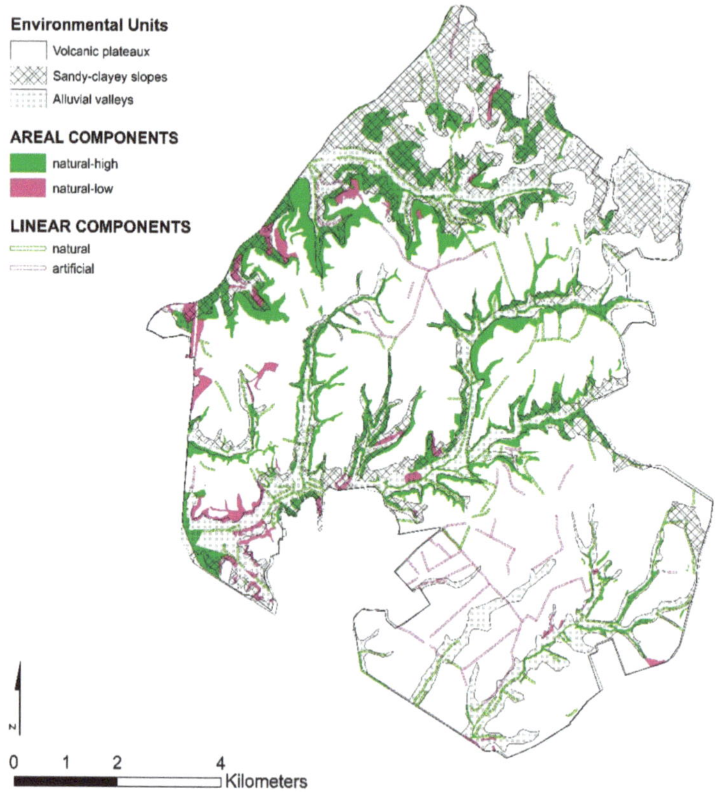

Figure 5. Spatial arrangement of suitable GI components in the environmental units of the Marcigliana Nature Reserve, distinguished according to structural features (areal or linear extent) and degree of naturalness.

3.2. *Current Ecological Connectivity*

Both structural connectivity features, in terms of absolute frequency of MSPA classes and NC, and functional connectivity features, in terms of IIC, for each of the four levels of investigation and for the three different EUNs at the Level 4, are reported in Table 2.

In the entire Reserve, and independently from the level of investigation, a conspicuous number of cores and NC is observed (with respect to NC_{min} = 1), denoting a high degree of habitat fragmentation. Moreover, the branches are always much more numerous than bridges, showing a high degree of discontinuity in existing corridors and further contributing to the observed NC. With respect to these general features, when linear elements are definitely taken into account (Levels 2 and 4 vs. Levels 1 and 3), the increase in connectivity is denoted by: (i) a higher number of continuous and discontinuous corridors (i.e., bridges and branches) and a consequent fewer NC; (ii) a higher number of cores, showing the potential role of linear elements as habitat providers themselves, even with a high degree of naturalness (Level 4); and (iii) a quadrupled values of the IIC. Alternatively, when the degree of naturalness of habitat patches and corridors is explicitly considered (Levels 3 and 4 vs. Levels 1 and 2), the effect of quality can be distinguished from that of quantity. In this case, the decrease in NC does not indicate a better structural connectivity, but rather the complete attrition of useful components for the dispersal of target species, also denoted by the halving of the IIC.

Table 2. Structural and functional connectivity features for the alternative levels of investigation. Note that, due to the "edge" parameter, many of the narrow forest ecosystems occurring along narrow slopes and valleys were eroded and fragmented. Therefore, the number of MSPA cores, branches, and bridges far exceeds the number of patches and linear elements in the original map.

Connectivity Feature	Level 1 (All Areal Components)	Level 2 (All Areal and Linear Components)	Level 3 (Natural-High Areal Components)	Level 4 (Natural-High Areal and Linear Components)	Level 4/VPL (Volcanic Plateaux)	Level 4/AV (Alluvial Valleys)	Level 4/SCS (Sandy-Clayey Slopes)
Number of MSPA CORES	300	332	265	281	194	275	136
Number of MSPA ISLETS	7	64	7	57	205	92	72
Number of MSPA EDGES	179	200	146	168	192	239	180
Number of MSPA LOOPS	0	12	1	8	19	30	22
Number of MSPA BRIDGES	119	194	105	161	56	141	42
Number of MSPA BRANCHES	1132	1506	969	1279	535	734	310
Number of Components (NC)	78	55	67	50	107	77	82
Integral Index of Connectivity (IIC)	0.004	0.017	0.002	0.008	0.00083	0.00503	0.00504

Finally, the comparison between the three different EUNs allowed ecological connectivity features to be spatially contextualized, and VPL to be recognized as the most critical EUN with respect to AV and SCS. Actually, in VPL: (i) the number of cores is not the highest but the islets are much more numerous, denoting a higher level of fragmentation and shrinkage in habitat patch dimension; (ii) the ratio between cores and bridges is higher (3.46 with respect to 1.95 in AV and 3.24 in SCS) as well as the NC, denoting a more marked isolation between residual habitats; and (iii) the IIC is six times lower than that of the other two EUNs, highlighting a low degree of connectivity also in functional terms.

3.3. Conservation Priorities

The ranking of adopted indicators, for the assignment of conservation priority scores to areal and linear GI components, is summarized in Table 3. The comprehensive conservation priority of each GI component, derived from the sum of partial indicator scores and ranked in 5 classes from 'very low' to 'very high', is instead represented in Figure 6.

Areal elements with a positive priority are 27. Notwithstanding those with maximum values are the largest ones, some medium-size (between 10 and 25 ha) and small-size patches (<10 ha) could be prioritized as well. Linear elements with a positive priority are 123 out of the 164 natural ones. For these GI components, 40 physiognomic-structural surveys were carried out. This surveys returned a prevalence of *Rubus ulmifolius* and *Prunus spinosa* shrub formations with oak specimens occurrence in 63% of the cases (25 linear elements with *Quercus cerris*, *Q. virgiliana*, and/or *Q. robur*). The Kendall Tau-b correlation showed a significant relationship (p-value = 0.052) between the abundance of the target species in linear elements and their conservation priority.

3.4. Restoration Priorities

By avoiding the encroachment on existing cultivated fields and the creation of connectivity loops, the conversion of *branches* into 20 new bridges was simulated. Notwithstanding the exiguous number of simulated new links, the conversion would lead to an ecological connectivity improvement of 79% in terms of IIC (from 0.008 to 0.014) and of 14% in terms of NC (from 50 to 43).

Table 3. Conservation priority scores assigned to GI components, at the level 4 stratified per EUN, according to individual indicators (a and b for areal components, and from b to f for linear components, respectively).

Conservation Priority Score	(a) Node Importance	(b) EUN Condition	(c) Link Removal	(d) Priority of the Connected Nodes	(e) Importance of Connection	(f) MSPA Class
5	dIIC > 11.57 (upper outliers)			two 'very high' priority nodes		
4	8.74 < dIIC < 11.56 (4th quartile)			at least one 'very high' priority node; two 'high' priority nodes		
3	5.05 < dIIC < 8.73 (3rd quartile)		dIIC > 13.59 (upper outliers)	at least one 'high' priority node; two 'medium' priority nodes		Bridge
2	2.30 < dIIC < 5.04 (2nd quartile)		1.30 < dIIC < 4.84 (from 1st to 4th quartile)	at least one 'medium' priority node		Branch
1	1.53 < dIIC < 2.29 (1st quartile)	VPL	dIIC < 1.00	two 'low' or 'very low' priority nodes	the link removal splits a component	Islet and Loop
Null (0)	dIIC < 1.00	SCS; AV			the link removal does not split any component	

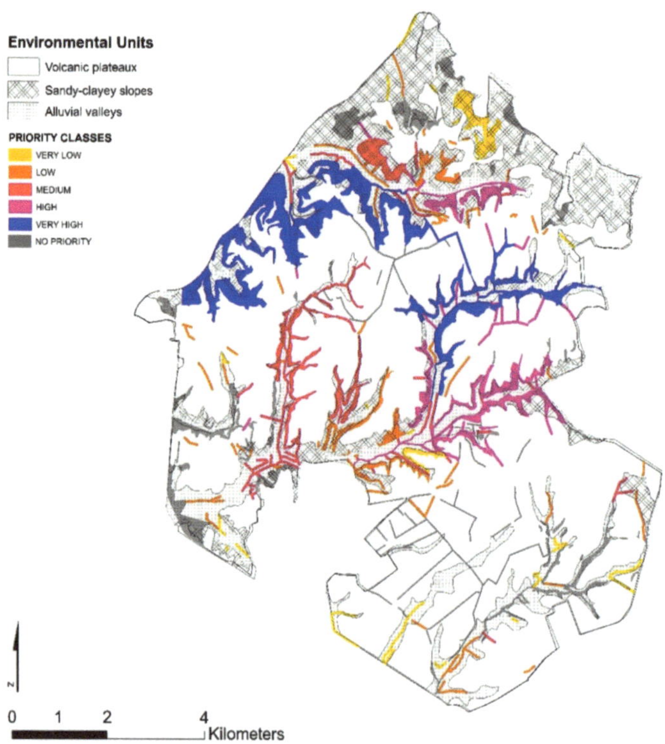

Figure 6. Distribution of comprehensive conservation priority of GI components in the Marcigliana Nature Reserve.

4. Discussion

A methodological approach was developed and tested for addressing the improvement of GI connectivity in peri-urban agricultural landscapes. The approach was first based on fine-scale environmental stratification into homogeneous EUNs, each supporting a unique type of PNV. Second, it was based on an in-depth definition of GI components, including linear woody elements, and, third, on the assessment of their naturalness.

With respect to approaches based on less detailed information [51], the greater mapping and assessment effort allowed some critical issues to be faced, especially pertaining to (i) a focused selection of target plant species to be favoured by GI connectivity, (ii) the reliability of structural and functional connectivity estimates, and (iii) the steering of conservation and restoration measure prioritization.

4.1. Strength and Weakness of Target Species Selection

The selection of target plant species was based upon the recognition of PNV types, so that not only limited dispersal ability but also representativeness of the varied ecological potential of the site has been considered. In the study area, different species of the genus *Quercus* comply with both these requirements. Their conservation and facilitation can thus actively contribute to boost native biodiversity, control biological invasions, facilitate ecological and biogeographic coherence of landscape management measures, and guarantee a high level of restoration success in a peri-urban rural landscape [52–54]. Moreover, even though these aspects have not been deepened and go beyond the objectives of the work, oak species are expected to play a crucial role for rural landscape resilience and agriculture sustainability as keystone components of mature vegetation communities [55,56]. Therefore, they should preferentially contribute to achieve GI multifunctionality with respect to species selected for their endemic, rare or threatened status and usually targeted for the exclusive objective of biodiversity conservation in ecological network design [57].

However, some factors may limit the restoration success for these target species, such as livestock overgrazing, intensive pruning, and shrub clearance [58–61]. These constraints should be carefully considered, especially in a prospective implementation phase, and eventually mitigated by coupling oak plantation and seeding with shrub restoration.

4.2. Strength and Weakness of Connectivity Assessment

As regards connectivity, the estimates performed at different levels of detail confirmed the significance of explicitly accounting for the occurrence of linear landscape elements in rural contexts, as matrix permeability enhancers [62–64]. Similarly, differences in estimates due to the varied naturalness of landscape mosaic components have been documented, complementing the evidence recently arising from broader scale investigations [7,65,66].

The limited number of alternative observations prevented however to test the statistical significance of such differences, so that more alternative settings and/or a comparison with similar case studies have to be explored for deepening knowledge in this respect. Moreover, the historical persistence of occurring hedgerows could be analyzed for strengthening the assessment of corridor effectiveness [67].

4.3. Strength and Weakness of Prioritization Procedure

For prioritization, all the collected information on environmental stratification and habitat and landscape condition was capitalized by means of an additive assessment, as already experimented but with different criteria and for different landscape contexts [68,69]. Accordingly, conservation and restoration measures were not only defined on the basis of connectivity metrics, but also differentiated accounting for the conservation status of the EUNs (i.e., the varying fragmentation degree due to differences in environmental suitability for intensive land uses), the naturalness of the occurring elements (avoiding a GI design just based on structural land cover information), and the current availability of ecological corridors for target species (in both structural and functional terms). Other authors already highlighted that ecological connectivity varies according to landscape types, without however incorporating such an information in the prioritization process [70,71]. Connectivity metrics alone, based on MSPA, Network Analysis and functional indices (e.g., IIC) have been instead commonly applied for setting habitat and corridor conservation priorities [46,72,73]. As an immediate advantage, the merge between environmental stratification and connectivity metrics mitigated the effect of patch area on node importance assessment [73,74], so that it was possible to include other nodes than the largest ones among the conservation priorities, bringing out their role as potential stepping stones. In spite of this benefit, implementation showed some limitations concerning EUNs affected by striking fragmentation. This is the case for the VPL unit in the southeastern sector of the study area, where only a few and low-priority conservation nodes could be identified and only 2 of the 20 restoration links were designed. Such a result suggests that conservation and restoration priorities should be framed in the first place on the difference between actual and potential cover of natural ecosystems, and only secondarily on the spatial pattern of remnants, as already proposed for the assessment of ecosystem conservation status at the national and regional level [42]. As regards the adopted connectivity indicators and with respect to consolidated practice [45,75], an approach not just based on node importance and link removal function allowed the contribution made by further elements of the landscape mosaic to be enhanced. Namely, branches and islets were explicitly included among the priorities so that their potential role as either stepping stones, discontinuous corridors, or habitat providers themselves [76–78] has been explicitly recognised while planning for conservation measures.

Some limitations could arise from the subjective choice of priority scores for each of the indicators, which however is often accepted as necessary in GI planning and could be eventually mitigated by including stakeholders and other disciplinary competences into the process [79]. The proposed restoration options are affected by a certain subjectivity as well. Nevertheless, these options comply with the evidence that new wooded links bring more benefits than converted ones and foster rodent dispersal, including that of *Apodemus* specimens [14,80–82]. Above all, however, the criteria for limiting as much as possible the consumption of productive space may facilitate, more than a more automatic but unfiltered least-cost path approach [83], the avoidance of potential conflicts with agricultural practices and the long-term persistence of planned interventions [49].

5. Conclusions

A set of criteria is presented for estimating and improving ecological connectivity at fine scales and that may be critical for planning effective GI in agricultural landscapes. The evidence provided by the implementation of these criteria in a peri-urban metropolitan sector emphasizes the usefulness of the ecological classification of land according to both the physical features of the environment and the biotic vegetation potential, and also provides a rationale for investing in detailed spatial representation and assessment of ecosystems. Notwithstanding the recognised limits, posed by the investigational character of the work but that can be quite easily disentangled in the case of a concrete GI deployment, it is hoped that the suggested approach will give useful hints for the requalification of transitional urban–rural areas and for the achievement of related sustainability goals, especially those prompted by the Green Infrastructure and Farm to Fork Strategies in Europe and by the Urban Green Strategy and the "Climate Decree" (national law decree n. 111/2019) in Italy.

Supplementary Materials: The following are available online at https://www.mdpi.com/2073-445X/10/8/807/s1, Table S1: adopted definitions for the concepts of ecosystem naturalness and ecological connectivity.

Author Contributions: Conceptualization, G.C. and S.V.; Methodology, S.V. and G.C.; Investigation, S.V. and G.C.; Resources, G.C.; Data Curation, S.V., G.C. and L.Z.; Writing—Original Draft Preparation, G.C. and S.V.; Writing—Review & Editing, S.V., G.C. and L.Z. All authors have read and agreed to the published version of the manuscript.

Funding: This research received no external funding.

Data Availability Statement: Some publicly available datasets were analyzed in this study, as reported in the reference section [40,41]. The new data were created in this study are available on request from the corresponding author.

Acknowledgments: The authors would like to thank the research unit on "Socio-ecological Systems, Landscape and Local Development"of the Complutense University of Madrid (UCM) for the support in applying the Kendall Tau-b statistic and for expert advice on linear element restoration criteria.

Conflicts of Interest: The authors declare no conflict of interest.

Abbreviations

Abbreviations

AV = Alluvial Valleys; EUN = Environmental Unit; GI = Green Infrastructure; IIC = Integral Index of Connectivity; MSPA = Morphological Spatial Pattern Analysis; NC = Number of Components; PNV = Potential Natural Vegetation; SCS = Sandy-Clayey Slopes; VPL = Volcanic Plateaux.

References

1. Baguette, M.; Blanchet, S.; Legrand, D.; Stevens, V.M.; Turlure, C. Individual dispersal, landscape connectivity and ecological networks. *Biol. Rev.* **2013**, *88*, 310–326. [CrossRef]
2. Wang, Z.; Wang, Z.; Zhang, B.; Lu, C.; Ren, C. Impact of land use/land cover changes on ecosystem services in the Nenjiang River Basin, Northeast China. *Ecol. Proc.* **2015**, *4*, 11. [CrossRef]
3. Honeck, E.; Moilanen, A.; Guinaudeau, B.; Wyler, N.; Schlaepfer, M.A.; Martin, P.; Sanguet, A.; Urbina, L.; von Arx, B.; Massy, J.; et al. Implementing Green Infrastructure for the Spatial Planning of Peri-Urban Areas in Geneva, Switzerland. *Sustainability* **2020**, *12*, 1387. [CrossRef]
4. Zhang, Z.; Sara, M.; Newell, J.; Lindquist, M. Enhancing Landscape Connectivity through Multi-functional Green Infrastructure Corridor Modeling and Design. *Urban For. Urban Green.* **2019**, *38*, 305–317. [CrossRef]
5. Bennett, A.F.; Radford, J.Q.; Haslem, A. Properties of land mosaics: Implications for nature conservation in agricultural environments. *Biol. Conserv.* **2006**, *133*, 250–264. [CrossRef]
6. La Point, S.; Balkenhol, N.; Hale, J.; Sadler, J.; van der Ree, R. Ecological connectivity research in urban areas. *Funct. Ecol.* **2015**, *29*, 868–878. [CrossRef]
7. Carlier, J.; Moran, J. Hedgerow typology and condition analysis to inform greenway design in rural landscapes. *J. Environ. Manag.* **2019**, *247*, 790–803. [CrossRef] [PubMed]
8. La Rosa, D.; Barbarossa, L.; Privitera, R.; Martinico, F. Agriculture and the city: A method for sustainable planning of new forms of agriculture in urban contexts. *Land Use Policy* **2014**, *41*, 290–303. [CrossRef]
9. Ochoa, C.Y.; Jiménez, D.F.; Olmo, R.M. Green Infrastructure Planning in Metropolitan Regions to Improve the Connectivity of Agricultural Landscapes and Food Security. *Land* **2020**, *9*, 414. [CrossRef]
10. Zeller, A.K.; Lewsion, R.; Fletcher, J.R.; Tulbure, G.M.; Jennings, M. Understanding the Importance of Dynamic Landscape Connectivity. *Land* **2020**, *9*, 303. [CrossRef]
11. De Blois, S.; Domon, G.; Bouchard, A. Landscape issues in plant ecology. *Ecography* **2002**, *25*, 244–256. [CrossRef]
12. Godfree, R.; Firn, J.; Johnson, S.; Knerr, N.; Stol, J.; Doerr, V. Why non-native grasses pose a critical emerging threat to biodiversity conservation, habitat connectivity and agricultural production in multifunctional rural landscapes. *Landsc. Ecol.* **2020**, *32*, 1219–1242. [CrossRef]
13. Uroy, L.; Ernoult, A.; Mony, C. Effect of landscape connectivity on plant communities: A review of response patterns. *Landsc. Ecol.* **2020**, *34*, 203–225. [CrossRef]
14. Damschen, E.I.; Brudvig, L.A.; Burt, M.A.; Fletcher, R.J.; Haddad, N.M.; Levey, D.J.; Orrock, J.L.; Resasco, J.; Tewksbury, J.J. Ongoing accumulation of plant diversity through habitat connectivity in an 18-year experiment. *Science* **2019**, *365*, 1478–1480. [CrossRef]
15. Schleicher, A.; Biedermann, R.; Kleyer, M. Dispersal traits determine plant response to habitat connectivity in an urban landscape. *Landsc. Ecol.* **2011**, *26*, 529–540. [CrossRef]
16. Thiele, J.; Kellner, S.; Buchholz, S.; Schirmel, J. Connectivity or area: What drives plant species richness in habitat corridors? *Landsc. Ecol.* **2018**, *33*, 173–181. [CrossRef]

17. EC (European Commission). Communication from the Commission to the European Parliament, the Council, the European Economic and Social Committee and the Committee of the Regions 'Green Infrastructure (GI)—Enhancing Europe's Natural Capital'. 2013. Available online: http://eur-lex.europa.eu/resource.html?uri=cellar:d41348f2-01d5-4abe-b817-4c73e6f1b2df.0014.03/DOC_1&format=PDF (accessed on 3 May 2021).
18. Ferrari, G.; Bocci, E.; Lepisto, E.; Cavallero, P.; Rombai, L. Territories and Landscapes: Place Identity, Quality of Life and Psychological Well-Being in Rural Areas. In *Italian Studies on Quality of Life, Social Indicators Research Series*; Bianco, A., Conigliaro, P., Gnaldi, M., Eds.; Springer: Cham, Switzerland, 2019; pp. 287–305.
19. Rescia, A.J.; Willaarts, B.A.; Schmitz, M.F.; Aguilera, P.A. Changes in land uses and management in two Nature Reserves in Spain: Evaluating the social–ecological resilience of cultural landscapes. *Landsc. Urban. Plan.* **2010**, *98*, 26–35. [CrossRef]
20. Blasi, C.; Capotorti, G.; Copiz, R.; Guida, D.; Mollo, B.; Smiraglia, D.; Zavattero, L. Classification and mapping of the ecoregions of Italy. *Plant. Biosyst.* **2014**, *148*, 1255–1345. [CrossRef]
21. Blasi, C.; Canini, L.; Capotorti, G.; Celesti, L.; Del Moro, M.A.; Ercole, S.; Filesi, L.; Fiorini, S.; Lattanzi, E.; Leoni, G.; et al. Flora vegetazione ed ecologia del paesaggio delle aree protette di RomaNatura. *Inf. Bot. Ital.* **2001**, *33*, 14–18.
22. Cavallo, A.; Di Donato, B.; Guadagno, R.; Marino, D. Cities, agriculture and changing landscapes in urban milieu: The case of Rome. *Rivista di Studi sulla Sostenibilità* **2015**, *1*, 79–97. [CrossRef]
23. Egidi, G.; Halbac-Cotoara-Zamfir, R.; Cividino, S.; Quaranta, G.; Salvati, L.; Colantoni, A. Rural in town: Traditional agriculture, population trends, and long-term urban expansion in metropolitan rome. *Land* **2020**, *9*, 53. [CrossRef]
24. Salvati, L.; Munafo, M.; Morelli, V.G.; Sabbi, A. Low-density settlements and land use changes in a Mediterranean urban region. *Landsc. Urban Plan.* **2012**, *105*, 43–52. [CrossRef]
25. Frondoni, R.; Mollo, B.; Capotorti, G. A landscape analysis of land cover change in the Municipality of Rome (Italy): Spatio-temporal characteristics and ecological implications of land cover transitions from 1954 to 2001. *Landsc. Urban Plan.* **2011**, *100*, 117–128. [CrossRef]
26. Zavattero, L.; Frondoni, R.; Capotorti, G.; Copiz, R.; Blasi, C. Towards the identification and mapping of traditional agricultural landscapes at the national scale: An inventory approach from Italy. *Landsc. Res.* **2021**. [CrossRef]
27. Blasi, C.; Zavattero, L.; Marignani, M.; Smiraglia, D.; Copiz, R.; Rosati, L.; Del Vico, E. The concept of land ecological network and its design using a land unit approach. *Plant. Biosyst.* **2008**, *142*, 540–549. [CrossRef]
28. Capotorti, G.; De Lazzari, V.; Alós Ortí, M. Local Scale Prioritisation of Green Infrastructure for Enhancing Biodiversity in Peri-Urban Agroecosystems: A Multi-Step Process Applied in the Metropolitan City of Rome (Italy). *Sustainability* **2019**, *11*, 3322. [CrossRef]
29. Andreasen, J.K.; O'Neill, R.V.; Noss, R.; Slosser, N.C. Considerations for the development of a terrestrial index of ecological integrity. *Ecol. Indic.* **2001**, *1*, 21–35. [CrossRef]
30. Ferrari, C.; Pezzi, G.; Diani, L.; Corazza, M. Evaluating Landscape Quality with Vegetation Naturalness Maps: An Index and Some Inferences. *Appl. Veg. Sci.* **2008**, *11*, 243–250. [CrossRef]
31. Farris, E.; Filibeck, G.; Marignani, M.; Rosati, L. The power of potential natural vegetation (and of spatial-temporal scale), a response to Carrión & Fernández (2009). *J. Biogeogr.* **2010**, *37*, 2211–2213.
32. Bennett, A.F. *Linkages in the Landscape: The Role of Corridors and Connectivity in Wildlife Conservation*; International Union for Conservation of Nature and Natural Resources: Cambridge, UK, 1999.
33. Calabrese, J.M.; Fagan, W.F. A comparison-shopper's guide to connectivity metrics. *Front. Ecol. Environ.* **2004**, *2*, 529–536. [CrossRef]
34. Capotorti, G.; Del Vico, E.; Anzellotti, I.; Celesti-Grapow, L. Combining the Conservation of Biodiversity with the Provision of Ecosystem Services in Urban Green Infrastructure Planning: Critical Features Arising from a Case Study in the Metropolitan Area of Rome. *Sustainability* **2017**, *9*, 10. [CrossRef]
35. Den Ouden, J.; Jansen, P.A.; Smit, R. Jays, mice and oaks: Predation and dispersal of *Quercus robur* and *Q. petraea* in north-western Europe. In *Seed Fate: Predation, Dispersal and Seedling Establishment*; Forget, P.M., Lambert, J.E., Hulme, P.E., van der Wall, S.B., Eds.; CABI Publishing: Wallingford, UK, 2005; pp. 223–240.
36. Sarrocco, S.; Battisti, C.; Brunelli, M.; Calvario, E.; Ianniello, L.; Sorace, A.; Tefofili, C.; Trotta, M.; Visentin, M.; Marco, E.; et al. L'avifauna delle aree naturali protette del comune di Roma gestite dall'ente romanatura. *Alula IX* **2002**, *9*, 3–31.
37. Capizzi, D.; Mortelliti, A.; Amori, G.; Colangelo, P.; Rondinini, C. *I Mammiferi del Lazio. Distribuzione, Ecologia e Conservazione*; Edizioni ARP: Rome, Italy, 2012; 266p.
38. Sunyer, P.; Muñoz, A.; Bonal, R.; Espelta, J.M. The ecology of seed dispersal by small rodents: A role for predator and conspecific scents. *Funct. Ecol.* **2013**, *27*, 1313–1321. [CrossRef]
39. Perea, R.; San Miguel, A.; Gil, L. Acorn dispersal by rodents: The importance of re-dispersal and distance to shelter. *Basic Appl. Ecol.* **2011**, *12*, 432–439. [CrossRef]
40. CIRBFEP (Centro di Ricerca Interuniversitario Biodiversità, Fitosociologia ed Ecologia del Paesaggio). Carta della Vegetazione reale della Provincia di Roma. 2013. Available online: http://websit.cittametropolitanaroma.it/BDV2014/Veget_Reale.aspx (accessed on 18 May 2021).
41. High Resolution Layers—Copernicus Land Monitoring Service. 2018. Available online: https://land.copernicus.eu/pan-european/high-resolution-layers/forests (accessed on 3 May 2021).

42. Capotorti, G.; Alós Ortí, M.M.; Anzellotti, I.; Azzella, M.M.; Copiz, R.; Mollo, B.; Zavattero, L. The MAES process in Italy: Contribution of vegetation science to implementation of European Biodiversity Strategy to 2020. *Plant. Biosyst.* **2015**, *149*, 949–953. [CrossRef]
43. Soille, P.; Vogt, P. Morphological segmentation of binary patterns. *Pattern Recognit. Lett.* **2009**, *30*, 456–459. [CrossRef]
44. Saura, S.; Pascual-Hortal, L. A new habitat availability index to integrate connectivity in landscape conservation planning: Comparison with existing indices and application to a case study. *Landsc. Urban Plan.* **2007**, *83*, 91–103. [CrossRef]
45. Saura, S.; Vogt, P.; Velázquez, J.; Hernando, A.; Tejera, R. Key structural forest connectors can be identified by combining landscape spatial pattern and network analyses. *For. Ecol. Manag.* **2011**, *262*, 150–160. [CrossRef]
46. Pascual-Hortal, L.; Saura, S. Integrating landscape connectivity in broad-scale forest planning through a new graph-based habitat availability methodology: Application to capercaillie (*Tetrao urogallus*) in Catalonia (NE Spain). *Eur. J. For. Res.* **2006**, *127*, 23–31. [CrossRef]
47. Saura, S.; Torné, J. Conefor 2.6 User Manual. 2012. Available online: http://www.conefor.org/files/usuarios/Manual_Conefor_26.pdf (accessed on 3 May 2021).
48. Kendall, M.G. A New Measure of Rank Correlation. *Biometrika* **1938**, *30*, 81–93. [CrossRef]
49. Benayas, J.M.R.; Altamirano, A.; Miranda, A.; Catalán, G.; Prado, M.; Lisón, F.; Bullock, J.M. Landscape restoration in a mixed agricultural-forest catchment: Planning a buffer strip and hedgerow network in a Chilean biodiversity hotspot. *Ambio* **2020**, *49*, 310–323. [CrossRef] [PubMed]
50. Brussaard, L.; Caron, P.; Campbell, B.; Lipper, L.; Mainka, S.; Rabbinge, R.; Babin, D.; Pulleman, M. Reconciling biodiversity conservation and food security: Scientific challenges for a new agriculture. *Curr. Opin. Environ. Sustain.* **2010**, *2*, 34–42. [CrossRef]
51. Van der Zanden, E.H.; Verburg, P.H.; Mucher, C.A. Modelling the spatial distribution of linear landscape elements in Europe. *Ecol. Indic.* **2013**, *27*, 125–136. [CrossRef]
52. Blasi, C.; Capotorti, G.; Marchese, M.; Marta, M.; Bologna, M.A.; Bombi, P.; Bonaiuto, M.; Bonnes, M.; Carrus, G.; Cifelli, F.; et al. Interdisciplinary research for the proposal of the Urban Biosphere Reserve of Rome Municipality. *Plant Biosyst.* **2008**, *142*, 305–312. [CrossRef]
53. Capotorti, G.; Bonacquisti, S.; Abis, L.; Aloisi, I.; Attorre, F.; Bacaro, G.; Balletto, G.; Banfi, E.; Barni, E.; Bartoli, F.; et al. More nature in the city. *Plant Biosyst.* **2020**, *154*, 1003–1006. [CrossRef]
54. Rodríguez, J.C.; Sabogal, C. Restoring Degraded Forest Land with Native Tree Species: The Experience of "Bosques Amazónicos" in Ucayali, Peru. *Forests* **2019**, *10*, 851. [CrossRef]
55. Bolliger, J.; Silbernagel, J. Contribution of connectivity assessments to Green Infrastructure (GI). *ISPRS Int. J. Geo-Inf.* **2020**, *9*, 212. [CrossRef]
56. Scherr, S.J.; McNeely, J.A. Biodiversity conservation and agricultural sustainability: Towards a new paradigm of "ecoagriculture" landscapes. *Philos. Trans. R. Soc. B Biol. Sci.* **2008**, *363*, 477–494. [CrossRef]
57. Ozinga, W.A.; Schaminée, J.H.J. Target species—Species of European concern; A database driven selection of plant and animal species for the implementation of the Pan European Ecological Network. In *Alterra-Report*; Alterra Wageningen UR: Wageningen, The Netherlands, 2005; Volume 1119.
58. Rousset, O.; Lepart, J. Shrub facilitation of *Quercus. humilis* regeneration in succession on calcareous grasslands. *J. Veg. Sci.* **1999**, *10*, 493–502. [CrossRef]
59. Smit, C.; Vandenberghe, C.; Ouden, J.D.; Müller-Schärer, H. Nurse plants, tree saplings and grazing pressure: Changes in facilitation along a biotic environmental gradient. *Oecologia* **2007**, *152*, 265–273. [CrossRef] [PubMed]
60. Muñoz, M.R.; Squeo, F.A.; León, M.F.; Tracol, Y.; Gutiérrez, J.R. Hydraulic lift in three shrub species from the Chilean coastal desert. *J. Arid Environ.* **2008**, *72*, 624–632. [CrossRef]
61. Pulido, F.; McCreary, D.; Cañellas, I.; McClaran, M.; Plieninger, T. Oak Regeneration: Ecological Dynamics and Restoration Techniques. In *Mediterranean Oak Woodland Working Landscapes: Dehesas of Spain and Ranchlands of California*; Campos, P., Huntsinger, L., Oviedo, J.L., Starrs, P.F., Diaz, M., Standiford, R.B., Montero, G., Eds.; Springer: New York, NY, USA, 2013; Volume 16, pp. 123–144.
62. Chardon, J.P.; Adriaensen, F.; Matthysen, E. Incorporating landscape elements into a connectivity measure: A case study for the Speckled wood butterfly (*Pararge aegeria* L.). *Landsc. Ecol.* **2003**, *18*, 561–573. [CrossRef]
63. León, M.C.; Harvey, C.A. Live fences and landscape connectivity in a neotropical agricultural landscape. *Agrofor. Syst.* **2006**, *68*, 15–26. [CrossRef]
64. Thiele, J.; Schirmel, J.; Buchholz, S. Effectiveness of corridors varies among phytosociological plant groups and dispersal syndromes. *PLoS ONE* **2018**, *13*, e0199980. [CrossRef]
65. Krosby, M.; Breckheimer, I.; Pierce, D.J.; Singleton, P.H.; Hall, S.A.; Halupka, K.C.; Gaines, W.L.; Long, R.A.; McRae, B.H.; Cosentino, B.L.; et al. Focal species and landscape "naturalness" corridor models offer complementary approaches for connectivity conservation planning. *Landsc. Ecol.* **2015**, *30*, 2121–2132. [CrossRef]
66. Cao, Y.; Yang, R.; Carver, S. Linking wilderness mapping and connectivity modelling: A methodological framework for wildland network planning. *Biol. Conserv.* **2020**, *251*, 108679. [CrossRef]
67. Lenoir, J.; Decocq, G.; Spicher, F.; Gallet-Moron, E.; Buridant, J.; Closset-Kopp, D. Historical continuity and spatial connectivity ensure hedgerows are effective corridors for forest plants: Evidence from the species–time–area relationship. *J. Veg. Sci.* **2021**, *32*, e12845. [CrossRef]

68. Hu, T.; Chang, J.; Liu, X.; Feng, S. Integrated methods for determining restoration priorities of coal mining subsidence areas based on green infrastructure: A case study in the Xuzhou urban area of China. *Ecol. Indic.* **2018**, *94*, 164–174. [CrossRef]
69. Norton, B.A.; Coutts, A.M.; Livesley, S.J.; Harris, R.J.; Hunter, A.M.; Williams, N.S. Planning for cooler cities: A framework to prioritise green infrastructure to mitigate high temperatures in urban landscapes. *Landsc. Urban Plan.* **2015**, *134*, 127–138. [CrossRef]
70. García-Feced, C.; Saura, S.; Elena-Rosselló, R. Improving landscape connectivity in forest districts: A two-stage process for prioritizing agricultural patches for reforestation. *For. Ecol. Manag.* **2011**, *261*, 154–161. [CrossRef]
71. Mimet, A.; Houset, T.; Julliard, R.; Simon, L. Assessing functional connectivity: A landscape approach for handling multiple ecological requirements. *Methods Ecol. Evol.* **2013**, *4*, 453–463. [CrossRef]
72. Aavik, T.; Holderegger, R.; Bolliger, J. The structural and functional connectivity of the grassland plant *Lychnis flos-cuculi*. *Heredity* **2014**, *112*, 471–478. [CrossRef] [PubMed]
73. Cui, N.; Feng, C.-C.; Wang, D.; Li, J.; Guo, L. The Effects of Rapid Urbanization on Forest Landscape Connectivity in Zhuhai City, China. *Sustainability* **2018**, *10*, 3381. [CrossRef]
74. Devi, B.S.; Murthy, M.S.R.; Debnath, B.; Jha, C.S. Forest patch connectivity diagnostics and prioritization using graph theory. *Ecol. Model.* **2013**, *251*, 279–287. [CrossRef]
75. Hernando, A.; Velázquez, J.; Valbuena, R.; Legrand, M.; García-Abril, A. Influence of the resolution of forest cover maps in evaluating fragmentation and connectivity to assess habitat conservation status. *Ecol. Indic.* **2017**, *79*, 295–302. [CrossRef]
76. Gutiérrez-Chacón, C.; Valderrama-A, C.; Klein, A.M. Biological corridors as important habitat structures for maintaining bees in a tropical fragmented landscape. *J. Insect Conserv.* **2020**, *24*, 187–197. [CrossRef]
77. Ascensão, F.; Santos-Reis, J.F.M. Highway verges as habitat providers for small mammals in agrosilvopastoral environments. *Biodivers. Conserv.* **2012**, *21*, 3681–3697. [CrossRef]
78. Benayas, J.M.R.; Bullock, J.M.; Newton, A.C. Creating woodland islets to reconcile ecological restoration, conservation and agricultural land use. *Front. Ecol. Environ.* **2008**, *6*, 329–336. [CrossRef]
79. Mikkonen, N.; Moilanen, A. Identification of top priority areas and management landscapes from a national natura 2000 network. *Environ. Sci. Policy* **2013**, *27*, 11–20. [CrossRef]
80. Wolff, J.O.; Sherman, P.W. *Rodent Societies: An. Ecological and Evolutionary Perspective*; University of Chicago Press: Chicago, IL, USA, 2008.
81. Ouin, A.; Paillat, G.; Butet, A.; Burel, F. Spatial dynamics of wood mouse (*Apodemus sylvaticus*) in an agricultural landscape under intensive use in the Mont Saint Michel Bay (France). *Agric. Ecosyst. Environ.* **2000**, *78*, 159–165. [CrossRef]
82. Dondina, O.; Saura, S.; Bani, L.; Mateo-Sánchez, M.C. Enhancing connectivity in agroecosystems: Focus on the best existing corridors or on new pathways? *Landsc. Ecol.* **2018**, *33*, 1741–1756. [CrossRef]
83. Lee, J.A.; Chon, J.; Ahn, C. Planning landscape corridors in ecological infrastructure using least-cost path methods based on the value of ecosystem services. *Sustainability* **2014**, *6*, 7564–7585. [CrossRef]

land

MDPI

Perspective

Developing a Metropolitan-Wide Urban Forest Strategy for a Large, Expanding and Densifying Capital City: Lessons from Melbourne, Australia

Martin Hartigan [1,2], James Fitzsimons [1,3,*], Maree Grenfell [2] and Toby Kent [2]

1 The Nature Conservancy, Suite 2-01, 60 Leicester Street, Carlton 3053, Australia; Martin.Hartigan@gww.com.au
2 Resilient Melbourne, City of Melbourne, GPO Box 1603, Melbourne 3001, Australia; Maree.Grenfell@melbourne.vic.gov.au (M.G.); toby@elliskent.com (T.K.)
3 School of Life and Environmental Sciences, Deakin University, 221 Burwood Highway, Burwood 3125, Australia
* Correspondence: jfitzsimons@tnc.org

Abstract: Urban forests provide many ecosystem services, such as reducing heat, improving air quality, treatment of stormwater, carbon sequestration, as well as biodiversity benefits. These benefits have resulted in increasing demand for urban forests and strategies to maintain and enhance this natural infrastructure. In response to a broader resilience strategy for Melbourne, Australia, we outline how a metropolitan-wide urban forest strategy (*Living Melbourne*) was developed, encompassing multiple jurisdictions and all land tenures. To this end, we mapped tree cover within the Melbourne metropolitan area, modelled potential habitat for some bird species, and investigated the role of tree cover for urban heat island mitigation. We outline the consultation and governance frameworks used to develop the strategy, the vision, goals and actions recommended, including canopy and shrub cover targets for different parts of the metropolitan area. The metropolitan-wide urban forest strategy acts as an overarching framework to guide local government authorities and various stakeholders towards a shared objective of increasing tree cover in Melbourne and we discuss the outcomes and lessons from this approach.

Keywords: canopy cover; vegetation mapping; connectivity; heat mapping; targets

1. Introduction

Nature is increasingly recognised as one of our most valuable assets to build resilience to 21st century urban challenges: the urban forest and the biodiversity that it supports can take pressure off our increasingly strained built infrastructure while supporting the liveability of our cities [1]. There is a growing understanding and demand for urban forests due to the multiple benefits that they provide. Urban forests provide many ecosystem services, such as reducing surface temperatures through shading, maintaining water quality through treatment of stormwater, providing carbon sequestration, and improving air quality through particulate matter removal [2]. In addition, the urban forest provides habitat for wildlife [3]. A recent and rapidly accumulating body of evidence demonstrates that experiencing nature in cities is critical to the health and well-being of the community [4,5]. In a context of rapid urbanisation, biodiversity conservation in towns and cities plays a significant role in minimising both the local extinction of species and maintaining the human experience of native plants and wildlife [1]. Although parks and reserves currently remain the focus for conserving urban nature, private gardens offer an extensive and undervalued resource for enhancing urban biodiversity [6].

Urban populations are growing [7], and many cities are densifying [8]. Densification can place pressure on canopy cover and other elements of vegetation in urban environments [9,10]. Acknowledging the many benefits of urban vegetation, and particularly trees, many city municipal governments have responded through developing urban forest strategies and targets [11]. It is atypical for urban forest strategies to be undertaken by coalitions of local government authorities (LGA). They are more typically undertaken by an individual LGA, as this is the extent of their jurisdiction (e.g., [12,13]).

Australia is no exception to the trends of both urban population growth and densification and responding through developing urban forest strategies. Australia is one of the most highly urbanised nations in the world, with almost 90% of the population living in urban areas in 2016 [14]. Municipal governments have been increasingly developing urban forest strategies [15], but as governance is typically fragmented in larger cities (through many local government entities), there is rarely coordination to develop whole-of-metropolitan urban forest strategies.

From 2012 to 2017, Melbourne, Australia's second largest city located in temperate south-eastern Australia, accommodated 87 percent of the total population growth of the State of Victoria [16]. A combination of natural increase and net immigration will make Melbourne Australia's largest city, with a projected population of 8 million people by 2051 [16]. To house this growing population, more than 1.6 million new dwellings will be needed, resulting in further density in existing areas and the addition of new suburbs on the urban fringe [17].

As with other Australian cities, Melbourne has a fragmented governance framework with no single metropolitan authority to coordinate action such as developing integrated urban forests. A key challenge, therefore, is achieving coordinated approaches across various jurisdictions and government entities. However, whilst challenging, it has successfully been achieved in some regions such as Chicago, USA where, for over 20 years, in excess of 250 partners, work together to improve conservation outcomes [18]. Here, we describe and discuss the process behind developing a metropolitan-wide urban forest strategy for Melbourne, the rationale behind various elements and outline lessons learnt, from the perspective of the authors of that strategy.

2. Background

2.1. Current Context for Melbourne

Melbourne sits at the junction of several bioregions, which are broad geographical regions composed of clusters of interacting ecosystems that share common physical and biological features such as climate, geology, landforms, soils and vegetation [19]. The different regions vary widely in canopy cover. This is a result of their natural attributes—which influence vegetation patterns—combined with the historical development and growth of Melbourne. The vegetation types that existed in Melbourne before European settlement included grasslands and grassy woodlands in the west, heaths and heathy woodlands in the south-east, and dry and damp forests in the east. In addition to these natural differences, Melbourne's development and growth since the 1830s has had an important influence on the shape and form of today's urban forest. Settlers' preferences for elevation, views, water and mature trees meant that Melbourne's early development moved outwards from the original European settlement on the banks of the Yarra River to the north-east, east and south-east, and tended to be on the hillier, treed terrain. The flatter northern and western areas—largely grassland plains—were considered less hospitable and desirable.

Melbourne is consistently ranked as one of the world's most liveable cities [20] and Victoria, the state for which it is the capital, has been known as 'Garden State'. Visitors to Melbourne (who contribute approximately AUD 8 billion to the city's annual economy) ranked parks and gardens as Melbourne's number-one unique attribute, and as the city's top 'must do' attraction [21]. Melbourne has one of the highest percentages of open green space of any city in the world [19]. The distribution and extent of native vegetation varies, with inner-city areas retaining less original native vegetation and having a larger proportion of introduced flora species than Melbourne's outer suburbs [22].

However, it was known that Melbourne was losing its canopy cover [23], and recent research found that between 2014 and 2018 there was a loss of canopy and shrub cover (combined) of 1.9% and tree cover alone of 0.7% across metropolitan Melbourne [24]. The loss is not uniform across metropolitan Melbourne, the eastern and southern regions showing greater rates of tree canopy loss (off a much higher base) resulting largely from densification of existing residential blocks (Figure 1a), and the western and northern regions of Melbourne showing a slight increase (off a low base) as a result of new developments on former farmland resulting in new streetscape, parkland and back/front yard plantings and during this period (Figure 1b).

Metropolitan Melbourne incorporates 32 LGAs. Local government is the third level of government in Australia and has a significant impact on the community. LGAs deliver more than 100 services including environmental management, managing land and community infrastructure. LGAs are responsible for a large range of capital and operational works such as waste management, road maintenance, drainage management, municipal building management and maintenance, social and community services, planning and building regulation, parks, open space and streetscape vegetation [25]. Fragmented governance hampers the ability to plan for and adequately protect a metropolitan-scale urban forest across different jurisdictions and land tenures. LGAs are the primary developers of urban forest strategies and are also major land managers of the urban forest itself, along with State government and statutory authorities (such as Parks Victoria and Melbourne Water) and private landholders.

As at 2019, 13 LGAs in Melbourne had or were developing an urban forest strategy and were making efforts to maintain or increase tree canopy cover (Table 1). Although some of these strategies focus narrowly on street trees and canopy on public land, others also consider vegetation on private land, have performance measures, are long term, and use a broad definition of an urban forest. Other LGAs have 'urban landscape', 'open space', or 'street tree' strategies which also contribute to the urban forest (Table 1). A range of Victorian Government documents provide support and guidance relating to parts of Melbourne's urban forest. These include *Plan Melbourne 2017–2050* [17], the *Victorian Climate Change Adaptation Plan 2017–2020* [26], *Protecting Victoria's Environment–Biodiversity 2037* [27], the *Victorian Public Health and Wellbeing Plan* [28] and the *Draft Metropolitan Open Space Strategy* [29]. However, a coordinated approach to planning Melbourne's urban forest across the entire metropolitan area, across different tenures, and encompassing native and exotic vegetation was lacking.

Some efforts have been made to coordinate across jurisdictions at a regional (sub-metropolitan) scale. For example, 'Greening the West' is a regional collaboration to help communities in Melbourne's west expand green spaces in parks, reserves, streetscapes, roofs and walls, backyards, car parks, sporting fields and waterways. A total of 23 organisations—LGAs, Victorian Government departments and agencies, water utilities and community groups—work successfully together to protect and enhance the urban forest, sharing knowledge and promoting and scaling up practical solutions in western Melbourne. The value of this collaboration is that, to date, Greening the West has generated AUD 30 million worth of green infrastructure projects in the Western Region of Melbourne [30].

2.2. Developing the 'Resilient Melbourne' Strategy

In 2014, Resilient Melbourne was established as part of the 100 Resilient Cities (100RC) initiative, pioneered by The Rockefeller Foundation. This global initiative aimed to help cities around the world become more resilient to the physical, social and economic challenges that are a growing part of the 21st century [31]. Melbourne was selected from 372 applicant cities around the world to be in the first wave of 33 cities to join the network. A key element of all cities as part of this program was the development of a resilience strategy.

a)

b)

Figure 1. (**a**) Historic vs. recent growth patterns in established urban areas in Melbourne's eastern region (upper images: higher residential density and single-dwelling renovation or construction of new dwellings; lower images: a redevelopment of four allotments with four detached dwellings to yield 12 dwellings) and (**b**) historic vs. current growth patterns in urban growth areas in Melbourne's northern region (note growth of street trees over an eight-year period). Year of photo under image.

Table 1. Municipal urban forest plans in Greater Melbourne, including current total tree canopy cover (as specified in individual local government plans or related documents), measurable canopy cover targets (public and private), overall canopy cover target and measurable tree diversity target (as at 2019).

Local Government	Urban Forest Strategy	Current Total Tree Canopy Cover	Measurable Canopy Cover Target/s (Public Land)	Measurable Canopy Cover Target/s (Private Land)	Overall Canopy Cover Target	Measurable Tree Diversity Target
City of Melbourne	Yes	11.0%	40% by 2040			No more than 5% species, 10% genus, 20% family
City of Port Phillip	Yes	22.0%	Yes, individual suburb targets by 2027			Increase diversity
City of Yarra	Yes	17.0%			21.25% by 2040	
City of Banyule	Yes	Parks and bushland reserves: 37.0%	Parks: 50%; Streets 75%	Increase by 20%		Increase diversity
City of Darebin	Yes	9.8%	At least 25% by 2028			
City of Moonee Valley	In development	11.0%				
City of Moreland	Yes	14.0%	21.3% by 2050	Maintain at current 9% by 2050	30.3% by 2050	Street tree planting guide: No more than 40% one family, 15% one genus and 5% one species
City of Whittlesea	In development	19.72%	Yes	Yes	Increase by 20% by 2040	
City of Knox	Street Tree Asset Management Plan	Streets: 22.0%	25% by 2030 in streetscapes			Improve diversity
City of Casey	2009 Revegetation strategy	Remnant vegetation: 7.0%			Native vegetation 30%	
City of Greater Dandenong	Yes	9.90%			15% by 2028	
City of Frankston	Urban Forest Policy	17.0%			20% by 2040	
City of Glen Eira	In development	12.5%			14% by 2040	
City of Monash	Monash Urban Landscape and Canopy Vegetation Strategy	22.0%			30% by 2040	
City of Stonnington	Yes	25.0%				No more than: 30% of any one family, 20% of any one genus, 10% of any one species
City of Brimbank	Yes	6.20%	50% canopy cover in urban parks and open spaces		30% by 2046	No more than 50% of the same family and introduce new families
City of Wyndham	Yes	9.0%	Streets: 25% by 2040 Open Space: 35% by 2040	Established areas: 15% by 2040; New areas: 10% by 2040		No more than 30% family, 20% genus, and 20% species

Following significant engagement across metropolitan Melbourne, including all LGAs, Victorian Government, and many other stakeholders, the *Resilient Melbourne* strategy was released in 2016 [32]—the first ever metropolitan-wide strategy in Australia that was led by local government. The Resilient Melbourne strategy recognised that a metropolitan-wide urban forest strategy would bring city-wide benefits that could not be achieved by individual LGAs in isolation and recommended the development of such a strategy as a flagship action (which was also considered the highest priority action for LGAs (Figure 2).

The Nature Conservancy, a global conservation NGO, worked in cities across Europe, North America and Africa, and Asia collaboratively with organisations and communities using strategic planning and science-based solutions make cities resilient, healthy and equitable (e.g., [33,34]). In 2014, The Nature Conservancy scoped the potential for a city-wide biodiversity strategy for Melbourne and found strong interest amongst stakeholders. The Nature Conservancy was a global platform partner with 100 Resilient Cities and in

2016 agreed to partner with Resilient Melbourne to lead the development of a metropolitan urban forest strategy which would become *Living Melbourne: Our Metropolitan Urban Forest* [35].

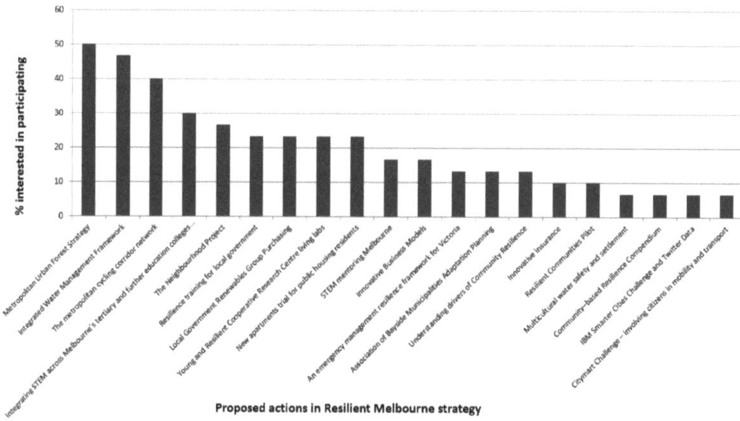

Figure 2. Local Government voting results on the most important *Resilient Melbourne* Strategy actions.

3. Developing a Metropolitan-Wide Urban Forest Strategy

Three major processes were run concurrently to gather information and ensure stakeholder engagement and ownership: (1) mapping of the urban forest (in the absence of a consistent metropolitan-wide dataset), (2) governance and advisory, and (3) broader engagement and co-design. These are outlined below.

3.1. Mapping and Modelling to Inform Priorities

3.1.1. Mapping Metropolitan Melbourne's Urban Forest

Mapping the extent of the urban forest is essential for its development, protection, maintenance and growth. Comprehensive and accurate maps give government authorities and other land managers the information they need to make informed decisions about improving, re-establishing and connecting natural environments [36]. Canopy cover is an important measure of the urban forest's ability to benefit the community and the environment [37]. One of the major inhibitors for developing metropolitan-wide strategy was a lack of consistent data across the region. Thus, developing a consistent, metropolitan-wide vegetation and canopy map for the *Living Melbourne* strategy was a critical step in assessing the current status of metropolitan Melbourne's urban forest, and for setting future targets and actions [12].

100 Resilient Cities (100RC) secured global platform partners to undertake the mapping of metropolitan vegetation and subsequent development of *Living Melbourne.* To better inform the strategic approach of connecting and enhancing our metropolitan urban forest, new mapping resources including imagery and mapping derivatives were developed. eCognition Essentials software provided by global software developer Trimble was used to map metropolitan Melbourne's urban forest. The two main sources of data were: (1) two-metre resolution multispectral satellite imagery provided by DigitalGlobe (November 2016–March 2017) and (2) Victorian Government Light Detection and Ranging (LiDAR) datasets (years 2005–2006, 2008–2009, 2011–2012, 2012–2013) and 1 m resolution Digital Terrain Model (DTM) and Digital Surface Model (DSM). Overall, the provided source data were well suited and of good quality (Morphum Environmental pers. comm. 2018). These data sets were prepared and formatted before being processed by eCognition Essentials.

The output provided a map of the distribution and height of vegetation across metropolitan Melbourne (Figures 3 and 4 and further details in [38]). The resultant mapping classified vegetation height into 5 Height Class categories: 1 = 0–0.3 m (grass), 2 = 0.3–3 m (shrub), 3 = 3–10 m (small tree), 4 = 10–15 m (tree), 5 = 15+ m (large trees). In general use, it is recommended that due to (a) the variations in the dates and times of the satellite imagery and LiDAR datasets, (b) the DTM and DSM resolution, and (c) the satellite imagery resolution that the vegetation data are best viewed at a scale of approximately 1:12,000. The overall accuracy of the metropolitan Melbourne urban vegetation height dataset is 82.0%, with Producer's Accuracy ranging from 74.3% (Height Class 4) to 90.1% (Height Class 1), and User's Accuracy ranging from 80.0% (Height Class 5) to 85.0% (Height Class 4). When the height classes are condensed down to belonging to Understory Vegetation (<3.0 m) or Tree Canopy (>3.0 m), the Overall Accuracy increases to 94.5%, with Producer's Accuracy and User's Accuracy for individual classes above 90%.

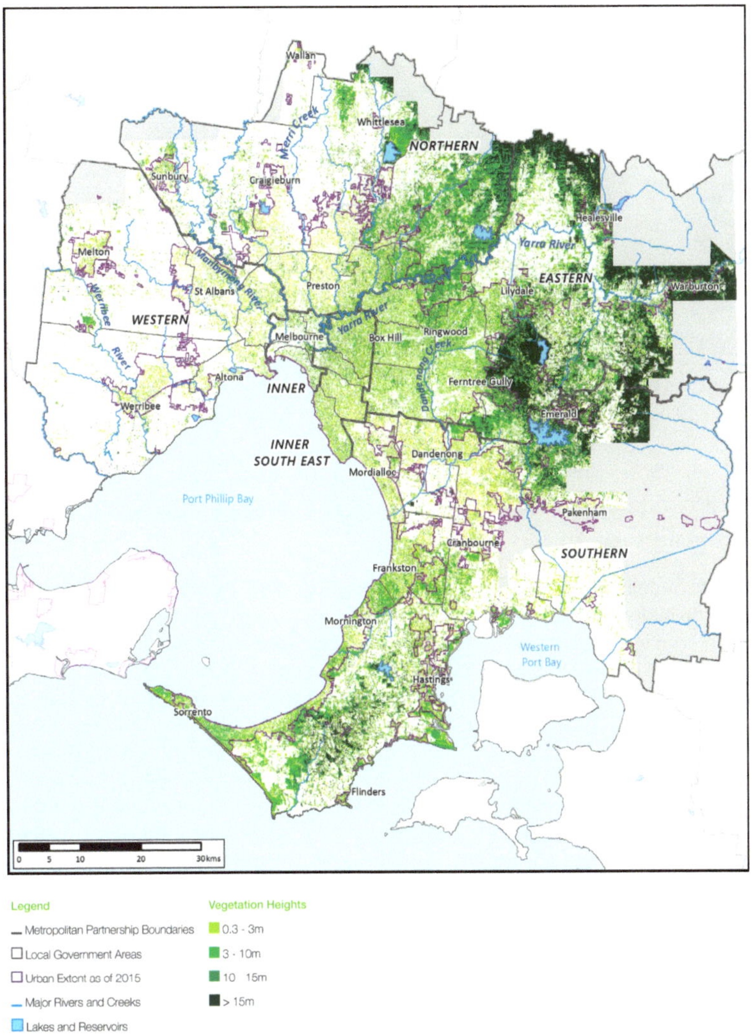

Figure 3. Metropolitan Melbourne's Urban Forest by vegetation height. Note grasslands not shown.

Figure 4. Canopy vegetation, overlayed on aerial imagery and centred on the suburb of Burnley, inner eastern Melbourne.

The metropolitan urban area has a total canopy cover of 15 percent. Table 2 provides a regional breakdown of the distribution of canopy cover across metropolitan Melbourne (Regions shown in Figure 3). Canopy is defined as vegetation above three metres in height. Canopy cover is highest in the Eastern (25%) and Inner South-East (22%) regions. The Southern (16%), Inner (13%) and Northern (12%) regions have less canopy cover. Canopy cover is lowest in the Western Region (4%).

3.1.2. Habitat Connectivity

Connectivity and the ability of animals to move, and plants to disperse, within or between patches of habitat is critical for conservation [39]. Many individual patches of habitat in urban areas are too small and widely dispersed to support viable populations [40]. It is therefore important that green spaces, such as gardens and public open space, are not viewed at the individual scale, but instead considered collectively as interconnected networks of green spaces across the urban landscape [41,42].

In Melbourne, like in other Australian cities, streetscapes with native trees support significantly more diverse and abundant populations of native birds than streets with mostly exotic trees [41,43]. Birds provide a useful candidate taxon for monitoring as they are relatively easy to detect and identify, census methods are well developed and formal and informal monitoring and databases in existence (e.g., [44]). Using the urban forest canopy cover mapping layer with bird atlas data (supplied by BirdLife Australia) and other important datasets (such as ecological vegetation classes), enabled the modelling of different levels of connectivity and landscape permeability for different bird species and bird groups (Figure 5).

Table 2. Canopy cover in urban Melbourne.

Metropolitan Region	Percentage of Land with Canopy Cover (of Trees 3 Metres High or Taller) *
Eastern	25
Inner South East	22
Southern	16
Inner	13
Northern	12
Western	4
Total metropolitan tree canopy cover	15

* Figures rounded to nearest whole number.

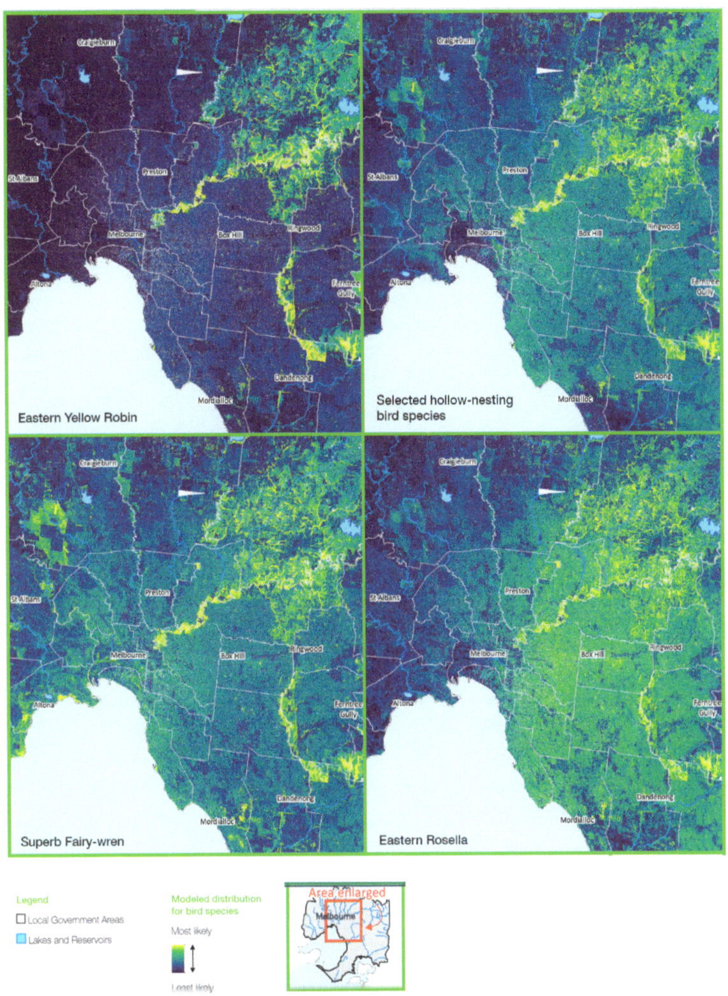

Figure 5. Modelled distribution for various bird species or bird groups in Melbourne's urban landscape, with suitability ranging from most likely (**yellow**) to least likely (**dark blue**).

The high suitability of particular riparian corridors (such as along the Yarra River and Dandenong Creek) for many of these bird species highlights the importance of these features for the persistence of some species in the urban landscape, and the importance of connectivity [40,45]. However, for species that are better able to exist in the urban environment, suitable habitat and connectivity is also provided by streetscapes and backyards.

Identifying a network of existing and potentially new habitat corridors at different scales for a range of species (and protecting and improving these corridors) will be an important step in creating an enhanced urban forest for Melbourne. For example, combining habitat models based on species records and known habitat preferences with canopy mapping can reveal areas for future corridor improvement (Figure 6).

3.1.3. Correlations between the Urban Forest and Heat Vulnerability

Urban forests cool surrounding environments. Built-up areas of cities can be as much as 7 °C warmer than surrounding areas [46]. This 'urban heat island effect' is caused by the heat-absorptive thermal mass of concrete, bitumen and bricks. The urban heat island disproportionately affects vulnerable people, including young children, the elderly, people who are unwell or socially isolated, and those who are financially disadvantaged [47].

In developing the *Living Melbourne* strategy, the land surface temperature was analysed using a variety of Australian Bureau of Statistics indices, such as the Socio-Economic Indexes for Areas, which ranks areas in Australia according to relative socio-economic advantage and disadvantage. This found a close correlation between "hot spots" in the landscape and vulnerable populations (Figure 7, Table 3).

Table 3. Mean Percent residential rental properties and mean average weekly household income (AUD) compared between urban heat island cool spots and hot spots across Melbourne's regions.

Region	Residential Rental Properties (%)		Average Weekly Household Income (AUD)	
	Cool Spots (Urban Heat Island < 0 °C)	**Hot Spots (Urban Heat Island > 10 °C)**	**Cool Spots (Urban Heat Island < 0 °C)**	**Hot Spots (Urban Heat Island > 10 °C)**
Eastern	14.2	18.3	1503.1	778.7
Inner	35.0	25.3	1188.8	350.3
Inner South-East	22.1	NA	2008.5	NA
Northern	16.0	23.3	2039.0	1250.8
Southern	19.4	26.6	1458.0	1347.8
Western	16.7	22.5	1284.0	1235.5
Total metropolitan area	18.9	23.5	1524.1	1228.1

The land surface temperature analysis was also undertaken to explore the effect of total vegetation cover and vegetation height. About half of metropolitan Melbourne was 5 °C above the city's estimated non-urban baseline temperature. This applies to about 80 percent of the Northern and Western regions. Canopy vegetation between three and 10 metres high was predominant across Melbourne, and vegetation at this height range provided for more cool spots (areas that are equal to or below their estimated non-urban baseline temperature) (Table 4). Overall, in cool spots there is more vegetation and far more canopy (Table 4). On average, hot spots had less than three percent canopy and no tall trees (trees greater than 15 metres high).

The highest numbers of hot spots occur in the north, west and south of Melbourne (Figure 7). The north and west also had far more hot spots than cool spots. The size of hot spots and cool spots varied—hot spots in the east were on average three hectares in area, compared to hot spots in the west averaging nearly 10 hectares in area. The opposite was true for the cool spots—these are larger in southern and eastern Melbourne (which is more heavily vegetated) and smaller in the north and west.

We also found, in most cases, a greater number of hot spots where the percentage of residential rental properties is higher and where weekly household income is lower (Table 3).

a) Current aerial photography of part of the Dandenong Creek Valley Parklands in eastern Melbourne.

b) Mapped canopy cover over the Dandenong Creek Valley and surrounding suburbs over aerial photography.

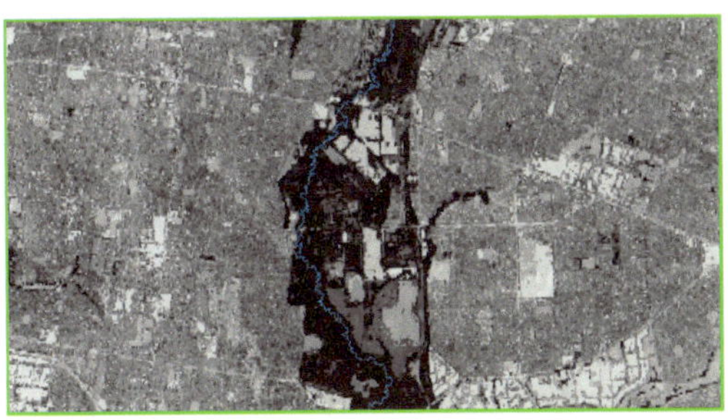

c) Likelihood of occurrence for a suite of hollow-nesting bird species in the Dandenong Creek Valley (black and dark grey areas are most likely).

d) Canopy cover (green) overlayed on hollow-nesting bird species suitability. The visible areas of darkest shading (black and dark grey) indicate areas without existing canopy that could be suitable for restoration and connectivity improvement for hollow-nesting bird species.

Figure 6. Canopy mapping and bird species habitat modelling used together, to reveal areas suitable for corridor improvement by expanding the urban forest: Dandenong Creek Valley Parklands in eastern Melbourne.

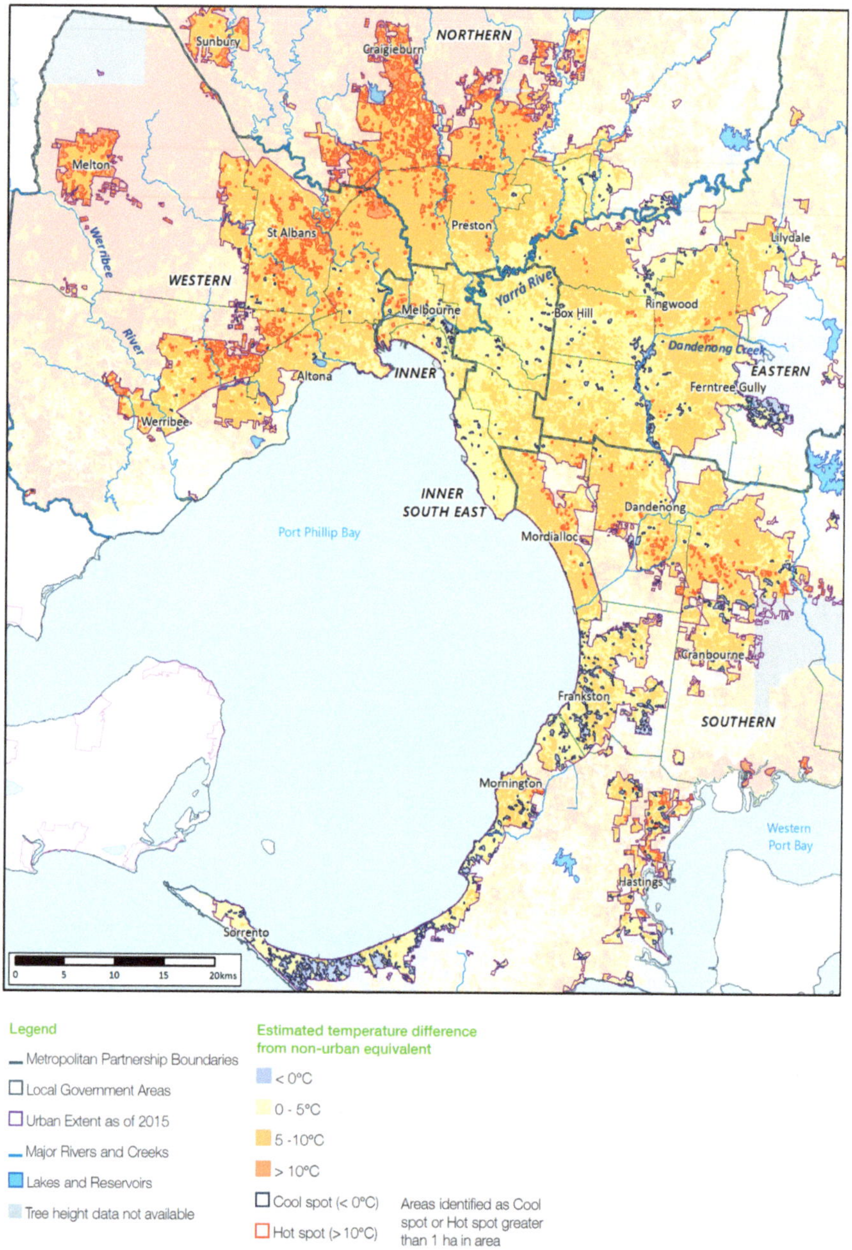

Figure 7. Urban heat islands across metropolitan Melbourne. Hot spots (>10 °C warmer than non-urban conditions) bordered in bright red. Cool spots (areas that are equal to or below their estimated non-urban baseline temperature =< 0 °C) bordered in dark blue.

Table 4. Number, average area and total area of urban heat island cool spots and hot spots across the regions, and total metropolitan area and mean percent cover of vegetation in different height classes compared between urban heat island cool spots and hot spots across the regions. NA = Not applicable.

Region	No. Features	Average Area (ha)	Total Area (ha)	Mean Percent Cover of Vegetation		
				Low Vegetation (<3 m)	Tree Canopy (>3 m)	Total Vegetation
Cool spot						
Eastern	165	10.6	1754.8	22.6	63.2	85.8
Inner	40	5.2	209.0	22.2	36.4	58.6
Inner South-East	64	3.9	252.5	24.3	37.1	61.4
Northern	47	5.6	261.4	11.9	72.2	84.1
Southern	380	11.9	4529.1	29.1	37.5	66.7
Western	54	3.2	174.0	19.1	2.9	22.0
Total metropolitan area	*750*	*9.6*	*7180.8*	*26.3*	*44.2*	*70.5*
Hot spot						
Eastern	55	3.0	167.0	14.6	2.1	16.7
Inner	20	5.0	99.8	2.2	0.3	2.5
Inner South-East	NA	NA	NA	NA	NA	NA
Northern	453	12.1	5476.0	22.7	1.6	24.3
Southern	243	6.6	1605.9	39.2	6.1	45.4
Western	401	12.7	5081.1	31.1	1.7	32.8
Total metropolitan area	*1172*	*10.6*	*12,429.8*	*28.0*	*2.2*	*30.2*

3.2. Governance Arrangements

The development of *Living Melbourne* was guided by two key bodies: the 'Senior Reference Group' and the 'Technical Advisory Group'. The Senior Reference Group's role was to oversee the progress of the strategy and consisted of senior representatives from local government sub-regions of metropolitan Melbourne and a senior representative from Melbourne Water, Parks Victoria, and the Department of Environment, Land, Water and Planning. The Technical Advisory Group, constituting specialists in state and local governments and academia, provided technical input, advice and recommendations to the *Living Melbourne* project team about the key issues, opportunities and barriers to meeting strategy objectives that should be considered.

3.3. Concurrent Consultation with Stakeholders

A critical part of the development of *Living Melbourne* was the collaborative approach that was taken by The Nature Conservancy and Resilient Melbourne, with approximately 65 other organisations and 250 individuals either attending workshops or providing feedback on the drafts of the strategy, in addition to the advice of a Senior Reference Group and a Technical Advisory Group. This process included a series of workshops to guide development of the strategy, incorporate stakeholder perspectives and review the strategy as it progressed. Four major workshops helped to:

- Establish the current status of the urban forest and associated management issues;
- Develop the vision and goals;
- Develop the critical strategic areas that the strategy would address;
- Identify technical evidence to guide the strategy;
- Frame and develop the draft strategy.

The stakeholders involved were metropolitan Melbourne's 32 LGAs, Victorian Government departments (e.g., the Department of Environment, Land, Water and Planning, Department of Health and Human Services, Department of Transport) and statutory agencies (e.g., Melbourne Water, retail water authorities), technical experts, land managers, policy makers, planners, developers and community representatives.

The workshops were complemented by formal and informal consultation with individuals, research organisations, individual stakeholder organisations, associated local government alliances and industry and relevant professional groups.

Each of the workshops had a critical mass of stakeholders present and many stakeholders attended all workshops. In addition to the strong workshop engagement, there was a co-design process including smaller working groups that helped define and delineate the vision, goals and actions arising from the workshop process.

Stakeholders contributed fundamental data that informed various elements of *Living Melbourne*. Critical data gathered from stakeholders included:

- Twenty-six LGAs contributed their tree asset inventories of street trees and open space trees. While these databases vary in size and scope, these datasets were used to help determine the monetised value of these assets including the economic savings in the form of pollution removal, carbon storage, carbon sequestration, and avoided water run-off. The datasets also helped calculate the cost of replacing these assets with similar trees based on size, species, health and location.
- BirdLife Australia supplied bird atlas data. Using the urban forest canopy cover mapping layer with bird atlas data and other important datasets (such as ecological vegetation classes), enabled the modelling of different levels of connectivity and landscape permeability for different bird species and bird groups.
- Thirty organisations, including 27 LGAs and three major public land managers (Parks Victoria, Melbourne Water and VicRoads) supplied their operational costs relating to the establishment, management and maintenance of vegetation.
- Tree purchasing, planting and maintenance costs were supplied by organisations such Council Arborists Victoria, Nursery and Garden Industry Victoria and public land management agencies to inform estimates of costs for future canopy targets.
- The Clean Air and Urban Landscapes (CAUL) Hub and RMIT University provided road partition features and valuable advice and guidance around the use of the "Melbourne Vegetation and land use cover 2014" [24,48]. This dataset was used to assist the urban heat island analysis.

First and second drafts of the strategy were circulated to stakeholders to seek feedback. In many cases, this feedback was followed up and discussed with the stakeholder to ensure that their views were clearly understood and where appropriate articulated in the strategy before the strategy was finalised. *Living Melbourne* was launched in mid-2019. Its key elements are outlined below.

4. Vision, Goals and Actions of *Living Melbourne*

Through the co-design process, a strategic vision was agreed of 'our thriving communities are resilient, connected through nature'. This vision is supported by three goals:

- Healthy people: Protect and increase access to nature, green space and canopy cover, to reduce heat exposure, and improve mental and physical wellbeing.
- Abundant nature: Protect and extend habitat connectivity and corridors to enhance biodiversity.
- Natural infrastructure: Protect and increase vegetation on public and private land, in order to cool urban areas, retain water in the soils, reduce flood risk and increase water and air quality.

These goals are supported by a program of six interrelated actions that together aim to support and coordinate action across the metropolitan area. These actions are outlined below:

Action 1: Protect and restore habitats, and increase ecological connectivity of all types between streetscapes, conservation reserves, riparian and coastal areas, open spaces and other green infrastructure across metropolitan Melbourne.

Urban areas can play a significant role in conserving biodiversity, but without conscious efforts to protect and enhance habitat and linking corridors, Melbourne's natural environment, which is 'fundamental to the health and wellbeing of every Victorian' [27] will continue to decline. *Living Melbourne* recommended restoring corridors and connectivity in Melbourne at different scales and with different approaches needed. The strategy emphasised restoring corridors on public land, with a focus on riverine corridors, and a priority for using indigenous plant species. In an urban setting, while indigenous plant species are usually best for native fauna (especially mobile fauna such as birds, bats and insects), non-indigenous native plant species can also provide important resources (such as food and shelter). Although introduced species of trees and shrubs typically offer fewer resources to fauna, they still provide ecosystem services and, in neighbourhoods where introduced trees dominate, the habitat value of these areas can be improved by increasing structure (such as by planting shrubs and native understorey). The strategy noted the importance of managing the different elements of the urban forest collectively, and building upon existing habitat and vegetation to form an interconnected matrix of green spaces across the urban landscape.

Action 1 comprised the following components:

1.1 Consolidate data, maps and other relevant information.

1.2 Assess the values and quality of information, to develop a list of priority areas for immediate protection.

1.3 Map existing and new areas for biodiversity connectivity at different scales, and prioritise areas for strengthening connectivity and biolinks, including responses to climate change, within each municipality and across the region.

1.4 Implement priorities for conservation, and secure and build habitat connectivity.

Action 2: Set urban canopy and understorey targets for each metropolitan region, and decide on a clear and consistent method for long-term monitoring and evaluation of the quality and extent of the urban forest.

Increasing canopy cover is a key performance measure for most urban forest strategies [12].

Action 2 comprised the following components:

2.1 Establish and implement urban forest greening targets including, as a minimum, 'tree canopy' and 'tree canopy and shrub' cover for each region.

2.2 Establish a measure of permeability across the regions, with the aim of implementing a permeability target for public and private land.

2.3 Establish a method for monitoring, evaluating and reporting on the improvement of the urban forest, including indicators and measures for quality and extent.

2.4 Develop a system for consistently collecting and analysing urban forest data, and coordinate the collection and publication of data in a publicly available, comparable database.

The strategy proposed the adoption of regional targets for canopy and vegetation across Melbourne that were based on a common analysis of metropolitan-scale vegetation, current literature and an analysis of the targets incorporated in existing urban forest strategies in 13 municipalities across Melbourne [35]. In addition, these targets were tested through two major rounds of consultation in December 2018 and February 2019 and a final strategy was circulated for endorsement in April 2019.

Research suggests that targets should be specific to each region and should take into account local conditions such as development density, land use, and climate [49]. After considering the baseline canopy and canopy and shrub cover for each region, the literature, current LGA urban forest strategies, and regional context, ambitious regional tree canopy cover targets of between 20 and 30 percent were proposed. Because native understorey and tree canopy cover of at least 30 percent benefits biodiversity [3], both tree canopy and shrub targets of between 30 and 50 percent by 2050 were proposed.

As a starting point for regional agreement, *Living Melbourne* proposed targets that:

- Were calculated based on vegetation in the existing urban area at 2015.
- Applied to all land (public and private) in each region.
- Recommended increases each decade.
- Were supported by principles, including no net loss of tree or shrub cover on public and private urban land in each metropolitan region.

Targets for 'canopy cover' and 'canopy and shrub' for each metropolitan region are proposed in Table 5 and Figure 8.

Table 5. Targets for 'tree canopy', and for 'canopy and shrubs', by region to 2050.

Region	Local Government Authorities	Existing 2015		Target 2030		Target 2040		Target 2050	
		%Tree Canopy	%Tree Canopy and Shrubs	%Tree Canopy	%Tree Canopy and Shrubs	%Tree Canopy	%Tree Canopy and Shrubs	%Tree Canopy	%Tree Canopy and Shrubs
Western	Brimbank, Hobsons Bay, Maribyrnong, Melton, Moonee Valley, Wyndham	4	15	9	20	14	25	20	30
Northern	Banyule, Darebin, Hume, Mitchell, Moreland, Nillumbik, Whittlesea	12	24	17	29	22	34	27	39
Inner	Melbourne, Port Phillip, Yarra	13	18	18	23	23	28	28	33
Southern	Casey, Frankston, Greater Dandenong, Kingston, Cardinia, Mornington Peninsula	16	34	21	39	26	44	30	50
Inner South-East	Bayside, Boroondara, Glen Eira, Stonnington	22	39	24	44	27	49	30	50
Eastern	Knox, Manningham, Maroondah, Monash, Whitehorse, Yarra Ranges	25	44	27	49	29	50	30	50

The 2015 baseline and targets for each of the six metropolitan regions vary, as they acknowledge the historical development of these regions and the ecological characteristics of their underlying bioregions. In addition, they acknowledge the ambitious canopy cover targets set by some metropolitan municipalities. For example, the Eastern Region of Melbourne was developed before the Western Region, and is largely in the Gippsland Plain bioregion, which before European settlement was mostly dry and damp forest. The result is a canopy cover of about 25 percent in 2015, a higher 2050 target, and less action required to achieve this target. By contrast, the Western Region is a newer urban growth area and is predominantly in the Victorian Volcanic Plain bioregion, which was originally a largely plains grassland landscape. The result is a 2015 canopy cover of only 4 percent, a lower 2050 target, and significantly more effort required to achieve this lower target. The strategy also sets out principles to guide the implementation of urban forest targets along with thresholds for the percentage of vegetation to be maintained or established on public and private land [35].

Action 3: Strengthen planning and development standards and relevant guidelines to increase the greening of the private realm.

Land zoned for residential, commercial, industrial, special or rural use contributes more than 66 percent of the existing tree canopy, with residential land alone contributing 58 percent. With a trend towards larger houses on smaller lots in greenfield developments, and higher-density development in urban infill areas, gardens have become smaller, and impervious surfaces have increased. The result is a rapid diminishing of the urban forest, as room for vegetation in private space shrinks. Protecting, maintaining and nurturing the trees and suitable understorey vegetation on private land is critical to maintaining and expanding the urban forest.

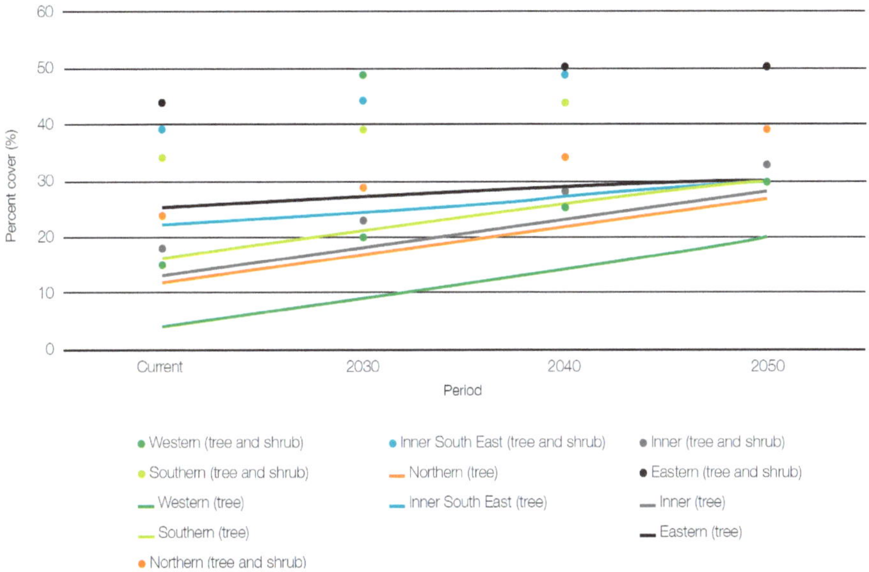

Figure 8. Commencement points and potential progress towards targets for each region. Tree canopy and tree and shrub cover targets: 2015(Current)–2050 for greater Melbourne region.

Action 3 comprised the following components:

3.1 Strengthen regulations to support greening in new subdivisions and developments—to benefit human health and wellbeing, and increase biodiversity.

3.2 Strengthen regulations to protect canopy trees.

3.3 Encourage private landholders to protect and enhance the urban forest and expand greening activities by offering incentives for planting, installing and maintaining natural infrastructure.

Action 4: Encourage collaboration between sectors and regions, to protect and expand the urban forest by strengthening existing regional partnerships, and establishing new ones, and by accelerating greening efforts on private land.

Although work is under way to protect and expand the urban forest across Melbourne, the fragmentation of these efforts within and between municipalities is one of the most significant barriers to reaching metropolitan Melbourne's urban forest goals.

Several successful alliances and cross-organisational governance agreements already existed in different parts of Melbourne. The strategy advocated that these should be the starting point for building further collaborations and alliances for greening across the city. Local communities can help urban forest efforts by gathering and sharing important data, so that investment in the urban forest is targeted wisely. To obtain specific tree information, such as species and numbers, the strategy advocated the involvement of citizen scientists and drawing on Aboriginal ecological knowledge.

Action 4 comprised the following components:

4.1 Capitalise on existing collaborations between local and state governments and the private sector.

4.2 Mobilise broad community support.

4.3 Support and develop existing and new methods to obtain and apply community knowledge.

4.4 Foster and promote urban forest champions, in both the public and private sectors.

Action 5: Equip practitioners to protect and enhance the urban forest by building on existing resources and creating a shared toolkit to facilitate implementation of best practices.

Helping practitioners identify, agree on and adopt best practice is central to the successful protection and expansion of our urban forest.

There is a growing list of open-source tools, resources and reference materials that practitioners can apply to their greening efforts, but new tools are needed. In particular, materials that will help involve the wider community and the land development industry, such as best practice guidelines and case studies; clear and agreed procedures to attract involvement by private and semi-public utility companies; and vegetation-management and associated technical training for staff of utility companies, to improve decision-making.

Managing, maintaining and expanding the urban forest requires significant capital and operational expenditure. The strategy recommended guidance on current and projected future costs to help all parties forecast future funding requirements more accurately, identify any unnecessary or unreasonable expenditures, and bring efficiencies in funding the urban forest.

Action 5 comprised the following components:

5.1 Build the capacity of public and private sector practitioners to protect, enhance and expand the urban forest.

5.2 Build on, and develop new tools for public sector land managers.

5.3 Build on, and develop new guidance materials for managing the capital and operational costs of urban forest endeavours.

Action 6: Establish a set of funding and financing options to suit different types of urban forest action.

Estimates undertaken as part of developing this strategy, drawn from existing government and peer-reviewed sources, suggest that Melburnians already enjoy benefits from nature valued at close to AUD 5 billion dollars per annum from ecosystem services [38]. Reaching the canopy and broader vegetation targets set out in Action 2 will require an estimated investment of AUD 1 billion over the next 30 years, with the bulk of this investment to be made in the decade to 2030. A range of financing tools will be required to achieve the goals of the *Living Melbourne* strategy. These sources range in scale and complexity from conventional government budget appropriations and philanthropy to public-private partnerships, ecosystem service payments, performance-based incentives and hybrid instruments that feature a range of revenue streams.

Action 6 comprised the following components:

6.1 Identify and secure long-term financing to realise the *Living Melbourne* vision.

6.2 Investigate and establish a targeted grants program to support innovation and action for greener neighbourhoods.

5. Discussion

5.1. Key Processes and Enabling Conditions That Facilitated Success

The development of *Living Melbourne* was supported by a high degree of stakeholder engagement throughout the process and has been described "as an instance of metropolitan governance in action" [50–52]. Approximately 65 organisations and over 250 individuals made contributions to the strategy. The efforts to involve and secure senior executive and political (mayor, CEO and executive officer level) buy-in during the development of the *Resilient Melbourne* strategy greatly enhanced the ability for subsequent engagement in *Living Melbourne*, as did the early identification of the metropolitan urban forest as a priority action (Figure 2). This senior stakeholder engagement was critical to the success of *Living Melbourne* as it provided the staff of stakeholder organisations, in particular LGAs, with a mandate to fully engage in the co-design process. The transparent co-design process provided stakeholders with confidence that *Living Melbourne* reflected their views.

The scheduling of the development of *Living Melbourne* benefitted from substantial urban greening research, policy, strategy and on-ground activity across metropolitan Melbourne (e.g., [15,53,54]). Examples of this activity included:

- The City of Melbourne had developed a world-class urban forest strategy.
- Thirteen LGAs (including City of Melbourne) had already developed (or were developing) municipal urban forest strategies and other metropolitan LGAs had relevant urban greening and open space related policies or strategies.
- Greening the West had been working as an alliance of organisations working toward a common agenda and was in the process of planting one million trees.
- The State Government had released several strategic plans, including *Plan Melbourne 2017–2050* [17] and *Biodiversity 2037* [27] that supported cooling and greening metropolitan Melbourne.
- Melbourne Water was developing the *Healthy Waterways Strategy 2018–2028* [55], a strategy shared by Melbourne Water, state and local governments, water corporations and the community. It covers the rivers, creeks, estuaries and wetlands of the Port Phillip and Westernport region, providing a single framework to protect and improve the waterways' environmental, social, economic and cultural values for the community.
- The Nature Conservancy had recently scoped the potential for a city-wide biodiversity strategy for Melbourne.

5.2. Challenges, Limitations and Critical Areas to Build on for Implementing the Strategy

Living Melbourne is largely positioned to support, expand and extend existing urban greening initiatives and provide enabling conditions to encourage better urban greening outcomes. As with any planning and strategy development process, there were time and resource constraints in which it was completed. There are five key areas where, had additional time and resources allowed, knowledge gaps might have been filled, allowing for an even richer and more directive strategy. They are each identified below.

1. A critical element of the strategy included the science associated with the vegetation mapping and the resultant use of the vegetation mapping in urban heat island analysis and models for some elements of biodiversity. Although it was considered that the input data to create the vegetation were of good quality and that there was a high level of accuracy, the large geographic area of metropolitan Melbourne meant that there were variations in the dates and times of the satellite imagery and LiDAR datasets, and the DTM and DSM resolution. Additional technical analysis would have enhanced the strategy, although these remain areas that could be readily undertaken using the base vegetation mapping. Most municipalities have biodiversity-related policy, strategy and or detailed plans. A granular analysis of a larger suite of metropolitan biodiversity values and threats, using available local government biodiversity-related strategy and existing Victorian Government datasets, could have provided more detailed direction for biodiversity benefits and habitat connectivity opportunities. Acknowledging the continuing need for this activity, Action 1 in the strategy provides a step-by-step approach for improving biodiversity and connectivity. The *Living Melbourne* canopy analysis, undertaken using the software eCognition Essentials (Trimble Geospatial), provided tenure-

neutral canopy mapping of 639,124 hectares of urban, peri-urban and rural land across the Melbourne metropolitan area. While the canopy mapping identified vegetation distribution and height, future analysis that could better differentiate different tree types (e.g., deciduous, eucalyptus, broadleaf trees, etc.) would be useful for finer scale conservation planning and assist the further prioritisation of on-ground urban greening locations. Sixteen local government street and park tree inventories were gathered during the development of the strategy and were most helpful to inform some research related to the financial value of such assets. These inventories could also be used to inform biodiversity connectivity mapping and identify where opportunities exist to provide good heat mitigation walking-corridors. In addition, such inventories, combined with the values available from the multi-spectral satellite imagery and canopy mapping might assist to validate remote species identification.

2. Collaboration was sought with land managers (including urban development industry representatives) and policy makers and not directly with the general public. It was agreed by the Senior Reference Group that LGAs, being the closest tier of government to the community should be relied upon to engage with relevant community stakeholders about *Living Melbourne*. Had significantly greater resources been available, direct community engagement at some level could have been advantageous. It should be acknowledged as part of the ongoing implementation of the strategy that public land management practitioners, in particular LGAs, will continue to closely engage with the public.
3. LGAs as key stakeholders and other major land management organisations (e.g., water authorities and public land managers) were specifically targeted as key endorsers as a subset of the wider consultation undertaken. Suggestions by some academic commentators [51,52] that the views of other key stakeholders (e.g., transport, health) were not sought is erroneous.
4. The rationale for the targets proposed in *Living Melbourne* is articulated above. An alternative target strategy would have been for all regions to reach 30% canopy by 2050 rather than the range of targets that were offered. This would have lent an increased element of regional equity for residents to benefit from ambitious canopy targets. It would also have required a much more aggressive canopy enhancement approach for some regions each decade. Most notably the Western Region, and to a lesser extent the Northern and Inner Regions, would have had to have increased canopy beyond the 5 percent increase per decade. It is uncertain whether this would be achievable.
5. Endorsing authorities were asked to support *Living Melbourne's* vision, goals and actions and commit to work in partnership with the other endorsing organisations towards its implementation. More could have been requested of endorsing organisations, such as a specific commitment of financial and other resources.

5.3. *Implementation Post Strategy Development*

In 2020, Resilient Melbourne's remit changed from metropolitan Melbourne to the City of Melbourne and has transitioned into the Climate Change and City Resilience Branch. Given this change, one of the key challenges for *Living Melbourne* will be to secure a new and innovative governance model to drive implementation of the strategy. To this end, a new governance model to implement the strategy is currently evolving which seeks to include and empower critical endorsing organisations and a broader stakeholder group.

In 2019, a discussion paper was developed by The Nature Conservancy in collaboration with Resilient Melbourne and the Victorian Government to provide a targeted evaluation of ways to fund urban forest investments in *Living Melbourne*. The paper is one evidence base for structuring funding and financing of *Living Melbourne* urban forest investments with investors.

At the time of the launch of *Living Melbourne*, Victoria's Department of Environment, Land, Water and Planning recognised, through *Plan Melbourne 2017–2050* [17] the need to create urban forests throughout metropolitan Melbourne by preparing new guidelines and regulations that support greening new subdivisions and developments via landscaping, green roofs, and increase the percentage of permeable site areas in developments.

In February 2020, following a stakeholder workshop in September 2019 and further feedback and consultation, 26 projects were agreed to take *Living Melbourne* forward over a three year period. As of July 2021, besides the two projects mentioned above, three additional projects have been completed and a further seven projects are being progressed, led by endorsing organisations or implementation partners and assisted by a range of stakeholders. The three projects completed under the umbrella of *Living Melbourne* include: (1) *Greener Spaces Better Places* led the development of the Urban Greeners' Resource Hub, a curated on-line collection of best practice tools, guides, resources and case studies to help urban greening professionals protect and enhance Australia's urban forests and green cover in towns and cities; (2) Living Melbourne, in association with Resilient Melbourne, provided a written submission to the Victorian Planning Authority regarding the alignment between the draft *Guidelines for Precinct Structure Planning in Melbourne's Greenfields* [56] and the vision, goals and action *Living Melbourne*; and (3) Melbourne Water completed modelling of Melbourne's greening and cooling irrigation demand out to 2070 under different climate scenarios.

6. Conclusions

Urban forests help mitigate the damage caused by several types of acute shocks and chronic stresses. Protecting and improving urban forests is an opportunity to unlock the economic, health and social dividends that strengthen the ability for cities to thrive. The development of a metropolitan-wide urban forest strategy for Melbourne was the first for Australia and one of the first at this scale globally. It is unlike many urban forest plans in that it is an enabling strategy rather than a strategy focussed on directing geospatial urban greening activity. *Living Melbourne* provides some of the evidence that could influence on-ground activity, for example urban heat island data modelled against canopy cover and vulnerable communities. However, it focuses on both identifying and recommending action to a range of themes that would significantly improve conditions for urban greening. This includes elements such as removing or amending regulatory barriers, increasing and consolidating existing tools and resources and, enhancing existing positive and successful initiatives. Being one of the first of its kind, it should be considered both a first step in expanding metropolitan Melbourne's urban forest, as well as playing an important role in encouraging the take up of urban forestry practice in many cities around the world. The lessons here (both key areas of success and challenges experienced) add to the growing body of literature on the characteristics and design of urban forest strategies and city resilience more broadly (e.g., [12,57,58]).

Author Contributions: Conceptualization, M.H., J.F., M.G., T.K.; methodology, M.H., J.F., M.G., T.K.; writing-original draft preparation, M.H., J.F., M.G., T.K.; writing-review and editing, M.H., J.F., M.G., T.K. All authors have read and agreed to the published version of the manuscript.

Funding: This research received no external funding.

Data Availability Statement: The new data were created in this study are available on request.

Acknowledgments: *Living Melbourne* was developed in collaboration with 32 local government authorities, Victorian State Government departments and agencies, non-government not-for-profit and private sector organisations and industry groups. Their buy-in and willingness to collectively co-design the strategy was critical to its development. Amy Hahs and Morphum Environmental Pty Ltd. undertook the majority of the canopy mapping. Hahs was also responsible for the detailed statistical analysis of the vegetation model against various climatic and socio-demographic variables and the detailed statistical analysis of the urban heat island mapping. We thank 100RC colleagues Sam Kernaghan and Henri Blas and The Nature Conservancy's Rob McDonald, Rich Gilmore, Bob

Moseley, Nate Peterson, Tim Boucher and Rebecca Keen for their contributions to *Living Melbourne*. Comments on earlier drafts from Hugh Possingham, Megan Good, Justine Hausheer and three anonymous referees improved the manuscript.

Conflicts of Interest: The authors declare no conflict of interest.

References

1. Oke, C.; Bekessy, S.A.; Frantzeskaki, N.; Bush, J.; Fitzsimons, J.A.; Garrard, G.E.; Grenfell, M.; Harrison, L.; Hartigan, M.; Callow, D.; et al. Cities should respond to the biodiversity extinction crisis. *NPJ Urban Sustain.* **2021**, *1*, 11. [CrossRef]
2. McDonald, R.; Kroeger, T.; Boucher, T.; Longzhu, W.; Salem, R.; Adams, J.; Bassett, S.; Edgecomb, M.; Garg, S. *Planting Healthy Air: A Global Analysis of the Role of Urban Trees in Addressing Particulate Matter Pollution and Extreme Heat*; The Nature Conservancy: Arlington, VA, USA, 2016.
3. Threlfall, C.G.; Mata, L.; Mackie, J.A.; Hahs, A.K.; Stork, N.E.; Williams, N.S.; Livesley, S.J. Increasing biodiversity in urban green spaces through simple vegetation intervention. *J. Appl. Ecol.* **2017**, *54*, 1874–1883. [CrossRef]
4. Shanahan, D.F.; Fuller, R.A.; Bush, R.; Lin, B.B.; Gaston, K.J. The health benefits of urban nature: How much do we need? *BioScience* **2015**, *65*, 476–485. [CrossRef]
5. Díaz, S.; Pascual, U.; Stenseke, M.; Martín-López, B.; Watson, R.T.; Molnár, Z.; Hill, R.; Chan, K.M.A.; Baste, I.A.; Brauman, K.A.; et al. Assessing nature's contributions to people. *Science* **2018**, *359*, 270–272. [CrossRef]
6. Mumaw, L.; Bekessy, S. Wildlife gardening for collaborative public–private biodiversity conservation. *Austral. J. Environ. Manag.* **2017**, *24*, 242–260. [CrossRef]
7. United Nations. *Revision of the World Urbanization Prospects: 2018*; United Nations Department of Economic and Social Affairs: New York, NY, USA, 2018.
8. Jim, C.Y.; Konijnendijk van den Bosch, C.; Chen, W.Y. Acute challenges and solutions for urban forestry in compact and densifying cities. *J. Urban Plann. Dev.* **2018**, *144*, 04018025. [CrossRef]
9. Haaland, C.; Konijnendijkvan den Bosch, C. Challenges and strategies for urban green-space planning in cities undergoing densification: A review. *Urban For. Urban Green.* **2015**, *14*, 760–771. [CrossRef]
10. Saunders, A.; Duncan, J.; Hurley, J.; Amati, M.; Caccetta, P.; Chia, J.; Boruff, B. Leaf my neighbourhood alone! Predicting the influence of densification on residential tree canopy cover in Perth. *Landsc. Urban Plan.* **2020**, *199*, 103804. [CrossRef]
11. Ordóñez, C.; Threlfall, C.G.; Livesley, S.J.; Kendal, D.; Fuller, R.A.; Davern, M.; van der Ree, R.; Hochuli, D.F. Decision-making of municipal urban forest managers through the lens of governance. *Environ. Sci. Pol.* **2020**, *104*, 136–147. [CrossRef]
12. Ordóñez, C.; Duinker, P. An analysis of urban forest management plans in Canada; Implications for urban forest management. *Landsc. Urban Plan.* **2013**, *116*, 36–47. [CrossRef]
13. Hinch, R.; Maxwell, E.; Chen, N. *A Summary of Current Urban Forest Plans and Metrics*; The Nature Conservancy: Arlington, VA, USA, 2018.
14. ABS. *3105.0.65.001-Australian Historical Population Statistics, 2016*; Australian Bureau of Statistics, Commonwealth of Australia: Canberra, Australia, 2019. Available online: https://www.abs.gov.au/AUSSTATS/abs@.nsf/mf/3105.0.65.001 (accessed on 31 October 2020).
15. Phelan, K.; Hurley, J.; Bush, J. Land-use planning's role in urban forest strategies: Recent local government approaches in Australia. *Urban Pol. Res.* **2019**, *37*, 215–226. [CrossRef]
16. Infrastructure Victoria. *Growing Victoria's Potential: The Opportunities and Challenges of Victoria's Population Growth*; Infrastructure Victoria: Melbourne, Australia, 2019.
17. DELWP. *Plan Melbourne 2017–2050*; Victorian State Government: Melbourne, Australia, 2017.
18. Heneghan, L.; Mulvaney, C.; Ross, K.; Umek, L.; Watkins, C.; Westphal, L.M.; Wise, D.H. Lessons learned from Chicago Wilderness—Implementing and sustaining conservation management in an urban setting. *Diversity* **2012**, *4*, 74–93. [CrossRef]
19. VEAC. *Metropolitan Melbourne Investigation Final Report*; Victorian Environmental Assessment Council: Melbourne, Australia, 2011.
20. Wahlquist, C. Melbourne 'World's Most Liveable City' for Seventh Year Running. *The Guardian*, 6 August 2017. Available online: https://www.theguardian.com/australia-news/2017/aug/16/melbourne-worlds-most-liveable-city-for-seventh-year-running (accessed on 31 October 2020).
21. Destination Melbourne. *Greater Melbourne's Destination Management Visitor Plan: Executive Summary*; Destination Melbourne: Melbourne, Australia, 2018.
22. Hahs, A.; McDonnell, M.; Holland, K.; Caryl, F. *Biodiversity of Metropolitan Melbourne*; Prepared for Victorian Environmental Assessment Council by Australian Research Centre for Urban Ecology; Royal Botanic Gardens: Melbourne, Australia, 2009. Available online: https://veac.vic.gov.au/investigations-assessments/previous-investigations/document/getDownload?fid=MjAx (accessed on 31 October 2020).
23. Moore, G.M. Taking It to the Streets: Celebrating a Twenty Year History of Treenet—Responding to the Urban Forest Challenge. In Proceedings of the Treenet: 18th National Street Tree Symposium, Adelaide, Australia, 7–8 September 2017. Available online: https://treenet.org/wp-content/uploads/2017/10/Taking-it-to-the-Streets-Responding-to-the-Urban-Forest-Challenge_GM-Moore.pdf (accessed on 31 October 2020).
24. Hurley, J.; Saunders, A.; Both, A.; Sun, C.; Boruff, B.; Duncan, J.; Amati, M.; Caccetta, P.; Chia, J. *Urban Vegetation Cover Change in Melbourne 2014–2018*; Centre for Urban Research, RMIT University: Melbourne, Australia, 2019.

25. Municipal Association of Victoria. 2019. Available online: http://www.mav.asn.au/ (accessed on 11 December 2019).
26. DELWP. *Victoria's Climate Change Adaptation Plan 2017–2020*; Victorian Government: Melbourne, Australia, 2016.
27. DELWP. *Protecting Victoria's Environment: Biodiversity 2037*; Victorian State Government: Melbourne, Australia, 2017.
28. Victorian Government. *Victorian Public Health and Wellbeing Plan 2015–2019*; Victorian Government: Melbourne, Australia, 2015.
29. DELWP. *Melbourne Open Space Strategy: Ensuring Liveability for Future Generations. Draft for Comment September 2019*; Victoria State Government: Melbourne, Australia, 2019.
30. Greening the West. About Greening the West. Available online: https://greeningthewest.org.au/about/ (accessed on 12 December 2019).
31. Zebrowski, C. Acting local, thinking global: Globalizing resilience through 100 Resilient Cities. *New Perspect.* **2020**, *28*, 71–88. [CrossRef]
32. Resilient Melbourne. *Resilient Melbourne Strategy*; City of Melbourne: Melbourne, Australia, 2016.
33. McDonald, R.I.; Colbert, M.; Hamann, M.; Simkin, R.; Walsh, B. *Nature in the Urban Century: A Global Assessment of Where and How to Conserve Nature for Biodiversity and Human Wellbeing*; The Nature Conservancy: Arlington, VA, USA, 2018. Available online: https://www.nature.org/content/dam/tnc/nature/en/documents/TNC_NatureintheUrbanCentury_FullReport.pdf (accessed on 29 September 2020).
34. McDonald, R.I.; Mansur, A.V.; Ascensão, F.; Colbert, M.; Crossman, K.; Elmqvist, T.; Gonzalez, A.; Güneralp, B.; Haase, D.; Hamann, M.; et al. Research gaps in knowledge of the impact of urban growth on biodiversity. *Nat. Sustain.* **2020**, *3*, 16–24. [CrossRef]
35. The Nature Conservancy and Resilient Melbourne. *Living Melbourne: Our Metropolitan Urban Forest*; The Nature Conservancy and Resilient Melbourne: Melbourne, Australia, 2019.
36. Yan, J.; Zhou, W.; Han, L.; Qian, Y. Mapping vegetation functional types in urban areas with WorldView-2 imagery: Integrating object-based classification with phenology. *Urban For. Urban Green.* **2018**, *31*, 230–240. [CrossRef]
37. City of Melbourne. *Urban Forest Strategy: Making a Great City Greener: 2012–2032*; City of Melbourne: Melbourne, Australia, 2012.
38. The Nature Conservancy and Resilient Melbourne. *Living Melbourne: Our Metropolitan Urban Forest Technical Report*; The Nature Conservancy and Resilient Melbourne: Melbourne, Australia, 2019.
39. Crooks, K.; Sanjayan, M. *Connectivity Conservation*; Cambridge University Press: New York, NY, USA, 2006.
40. Palmer, G.C.; Fitzsimons, J.A.; Antos, M.J.; White, J.G. Determinants of native avian richness in suburban remnant vegetation: Implications for conservation planning. *Biol. Conserv.* **2008**, *141*, 2329–2341. [CrossRef]
41. White, J.G.; Antos, M.J.; Fitzsimons, J.A.; Palmer, G.C. Non-uniform bird assemblages in urban environments: The influence of streetscape vegetation. *Lands. Urban Plan.* **2005**, *71*, 123–135. [CrossRef]
42. Callaghan, C.T.; Major, R.E.; Lyons, M.B.; Martin, J.M.; Kingsford, R.T. The effects of local and landscape habitat attributes on bird diversity in urban greenspaces. *Ecosphere* **2018**, *9*, e02347. [CrossRef]
43. Champness, B.S.; Palmer, G.C.; Fitzsimons, J.A. Bringing the city to the country: Relationships between streetscape vegetation type and bird assemblages in a major regional centre. *J. Urban Ecol.* **2019**, *5*, juz018. [CrossRef]
44. Fraixedas, S.; Lindén, A.; Piha, M.; Cabeza, M.; Gregory, R.; Lehikoinen, A. A state-of-the-art review on birds as indicators of biodiversity: Advances, challenges, and future directions. *Ecol. Indic.* **2020**, *118*, 106728. [CrossRef]
45. Fitzsimons, J.A.; Antos, M.J.; Palmer, G.C. When more is less: Urban remnants support high bird abundance but diversity varies. *Pac. Conserv. Biol.* **2011**, *17*, 97–109. [CrossRef]
46. Coutts, A.M.; Tapper, N.J.; Beringer, J.; Loughnan, M.; Demuzere, M. Watering our cities: The capacity for water sensitive urban design to support urban cooling and improve human thermal comfort in the Australian context. *Prog. Phys. Geogr. Earth Environ.* **2012**, *37*, 2–28. [CrossRef]
47. Tapper, N.; Loughnan, M.; Nicholls, N. Towards management of urban heat stress in urban environments: Recent developments in Melbourne, Victoria. *Emerg. Med. Australas.* **2010**, *22*, A11–A12.
48. DELWP. *Melbourne Vegetation and Land Use Cover 2014*; Department of Environment, Land, Water and Planning: Melbourne, Australia, 2018. Available online: https://services.land.vic.gov.au/SpatialDatamart/index.jsp (accessed on 12 December 2020).
49. Leahy, I. Why We No Longer Recommend a 40 Percent Urban Tree Canopy Goal. *Loose Leaf: The Official Blog of American Forests*. 2017. Available online: https://www.americanforests.org/blog/no-longer-recommend-40-percent-urban-tree-canopy-goal/ (accessed on 26 September 2018).
50. Coenen, L.; Davidson, K.; Frantzeskaki, N.; Grenfell, M.; Håkansson, I.; Hartigan, M. Metropolitan governance in action? Learning from metropolitan Melbourne's urban forest strategy. *Aust. Plan.* **2020**, *56*, 144–148. [CrossRef]
51. Bush, J.; Coffey, B.; Fastenrath, S. Governing urban greening at a metropolitan scale: An analysis of the *Living Melbourne* strategy. *Aust. Plan.* **2020**, *56*, 95–102. [CrossRef]
52. Fastenrath, S.; Bush, J.; Coenen, L. Scaling-up nature-based solutions. Lessons from the Living Melbourne strategy. *Geoforum* **2020**, *116*, 63–72. [CrossRef]
53. Dobbs, C.; Kendal, D.; Nitschke, C. The effects of land tenure and land use on the urban forest structure and composition of Melbourne. *Urban For. Urban Green.* **2013**, *12*, 417–425. [CrossRef]
54. Gulsrud, N.M., Hertzog, K.; Shears, I. Innovative urban forestry governance in Melbourne?: Investigating "green placemaking" as a nature-based solution. *Environ. Res.* **2018**, *161*, 158–167. [CrossRef]
55. Melbourne Water. *Healthy Waterways Strategy 2018–2028*; Melbourne Water: Melbourne, Australia, 2018.

56. VPA. *Guidelines for Precinct Structure Planning in Melbourne's Greenfields: Draft for Public Engagement September 2020*; Victorian Planning Authority: Melbourne, Australia, 2020.
57. Ostoić, S.K.; Salbitano, F.; Borelli, S.; Verlič, A. Urban forest research in the Mediterranean: A systematic review. *Urban Forest. Urban Green.* **2018**, *31*, 185–196. [CrossRef]
58. Fastenrath, S.; Coenen, L.; Davidson, K. Urban resilience in action: The Resilient Melbourne Strategy as transformative urban innovation policy? *Sustainability* **2019**, *11*, 693. [CrossRef]

land

MDPI

Article

Urban Green Spaces Restoration Using Native Forbs, Site Preparation and Soil Amendments—A Case Study

Jaime Aguilar Rojas, Amalesh Dhar and M. Anne Naeth *

Department of Renewable Resources, University of Alberta, 751 General Services Building, Edmonton, AB T6G 2H1, Canada; aguilarr@ualberta.ca (J.A.R.); amalesh@ualberta.ca (A.D.)
* Correspondence: anne.naeth@ualberta.ca; Tel.: +1-780-492-9539

Abstract: Restoration of urban green spaces with native flora is especially important for promoting various ecosystem services. Although there have been years of research on land reclamation, ecological restoration and plant establishment, there is a lack of knowledge on how to reintegrate the native ecological component, specifically forb species in urban green spaces. We evaluated the restoration potential of 24 native forbs using different site preparation (herbicide, tillage, herbicide with tillage and control) and soil amendment (100% compost, 50% compost with 50% topsoil, 20% compost with 80% topsoil and control) treatments in a recreational park in Edmonton, Alberta, Canada. Soil texture and nutrients generally increased with increased compost application rate; some declined within a year, others increased. Based on survival and growth analysis, the forb species with the highest potential for use in urban green spaces were *Penstemon procerus*, *Fragaria virginiana*, *Heuchera cylindrica*, *Agastache foeniculum*, *Antennaria microphylla*, *Mentha arvensis* and *Geum aleppicum*. Native forb species response was more prominent with soil amendment than site preparation. Treatments with greater amounts of compost had greater survival, growth, species richness, cover and noxious weed cover than control treatments. This study suggests amendment of soil with compost can positively influence forb species restoration in urban green spaces; under some conditions site preparation may be required.

Keywords: compost; ecological restoration; herbicide; plant community; tillage; urban ecology

1. Introduction

In the past, urban sustainability efforts mostly focused on engineered buildings, road networks and parks [1], while only modest attention was given to the green spaces that intermingle with urban structures [2,3]. Building a sustainable society in urban areas with appropriate management of green spaces (gardens and parks) is necessary [4–6] as they provide various environmental, economic and quality-of-life benefits [7]. Environmental benefits include increased biodiversity and wildlife use, soil stabilization, improved ground water recharge, windbreaks for snow capture and dust reduction, reduction in atmospheric greenhouse gases and cleaner air [3,5,8,9]. Economic benefits include significant reductions in maintenance costs such as mowing, irrigation and herbicide use. Quality-of-life benefits include landscape beautification, increased green and shady areas for recreation, increased community awareness of environmental issues and noise reduction by mature plantings [10,11].

To beautify urban gardens and parks, non-native garden flora is frequently planted, the most common pathway for alien species introductions worldwide [12,13]. In Europe, over 80,000 plant taxa are found in botanical gardens; 783 of these are alien species that have been introduced from other parts of the world and can be found in city parks and recreation areas [14]. Many of these invaders can easily escape and establish outside of their planted areas without human assistance and become problematic for native biodiversity [12,15]. Eradication of these alien species is difficult and expensive, thus preventing them by planting native species of regional provenance in urban areas may be a good management option. A successful native species restoration strategy in urban green areas can significantly reduce city management costs, promote preservation of local species, restore environmental services and encourage more community members to embrace native species as a desirable strategy to follow [3,16]. Therefore, cities such as Edmonton proposed to transform urban habitats into habitats suitable for native plants found in the area [17].

Restoration with native species can prevent new alien species invasions, reduce soil compaction and increase soil organic matter and microbial activity [5,18,19]. Compacted soils can restrict root growth, which can limit successful plant establishment and long-term development [20]. Restoration with native species can reduce soil compaction through root expansion, increased biological activity and frost heave, consequently increasing infiltration and percolation [16,21]. Naturalized sites retain leaf litter and woody debris, which decompose, adding organic material which can increase plant available soil water [22]. Planting native forbs (wildflowers) in addition to trees and shrubs is a relatively new approach in landscape architecture that is gaining momentum among urban planners and landscapers and is recommended in many studies [5,23–25]. Adding native forbs for restoration of urban green spaces promotes native biodiversity and creates attractive flowering vegetation for recreational enjoyment and education [25]. Although there are considerable possibilities for native forbs to be used in urban green space restoration, scientific research on methods for using native forbs is scarce. The huge variety of forbs complicates their use due to lack of knowledge about them as individual species in urban green space restoration. Current species selection is usually based on visual appearance and plant material availability. However, successful restoration requires use of plants that are competitive, hardy and resilient in a highly competitive urban area with non-native species that are often present in urban green spaces [17]. Native forb response to urban conditions and best introduction techniques thus need to be better understood. The objective of our study was to assess the effects of site preparation and soil amendment on the survival and growth of 24 native forb species and on plant community development. The outcome of this study helps us to predict which combinations of plant species, soil preparation techniques and amendments have the greatest potential for urban green space restoration and provides the ground for further detailed study in urban restoration and green space management.

2. Materials and Methods

2.1. Study Area and Experimental Design

The study was conducted in a prominent recreational park in the City of Edmonton, Alberta, Canada (53°34′19″ N and 113°31′10″ W). Mean annual temperature is 4.2 °C, mean growing season temperature from May to October is 13.0 °C and winter temperature from November to April is −4.6 °C. Mean annual precipitation is 348 mm, with 284 mm of rain from June to October [26]. The area is flat with a gentle slope to the southwest. Immediately surrounding the roundabout is asphalt, then buildings, small canopy trees and open lawn areas. Traffic conditions are moderately high for vehicles near the roundabout; pedestrian traffic is mostly concentrated on walking paths.

2.2. Experimental Design and Treatments

The experiment used a complete randomized design with four replicates 50 m from each other. Each experimental plot was 10 m × 10 m, divided into sixteen 2.5 m × 2.5 m subplots, covering an area of 6.25 m^2 each (Figure A1). Site preparation treatments were randomly assigned vertically, and amendment treatments were applied randomly horizontally to the experimental plot (10 m × 10 m). Thus, there were 4 site preparation treatments × 4 amendment treatments × 4 replicates for a total of 64 plots. Plots were approximately 50 m from any roads and 10 m from all walking paths to reduce the traffic effect. There were 30 cm buffer zones between the subplots to reduce the potential neighbor effects.

Four site preparations and four soil amendment treatments were applied in the study area. Four site preparation techniques were soil tillage, foliar herbicide application, tillage plus herbicide and no site preparation (control) to remove existing vegetation, which consisted of lawn grass and some common annual weeds. Soil was tilled in June to a 15 cm depth with a rear tined, hydraulic drive, rototiller; first in one direction, then crossed perpendicularly. Glyphosate foliar herbicide Transorb™ was applied as a 1% solution (540 g/L glyphosate) 2 weeks prior to site tillage. Glyphosate has been predominantly used for controlling weeds in North America due to its effectiveness, non-selective nature, little or no soil residue and relatively low cost. Therefore, to control the competitive weeds, the practice of herbicide use prior to revegetation with native species is common in North American urban areas for reducing competition, although its use was questioned by many international agencies due to its toxicity and environmental safety. Some alternatives to glyphosate such as other chemicals Diquat (Reward™), pelargonic acid (Scythe™), glufosinate (Finale™); manual removal, fire, steam, hot foam and weeding were recommended for different jurisdictions and countries.

Four soil amendment treatments were 100% compost, 50% topsoil with 50% compost, 80% topsoil with 20% compost and a control (0% compost with 0% topsoil). Compost was 20% wood chips and 80% compost by volume, a standard mix used by the City of Edmonton. Topsoil was Ah horizon from development on previous agricultural land and clay loam to clay to silty clay loam in texture. Topsoil and compost were mixed in their treatment proportions, then applied using a mini steer loader. Amendment mixes were added to the surface of each subplot and spread by hand with shovels to a 15 cm deep layer.

2.3. Planting and Plot Management

Twenty-four native forb species from 12 families were selected for urban green space restoration recommended by the City of Edmonton. Forbs species were small with a shallow root system and selection was based on the visual appearance (flower color, shape and longevity), availability, geographic distribution (species that are adapted within the same geographic location) and growing conditions (water stress tolerant, frequent disturbance tolerant and ability to grow in a wide range of soil types) [17] (Table 1). All planting stock was procured from the City of Edmonton nursery and planted on July 8 and 9. In each treatment unit (subplot), one plant of each of the 24 forbs was planted with equal spacing. In total, 1536 plants (4 site preparation × 4 amendment × 4 replicates × 24 plants) were planted in the study area. Plants were watered 24 to 48 h after planting; then every 2 to 3 days for the next two weeks, twice per week for the next four weeks, then once per week until the end of the growing season. Manual weeding was conducted within 2 m from the edge of research plots as a weed control buffer zone.

2.4. Vegetation Assessments

Plant survival assessments were conducted in August and October of 2014, and June and August of 2015. Live and dead planted forbs were counted. In June and August 2015, planted forb-species spread was measured for each seedling. Diameter of forbs from tip to tip was determined with a tape measure. For species with cluster growth habits, the tape was placed on the farthest tip of one individual then pulled to the tip of the farthest individual of the cluster. Forbs were considered clusters when several of the same species were fewer than 5 cm apart with no vegetation between them. Other than planted forbs, vegetation cover was assessed in August 2014 and 2015, in three randomly located 1 m × 0.1 m quadrats inside each treatment. In total, 192 quadrats (4 site preparation × 4 amendment × 4 replication × 3 quadrat) were established and ocularly assessed for percent of live vegetation, bare ground, litter and other (rocks, trash and feces) cover. Total number of sample plots was considered adequate as species numbers reached a plateau for all treatment plots (Figure A2). Live vegetation was assessed on an individual species basis for both planted and naturally occurring species. Plant identification and nomenclature followed Moss [27].

Table 1. Planted native forb species.

Common Name	Scientific Name	Family
Black-eyed Susan	*Rudbeckia hirta* L.	Asteraceae
Dotted blazing star	*Liatris ligulistylis* A. Nels. K. Schum.	Asteraceae
Hairy false golden aster	*Heterotheca villosa* Pursh Shinners	Asteraceae
Little-leaf pussytoes	*Antennaria microphylla* Rydb.	Asteraceae
Prairie sagewort	*Artemisia frigida* Willd	Asteraceae
White prairie aster	*Symphyotrichum falcatum* Lindl. G.L. Nesom	Asteraceae
Harebell	*Campanula rotundifolia* L.	Campanulaceae
Bunchberry	*Cornus canadensis* L.	Cornaceae
Giant hyssop	*Agastache foeniculum* Pursh ktze.	Lamiaceae
Wild mint	*Mentha arvensis* L.	Lamiaceae
Prairie onion	*Allium textile* A. Nels. and J. F. Macbr.	Liliaceae
Yellow buckwheat	*Eriogonum flavum* Nutt.	Polygonaceae
Canada anemone	*Anemone canadensis* L.	Ranunculaceae
Long-fruited anemone	*Anemone cylindrica* Gray	Ranunculaceae
Prairie crocus	*Pulsatilla patens* L.	Ranunculaceae
Tall larkspur	*Delphinium elatum* L.	Ranunculaceae
Veiny meadow	*Thalictrum venulosum* Trel.	Ranunculaceae
Prairie cinquefoil	*Potentilla arguta* Pursh	Rosaceae
Three-flowered avens	*Geum aleppicum* Jacq.	Rosaceae
Wild strawberry	*Fragaria virginiana* Dcne.	Rosaceae
Northern bedstraw	*Galium boreale* L.	Rubiaceae
Round-leaved alumroot	*Heuchera cylindrica* Douglas ex Hook.	Saxifragaceae
Slender penstemon	*Penstemon procerus* Dougl. Ex Graham	Scrophulariaceae
Early blue violet	*Viola adunca* Sm.	Violaceae

2.5. Soils Sampling and Analyses

Soils were sampled in July of each study year from the plots to determine original soil conditions and changes with amendment treatments. One sample from each amended treatment in each plot was collected using 15 cm augers (total 16 soil samples). Collected samples were stored in ziploc plastic bags and sent to a commercial laboratory for analysis. Chloride in saturated paste was determined colorimetrically by auto-analyzer [28]. Inorganic and organic carbon were determined by carbon dioxide loss [29] and total carbon by combustion methods [30]. Cation exchange capacity was determined through ammonium acetate extraction [31]; ammonium by potassium chloride extraction; nitrate nitrogen colorimetrically in calcium chloride solution [32]; total nitrogen by combustion [33]; available phosphorus and potassium by modified Kelowna process [34]; sodium adsorption ratio, calcium, magnesium, sodium, potassium and sulfate in saturated paste by inductively coupled plasma; electrical conductivity and pH by meters [35]. Soil particles (sand, silt and clay) were determined by pipette method after removal of organic matter and carbonate [36].

2.6. Statistical Analyses

All statistical analyses were conducted using R version 4.0.3 [37] and significance level for analysis was $\alpha = 0.05$. In most cases, data from the last monitoring date of year 2 were statistically analyzed to evaluate overall performance of species at the end of the experiment. Chi-square analysis was used to identify effects of site preparation and soil amendment treatments on species survival. In a classical ecological experiment, replication of the treatments is prerequisite to test the hypothesis [38]. According to Oksanen [39] experiments, unreplicated or low-replicated treatments may also be the only or best option when (i) gross effects of treatments are anticipated, (ii) the experiment is conducted appropriately at large scales, (iii) only a rough estimate of effect is required and (iv) if the cost of replication is high. We conducted a study with low replication for individual species as the goal was to determine a rough estimate of effect for developing a foundation for future in-depth work, while minimizing the cost and labor requirements. Due to small numbers per species, statistical analysis was conducted on species grouped by family. Chi-square criteria were applied to groups and analyses were conducted only if assumptions were met (<20% of expected frequencies <5). Soil preparation and amendment effects were analyzed per species with one-way analysis of variance (ANOVA). Shapiro-Wilk test was used to determine normality of distribution and Levene's test for homogeneity of variance assessments. For significant factors, an HSD Tukey's test was applied for pairwise comparison. All statistical analyses were conducted using package 'stats' version 4.2.0 [37].

3. Results

3.1. Forb Survival Response to Treatments

Regardless of site preparation and soil treatment, top surviving and performing forbs species were *Penstemon procerus* (96.9%), *Fragaria virginiana* (95.3%), *Agastache foeniculum* (92.2%), *Antennaria microphylla* (92.1%), *Heuchera cylindrica* (89.1%), *Geum aleppicum* (89.0%) and *Mentha arvenses* (89.6%) at the end of the two-year experiment (Figure 1a,b). Survival was generally high at the first monitoring in August of year one then decreased with time, with fewer than 35% of the plants surviving by the end of the experiment for *Cornus canadensis* (0%), *Anemone cylindrica* (20.3%), *Pulsatilla patens* (21.8%), *Eriogonum flavum* (23.4%), *Allium textile* (29.7%), *Viola adunca* (34.4%) and *Liatris ligulistylis* (35.9%) (Figure 1a,b). *Cornus canadensis* was the only species that did not survive by the end of year two.

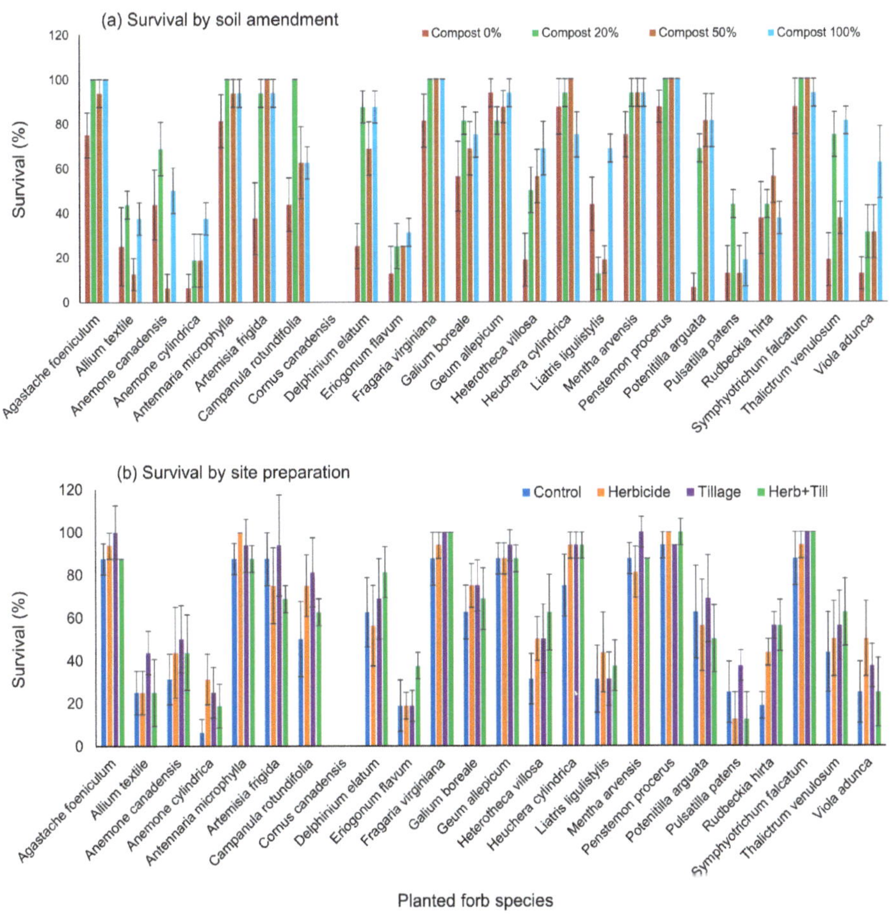

Figure 1. Mean (±SE) survival percent of planted forb by species at the end of monitoring dates in relation to (**a**) soil amendment and (**b**) site preparation. Herb + Till = Herbicide with tillage.

When species were analyzed grouped by family, a significant effect of soil amendment treatment on survival was found for Asteraceae, Ranunculaceae out of 12 families (Figure 2). Forb survival was significantly the lowest in compost 0% (unamended) for Asteraceae and the greatest in compost 100% (Figure 2). For Ranunculaceae, survival was significantly lower in compost 0% and compost 20%. Site preparation and interactions with amendment treatments did not significantly affect family survival.

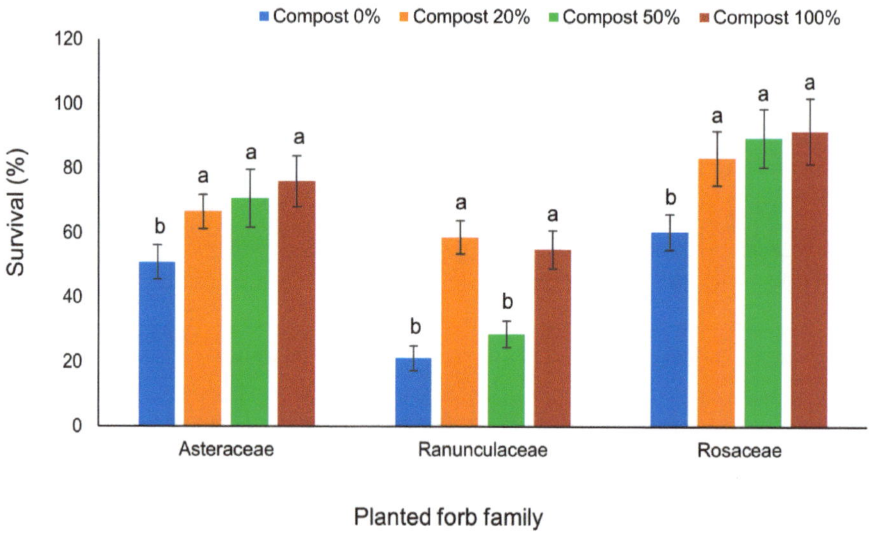

Figure 2. Mean (±SE) survival percent grouped by family and soil amendment. Different letters within species indicate significance differences at $\alpha = 0.05$.

3.2. Spread of Planted Forb Species

Spread of *Thalictrum venulosum* responded significantly ($p = 0.008$) to site preparation treatment; rate of spread was significantly higher with herbicide alone (25 cm) than herbicide–tillage together (14.8 cm) and tillage alone (12.9 cm), and statistically similar to untreated (19.4 cm) (data not shown). Soil amendment had a significant effect on spread for 9 of the 24 evaluated forb species (Table 2). *Fragaria virginiana, Penstemon procerus, Delphinium elatum, Symphyotrichum falcatum, Heuchera cylindrical, Antennaria microphylla, Rudbeckia hirta, Geum aleppicum* and *Mentha arvensis* had significantly greater spread in compost 100% than with no compost and more variable responses with the other two compost treatments (Table 2). The rate of spread was 6 to 128 cm across compost treatments, whereas in no compost 15 species had <10 cm spread (Table 2).

Table 2. Mean (±SE) spread (cm) by planted forb species in response to soil amendment treatments in year 2. Different letters within rows denote significant differences among treatments for species at α = 0.05.

Species	Compost 0%	Compost 20%	Compost 50%	Compost 100%
Agastache foeniculum	11.4 (1.2)	72.0 (4.3)	46.5 (3.4)	68.8 (7.2)
Allium textile	6.5 (0.9)	9.6 (1.9)	7.5 (0.5)	8.2 (2.3)
Anemone canadensis	8.7 (1.0)	16.2 (1.1)	17.0 (NA)	24.3 (1.4)
Anemone cylindrica	5.0 (NA)	11.7 (2.5)	9.7 (1.9)	18.7 (1.8)
Antennaria microphylla	20.7 (1.8) b	27.3 (2.3) ab	21.6 (1.1) b	30.7 (2.4) a
Artemisia frigida	16.2 (1.9)	128.0 (3.9)	118.7 (6.7)	99.2 (10.8)
Campanula rotundifolia	7.4 (0.7)	25.4 (3.7)	15.8 (1.7)	26.4 (2.9)
Delphinium elatum	5.3 (0.3) b	16.4 (1.8) a	16.4 (1.1) a	21.0 (1.9) a
Eriogonum flavum	8.5 (0.9)	14.0 (1.2)	9.0 (0.2)	12.4 (0.6)
Fragaria virginiana	9.5 (0.8) b	18.0 (1.7) a	16.6 (1.0) a	19.0 (1.2) a
Galium boreale	8.8 (1.1)	24.2 (2.6)	19.9 (1.0)	21.3 (2.6)
Geum aleppicum	12.1 (1.0) b	22.5 (1.7) a	18.1 (1.2) ab	21.0 (2.1) a
Heterotheca villosa	16.3 (2.1)	37.6 (5.5)	21.6 (3.3)	38.6 (4.9)
Heuchera cylindrica	9.4 (0.6) b	24.0 (1.2) a	19.8 (1.8) a	23.6 (1.6) a
Liatris ligulistylis	9.6 (1.0)	13.5 (0.5)	10.0 (1.6)	14.3 (0.9)
Mentha arvensis	9.7 (2.2) b	67.5 (10.3) a	39.1 (6.0) ab	52.2 (7.8) a
Penstemon procerus	18.9 (1.8) c	47.1 (2.7) a	33.3 (3.4) b	49.8 (2.5) a
Potentilla arguta	5.0 (NA)	25.7 (1.7)	22.6 (1.4)	30.2 (2.2)
Pulsatilla patens	2.5 (0.2)	8.1 (0.7)	6.0 (1.1)	10.0 (0.9)
Rudbeckia hirta	21.7 (2.7) b	39.1 (3.9) ab	38.4 (2.9) ab	46.8 (4.1) a
Symphyotrichum falcatum	21.8 (2.5) b	55.3 (4.6) a	60.1 (4.4) a	62.7 (5.6) a
Thalictrum venulosum	7.3 (1.3)	19.8 (2.2)	13.8 (2.9)	19.7 (1.3)
Viola adunca	7.5 (0.2)	11.2 (1.3)	10.6 (2.4)	12.6 (1.4)

3.3. Species Cover, Composition and Richness

Other than planted forbs, cover by plant categories followed similar trends for most soil preparation and amendment treatments, with a few exceptions (Figure 3). The untreated control, herbicide and tillage together and tillage only treatments had greater cover of native species, and the herbicide–tillage together treatment had greater bare ground than other treatments (Figure 3). Planted forb-species cover was significantly higher in compost treatments than in compost 0% at the end of year two.

A total of 28 plant species other than the planted forbs were identified across the plots (Table A1). There were 9 native, 15 non-native, 3 noxious (*Cirsium arvense* (L.) Scop. (Canada thistle), *Sonchus arvensis* L. (perennial sow thistle), *Tripleurospermum perforatum* (Mérat) M. Lainz (scentless chamomile)) and one prohibited noxious (*Potentilla recta* L. (sulphur cinquefoil)) species. Among the non-native species, *Festuca rubra* L. (creeping red fescue), *Polygonum convolvulus* L. (wild buckwheat) and *Taraxacum officinale* F.H. Wigg. (common dandelion) were the most common species. *Festuca rubra* and *Taraxacum officinale* were found on all site preparation treatments with compost 0% and *Polygonum convolvulus* was found on all site preparation treatments with compost 20%. The noxious species *Cirsium arvense* was found on almost 50% of the plots, being more frequent in the compost 100% treatment. Species richness excluding planted forbs differed with soil amendment but not site preparation treatments. Compost 0% had significantly greater overall species richness (R: 8.6; $p < 0.001$), native (R: 4; $p < 0.021$) and non-native (R: 3.5; $p < 0.045$) species richness than all soil amendments.

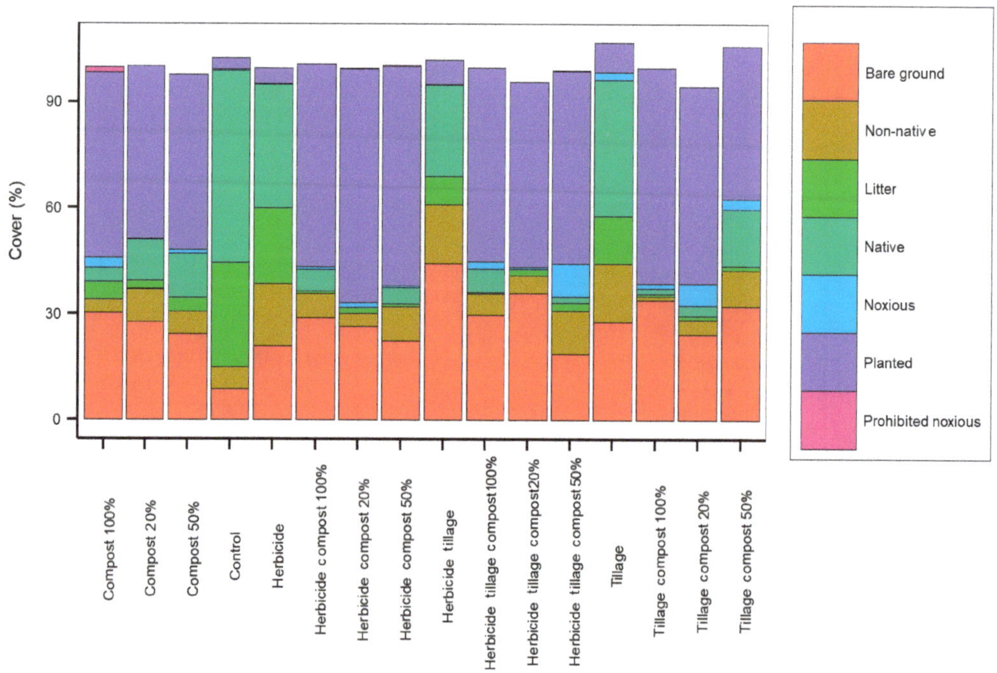

Figure 3. Percent cover by category for site preparation and soil amendment treatments.

3.4. Soil Response to Treatments

Most soil properties did not differ with year and soil amendment treatments. Soil nutrients generally increased with compost application: some declined slightly (total nitrogen, nitrate, total carbon, total organic carbon, ammonium, phosphorus, copper and zinc) and some increased slightly (sodium adsorption ratio, calcium, potassium, sodium and sulphate) within a year, being the highest and steadiest in both years with 100% compost (Table 3). Soil pH was acidic and increased with 100% compost (mean 5.7). Sodium adsorption ratio was very low across all amendment treatments (mean 0.5).

Table 3. Mean (±SE) soil properties by soil amendment treatments. Different letters indicate significant differences among amendment treatments in individual years at α = 0.05. EC = Electrical Conductivity, CEC = Cation Exchange Capacity, SAR = Sodium Adsorption Ratio and TOC = Total Organic Carbon.

Properties	Compost 0%		Compost 20%		Compost 50%		Compost 100%	
	Year 1	Year 2	Year 1	Year 2	Year 1	Year 2	Year 1	Year 2
pH	6.5 (0.1)	6.9 (0.1)	6.2 (0.2)	6.3 (0.1)	6.0 (0.1)	6.2 (0.1)	5.8 (0.1)	5.7 (0.1)
EC (dS m^{-1})	1.1 (0.1) c	1.2 (0.1) y	2.8 (0.3) b	2.2 (0.3) x	3.2 (0.3) a	2.0 (0.4) x	5.6 (0.6) a	2.1 (0.4) x
CEC (meq 100 g^{-1})	33.8 (1.5) b	35.7 (5.2) z	39.8 (4.0) b	41.5 (5.0) yz	43.1 (5.9) b	52.6 (4.7) y	61.6 (4.2) a	71.8 (5.2) x
SAR	0.7 (0.1)	1.0 (0.1)	0.4 (0.0)	0.5 (0.1)	0.4 (0.0)	0.5 (0.0)	0.3 (0.1)	0.3 (0.0)
Sodium (mg L^{-1})	26.4 (2.8)	56.4 (4.9) y	31.0 (4.8)	41.7 (6.2) xy	34.0 (6.2)	35.6 (8.5) x	28.3 (12)	22.5 (3.1) x
Total Carbon (%)	4.3 (0.4) c	3.4 (0.3) y	5.6 (1.7) c	5.1 (0.5) y	9.4 (1.6) b	6.9 (1.2) y	23.2.3 (2.5) a	20.86 (1.9) x
Total Nitrogen (%)	0.3 (0.0) c	0.3 (0.0) y	0.5 (0.1) c	0.5 (0.1) y	1.4 (0.1) b	0.6 (0.1) y	6.4 (0.8) a	1.5 (0.1) x
TOC (%)	4.3 (0.4) c	3.2 (0.3) y	5.2 (0.7) c	5.1 (0.5) y	8.3 (0.3) b	6.5 (1.2) y	22.5 (2.3) a	20.7 (1.4) x
Ammonium (mg L^{-1})	4.3 (1.8) c	2.3 (0.9) y	8.8 (3.5) bc	2.6 (0.3) y	17.4 (5.1) b	14.9 (4.6) x	70.4 (14.2) a	23.7 (6.4) x
Nitrate (mg L^{-1})	19.9 (3.3) b	2.9 (0.9) z	117.0 (23.2) a	56.4 (9.1) y	104.0 (15.8) a	55.1 (8.0) y	139.8 (22.3) a	91.6 (9.0) x
Phosphate (mg L^{-1})	29.9 (6.1) c	8.2 (1.8) z	251.0 (65.2) b	212.0 (41.2) y	470.2 (82.4)b	412.0 (79.5) y	2580.7 (179.3) a	1550.0 (154.2) x
Potassium (mg L^{-1})	176.5 (37.4) c	121.3 (23.8) z	231.0 (19.9) c	186.3 (14.1) y	330.7 (29.0) b	290.0 (62.3) y	1100.0 (58.6) a	1050.8 (100.8) x
Sulfate (mg L^{-1})	44.1 (2.2) c	82.1 (3.8) z	154.0 (23.6) b	129.4 (23.9) y	231.5 (32.4) ab	178.8 (31.1) y	372.7 (85.7) a	274.1 (61.1) x
Calcium (mg L^{-1})	111.4 (13.5) b	170.0 (15.8) y	340.1 (53.1) a	379.7 (48.2) x	396.8 (75.7) 4	328.0 (61.0) x	306.5 (45.4) a	299.8 (63.3) x
Chloride (mg L^{-1})	30.5 (3.8)	36.5 (4.3)	25.6 (5.0)	30.8 (5.7)	26.2 (7.7)	34.5 (4.9)	24.1 (6.5)	20.0 (3.1)
Copper (mg L^{-1})	16.6 (0.5) c	0.9 (0.1) z	44.0 (3.5) b	2.9 (0.4) y	55.5 (5.9) b	5.1 (1.9) y	308.0 (5.9) a	40.6 (2.7) x
Magnesium (mg L^{-1})	27.0 (4.2) c	39.5 (8.6) y	77.6 (15.1) b	89.7 (14.8) x	108.6 (21.4) a	91.5 (19.2) x	126.4 (19.4) a	114.0 (22.2) x

4. Discussion

Native forb species planted in green areas and exposed to urban disturbance and restoration treatments behaved quite differently. The limited impact of site preparation treatments in our study supports the results of Buonopane et al. [40] who found no differences in vegetation cover, germinant density or species richness between herbicide and non-herbicide plots in any group, including noxious weeds. Amendment with compost was a useful treatment for forb survival and spread in our study, similar to other studies that found a positive relationship between compost and forb survival [41,42]. Marrs and Gough [41] found floristic composition of wildflower meadows was controlled by soil fertility. Bretzel et al. [42] reported that the wildflower diversity index was related to cation exchange capacity and carbon–nitrogen ratio.

Native forbs used in our experiment were small with a shallow root system, and when planted in the upper 15 cm of soil that had been structurally altered and amended with compost, they had a new growing medium. Even small changes in nutrients in amended substrates may have impacted tiny plants at a vulnerable time when they needed nutrition. However, soil preparation and amendment application combinations were expected to influence soil water dynamics, indirectly determining stress and winterizing conditions. Site preparation techniques can alter soil water availability in the soil profile, and strategic plant treatments can increase revegetation success [43].

Although soil amendments resulted in a greater proportion of desired planted species cover, it exposed the site to invasion by non-native, noxious and prohibited noxious weed species. This finding is consistent with Skrindo and Pedersen [44], who found using topsoil as an amendment to restore a roadside in Norway increased vegetation cover from one year to the next for species such as *Cirsium arvense*. The loss of ecological memory in urban settings is thought to facilitate the establishment of alien or non-native invasive or weed species in recently disturbed urban environments, as these species have very high seed output, phenotypic and germination plasticity, adaptations for short- and long-distance dispersal, small seed size and high seed longevity [12,45]. Thus, these species are often difficult to control in newly naturalized landscapes, where they can quickly dominate and outcompete desired species [45]. Without management intervention such as native seeding, common seed bank species, especially exotic and noxious plants, may exclude or inhibit desirable later successional species until resources are made available by their damage or death [46].

Weed management in our study played a key role in assemblage of plant communities. Targeted hand weeding benefitted planted forbs, especially in amended plots where forbs grew larger. Weed management is a necessary tool to build plant communities rather than simply for containment and eradication of undesired species. Plant community weed management opens the possibility of using competitive native species to shift the plant community to a more desirable state and reduce weed management in the long term. Weed control can be complex for native forbs as they tend to be more sensitive to chemical control than other species [47]. There are few selective herbicides targeted to weeds that do not also kill the native forbs. Manually weeding the sites is an efficient but time-consuming practice and requires good plant identification skills. This type of manual weeding would need to be implemented early in the restoration program and continue at least beyond two years.

Due to the elevated level of exposure of the research site, it appeared that using native forbs was a great way to raise ecological awareness and involvement of the local community in citizen science [4,24,48]. People are often interested in wildflowers when they are in urban green spaces which opens up the possibility to integrate common citizens in maintenance and weed management strategies associated with naturalization, potentially reducing costs and creating a common goal among the community members [25,49]. Native forbs constitute part of our natural heritage and should be protected and preserved. This experiment confirmed that native forb species remain resilient in their endemic environments.

Human landscape modifications may provide opportunity for evolutionary adjustment, for growth, maturation and adaptation to new conditions.

Findings from this two-year study provide documented insight on how site preparation and soil amendment techniques can be used to improve the success of restoration with a relatively large number of native forb species. The outcomes of this study can provide a foundation for future work, including longer-term seedling establishment.

5. Conclusions

Soil amendment with compost was more influential than site preparation treatments for restoration of forb species in an urban green area as it had a direct positive impact on survival and growth of planted forbs. Treatments with greater amounts of compost had greater survival, growth, species richness, cover and noxious weed cover than control treatments. Soil amendment had a concurrent negative impact by increasing noxious weeds. Although site preparation treatments had little influence on survival of planted forbs, they could provide more benefits when combined with appropriate weed management that controls competition from baseline vegetation. Of 24 forb species, *Penstemon procerus, Fragaria virginiana, Heuchera cylindrica, Agastache foeniculum, Antennaria microphyla, Mentha arvensis* and *Geum aleppicum* showed the greatest potential for establishment under the management approach used in this study. These species are highly recommended for future use in restoration for the City of Edmonton and similar urban centers. *Cornus canadensis, Pulsatilla patens* and *Liatris ligulistylis* are not recommended for use due to their poor performance. Allium textile, *Eriogonum flavum, Viola adunca, Potentilla arguta, Heterotheca villosa, Anemone cylindrica, Rudbeckia hirta, Thalictrum venulosum* and *Anemone canadensis* need further study but may have potential for use in urban restoration programs. Since the results of this investigation are based on low replication, we recommend that urban planners and practitioners use our results but do so with caution as they may be site specific.

Author Contributions: J.A.R. collected and analyzed data and wrote the thesis; A.D. analyzed data, reviewed, edited and significantly modified the manuscript from the thesis; M.A.N. conceptualized the experiment and procured funding, developed the experimental design, supervised all the work and reviewed and edited the manuscript. All authors have read and agreed to the published version of the manuscript.

Funding: This research was funded by the City of Edmonton and the Land Reclamation International Graduate School (LRIGS) through the NSERC CREATE program.

Institutional Review Board Statement: Not applicable.

Informed Consent Statement: Not applicable.

Data Availability Statement: The data presented in this study are available on request from the corresponding author. The data are not publicly available due to copyright issues.

Acknowledgments: We thank Travis Kennedy, Karolina Peret, Nicole Fraser, Lesley Ravell, Megan Egler, Dustin Bilyk, Kelly Bakken, Brent Hamilton, Danny Petryliak and the spraying and planting crews of the City of Edmonton for their support to complete this project.

Conflicts of Interest: The authors declare no conflict of interest.

Appendix A

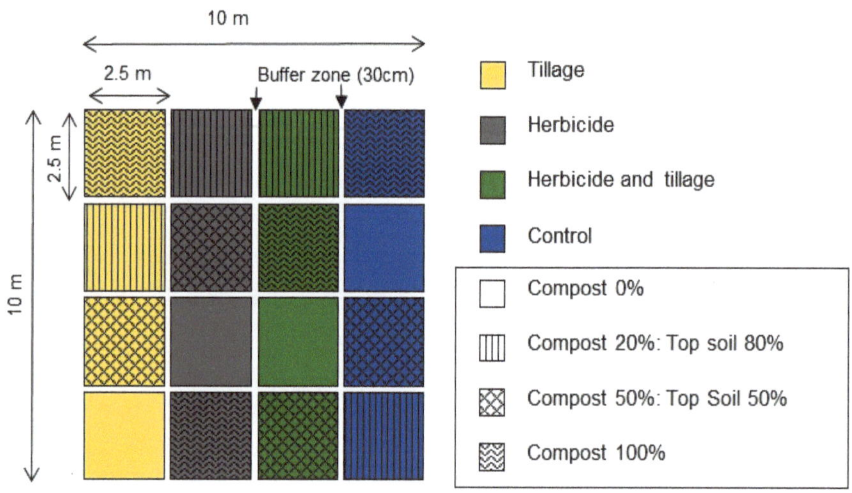

Figure A1. Plot design and soil preparation treatments (colored boxes) randomly applied in columns. Amendment treatments (patterns) randomly distributed within columns.

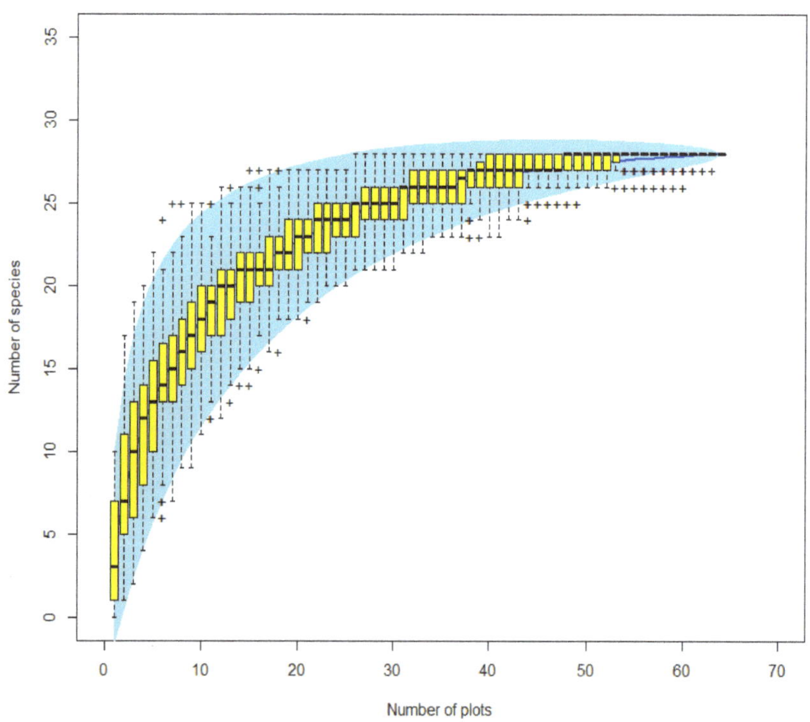

Figure A2. Species accumulation curves in reclamation well sites showing number of species (mean ± SD) versus number of plots.

Table A1. List of species found in the study site.

Plant Category	Common Name	Botanical Name
Native	Milkvetch	*Astragalus* L.
	Plains rough fescue	*Festuca hallii* (Vasey) Piper
	Mountain fescue	*Festuca saximontana* Rydb.
	Foxtail barley	*Hordeum jubatum* L.
	Western wheatgrass	*Pascopyrum smithii* (Rydb.) Á. Löve
	Alpine bluegrass	*Poa alpina* L.
	Wood bluegrass	*Poa nemoralis* L. subsp. interior (Rydb.) W.A. Weber
	Bluebunch wheatgrass	*Pseudoroegneria spicata* (Pursh) Á. Löve
	Prairie thermopsis	*Thermopsis rhombifolia* (Nutt. ex Pursh) Nutt. Ex Richardson
Non Native	Common oat	*Avena sativa* L.
	Smooth brome	*Bromus inermis* Leyss.
	Lamb's quarters	*Chenopodium album* L.
	Quackgrass	*Elymus repens* (L.) Gould
	Creeping red fescue	*Festuca rubra* L.
	Yellow sweet clover	*Melilotus officinalis* (L.) Lam.
	White man's foot	*Plantago major* L.
	Kentucky bluegrass	*Poa pratensis* L.
	Wild buckwheat	*Polygonum convolvulus* L.
	Prickly sow-thistle	*Sonchus asper* (L.) Hill
	Chickweed	*Stellaria media* (L.) Vill.
	Common dandelion	*Taraxacum officinale* F.H. Wigg.
	Alaska clover	*Trifolium hybridum* L.
	White clover	*Trifolium repens* L
	Rapeseed	*Brassica napus* L.
Noxious	Canada thistle	*Cirsium arvense* (L.) Scop.
	Perennial sow thistle	*Sonchus arvensis* L.
	Scentless chamomile	*Tripleurospermum perforatum* (Mérat) M. Lainz
Prohibited Noxious	Slphur cinquefoil	*Potentilla recta* L.

References

1. Benedict, M.A.; McMahon, E.T. Green infrastructure: Smart conservation for the 21st century. *Renew. Resour. J.* **2002**, *20*, 12–17.
2. Grimm, N.B.; Faeth, S.H.; Golubiewski, N.E.; Redman, C.L.; Wu, J.; Bai, X.; Briggs, J.M. Global change and the ecology of cities. *Science* **2008**, *319*, 756–760. [CrossRef] [PubMed]
3. Chiesura, A. The role of urban parks for the sustainable city. *Landsc. Urban Plan.* **2004**, *68*, 129–138. [CrossRef]
4. Aronson, M.F.; Lepczyk, C.A.; Evans, K.L.; Goddard, M.A.; Lerman, S.B.; MacIvor, J.S.; Vargo, T. Biodiversity in the city: Key challenges for urban green space management. *Front. Ecol. Environ.* **2017**, *15*, 189–196. [CrossRef]
5. Rojas, J.A.; Dhar, A.; Naeth, M.A. Urban Naturalization for Green Spaces Using Soil Tillage, Herbicide Application, Compost Amendment and Native Vegetation. *Land* **2021**, *10*, 854. [CrossRef]
6. Bartoli, F.; Savo, V.; Caneva, G. Biodiversity of urban street trees in Italian cities: A comparative analysis. *Plant Biosyst.* **2021**, 1–14. [CrossRef]
7. Song, P.; Kim, G.; Mayer, A.; He, R.; Tian, G. Assessing the Ecosystem Services of Various Types of Urban Green Spaces Based on i-Tree Eco. *Sustainability* **2020**, *12*, 1630. [CrossRef]
8. Millard, A. Indigenous and spontaneous vegetation: Their relationship to urban development in the city of Leeds, UK. *Urban For. Urban Green.* **2004**, *3*, 39–47. [CrossRef]
9. Lin, B.B.; Meyers, J.; Beaty, M.; Barnett, G.B. Urban green infrastructure impacts on climate regulation services in Sydney, Australia. *Sustainability* **2016**, *8*, 788. [CrossRef]
10. Lindemann-Matthies, P.; Brieger, H. Does urban gardening increase aesthetic quality of urban areas? A case study from Germany. *Urban For. Urban Green.* **2016**, *17*, 33–41. [CrossRef]
11. Hwang, Y.H.; Yue, Z.E.J.; Ling, S.K.; Tan, H.H.V. It's ok to be wilder: Preference for natural growth in urban green spaces in a tropical city. *Urban For. Urban Green.* **2019**, *38*, 165–176. [CrossRef]
12. Mayer, K.; Haeuser, E.; Dawson, W.; Essl, F.; Kreft, H.; Pergl, J.; Pysek, P.; Weigelt, P.; Winter, M.; Lenzner, B.; et al. Naturalization of ornamental plant species in public green spaces and private gardens. *Biol. Invasions* **2017**, *19*, 3613–3627. [CrossRef]
13. Çoban, S.; Yener, S.D.; Bayraktar, S. Woody plant composition and diversity of urban green spaces in Istanbul, Turkey. *Plant Biosyst.* **2021**, *155*, 83–91. [CrossRef]

14. Dullinger, I.; Wessely, J.; Bossdorf, O.; Dawson, W.; Ess, L.F.; Gattringer, A.; Klonner, G.; Kuttner, M.; Moser, D.; Pergl, J.; et al. Climate change will increase naturalization risk from garden plants in Europe. *Glob. Ecol. Biogeogr.* **2017**, *26*, 43–53. [CrossRef]
15. Vila, M.; Espinar, J.L.; Hejda, M.; Hulme, P.E.; Jarosik, V.; Maron, J.L.; Pergl, J.; Schaffner, U.; Sun, Y.; Pysek, P. Ecological impacts of invasive alien plants: A meta-analysis of their effects on species, communities and ecosystems. *Ecol. Lett.* **2011**, *14*, 702–708. [CrossRef]
16. Savard, J.-P.L.; Clergeau, P.; Mennechez, G. Biodiversity concepts and urban ecosystems. *Landsc. Urban Plan.* **2000**, *48*, 131–142. [CrossRef]
17. City of Edmonton. Environmental Stewardship Naturalization. 2021. Available online: https://www.edmonton.ca/city_government/environmental_stewardship/naturalization (accessed on 20 February 2022).
18. Pavao-Zuckerman, M.A. The nature of urban soils and their role in ecological restoration in cities. *Restor. Ecol.* **2008**, *16*, 642–649. [CrossRef]
19. Schaefer, V. Alien Invasions, ecological restoration in cities and the loss of ecological memory. *Restor. Ecol.* **2009**, *17*, 171–176. [CrossRef]
20. Millwood, A.A.; Paudel, K.; Briggs, S.E. Naturalization as a strategy for improving soil physical characteristics in a forested urban park. *Urban Ecosyst.* **2011**, *14*, 261–278. [CrossRef]
21. Alakukku, L. Persistence of soil compaction due to high axle load traffic. II. Long-term effects on the properties of fine-textured and organic soils. *Soil Tillage Res.* **1996**, *37*, 223–238. [CrossRef]
22. Gomez, A.; Powers, R.F.; Singer, M.J.; Horwath, W.R. Soil compaction effects on growth of young ponderosa pine following litter removal in California's Sierra Nevada. *Soil Sci. Soc. Am. J.* **2002**, *66*, 1334–1343. [CrossRef]
23. Bretzel, F.; Malorgio, F.; Carrai, C.; Pezzarossa, B. Wildflower plantings to reduce the management costs of urban gardens and roadsides. *Acta Hortic.* **2009**, *813*, 263–269. [CrossRef]
24. Muzafar, I.; Khuroo, A.A.; Mehraj, G.; Hamid, M.; Rashid, I.; Malik, A.H. Floristic diversity along the roadsides of an urban biodiversity hotspot in Indian Himalayas. *Plant Biosyst.* **2019**, *153*, 222–230. [CrossRef]
25. Itani, M.; Al Zein, M.; Nasralla, N.; Talhouk, S.N. Biodiversity conservation in cities: Defining habitat analogues for plant species of conservation interest. *PLoS ONE* **2020**, *15*, e0220355. [CrossRef]
26. Environment Canada. Canadian Climate Normals 1981–2010. Bindloss East Station Data. 2021. Available online: https://climate.weather.gc.ca/climate_normals/index_e.html (accessed on 2 March 2022).
27. Moss, E.H. *Flora of Alberta*, 2nd ed.; Packer, J.G., Ed.; University Press Inc: Toronto, ON, Canada, 1994; p. 687.
28. Hendershot, W.H.; Lalande, H.; Duquette, M. Ion exchange and exchangeable cations. In *Soil Sampling and Methods of Analysis*; Carter, M.R., Gregorich, E.G., Eds.; Canadian Society of Soil Science: Boca Raton, FL, USA, 2008; pp. 199–201.
29. Loeppert, R.H.; Suarez, D.L. Carbonate and gypsum. In *Methods of Soil Analysis Part 3—Chemical Methods*; Sparks, D.L., Page, A.L., Helmke, P.A., Loeppert, R.H., Eds.; Soil Science Society of America, American Society of Agronomy: Madison, WI, USA, 1996; pp. 437–474.
30. Nelson, D.W.; Sommers, L.E. Total carbon, organic carbon, and organic matter. In *Methods of Soil Analysis Part 3—Chemical Methods*; Sparks, D.L., Page, A.L., Helmke, P.A., Loeppert, R.H., Eds.; Soil Science Society of America, American Society of Agronomy: Madison, WI, USA, 1996; pp. 961–1010.
31. Chapman, H.D. Cation-exchange capacity. In *Methods of Soil Analysis*; Black, C.A., Ed.; Soil Science Society of America, American Society of Agronomy: Madison, WI, USA, 1965; pp. 891–901.
32. Maynard, D.G.; Kalra, Y.P.; Crumbaugh, J.A. Nitrate and exchangeable ammonium nitrogen. In *Soil Sampling and Methods of Analysis*; Carter, M.R., Gregorich, E.G., Eds.; Canadian Society of Soil Science: Boca Raton, FL, USA, 2008; pp. 71–80.
33. Bremner, J.M. Nitrogen—Total. In *Methods of Soil Analysis Part 3—Chemical Methods*; Sparks, D.L., Page, A.L., Helmke, P.A., Loeppert, R.H., Eds.; Soil Science Society of America, American Society of Agronomy: Madison, WI, USA, 1996; pp. 1085–1121.
34. Ashworth, J.; Mrazek, K. Modified Kelowna test for available phosphorus and potassium in soil. *Commun. Soil Sci. Plant. Anal.* **1995**, *26*, 731–739. [CrossRef]
35. Miller, J.J.; Curtin, D. Electronical conductivity and soluble ions. In *Soil Sampling and Methods of Analysis*, 2nd ed.; Carter, M.R., Gregorich, E.G., Eds.; Canadian Soil Science Society: Madison, WI, USA; CRC Press and Taylor and Francis Group: Boca Raton, FL, USA, 2007; pp. 153–166.
36. Burt, R. Soil Survey Field and Laboratory Methods Manual. Soil Survey Investigations Report No. 51 Version 2.0. US Department of Natural Resources. 2014. Available online: http://www.nrcs.usda.gov/Internet/FSE_DOCUMENTS/stelprdb1244466.pdf (accessed on 1 February 2022).
37. R Core Team. *R: A Language and Environment for Statistical Computing*; R Foundation for Statistical Computing: Vienna, Austria, 2021; Available online: http://www.Rproject.org/ (accessed on 15 July 2021).
38. Hurlbert, S.H. Pseudoreplication and the design of ecological field experiments. *Ecol. Monogr.* **1984**, *54*, 187–211. [CrossRef]
39. Oksanen, L. Logic of experiments in ecology: Is pseudo-replication a pseudoissue? *Oikos* **2001**, *94*, 27–38. [CrossRef]
40. Buonopane, M.; Snider, G.; Kerns, B.K.; Doescher, P.S. Complex restoration challenges: Weeds, seeds, and roads in a forested wildland urban interface. *For. Ecol. Manag.* **2013**, *295*, 87–96. [CrossRef]
41. Marrs, R.H.; Gough, M.W. Soil fertility—A potential problem for habitat restoration. In *Biological Habitat Reconstruction*; Buckley, G.P., Ed.; Belhaven Press: London, UK, 1989; pp. 29–44.

42. Bretzel, F.; Vannucchi, F.; Romano, D.; Malorgio, F.; Benvenuti, S.; Pezzarossa, B. Wildflowers: From conserving biodiversity to urban greening: A review. *Urban For. Urban Green.* **2016**, *20*, 428–436. [CrossRef]
43. Ruthrof, K.X.; McHenry, M.P.; Hardy, G.E.S.J.; Matusick, G.; Fontaine, J.B.; Buizer, M. Linking restoration outcomes with mechanism: The role of site preparation, fertilisation and revegetation timing relative to soil density and water content. *Plant Ecol.* **2013**, *214*, 987–998. [CrossRef]
44. Skrindo, A.B.; Pedersen, P.A. Natural revegetation of indigenous roadside vegetation by propagules from topsoil. *Urban For. Urban Green.* **2004**, *3*, 29–37. [CrossRef]
45. Fortuna-Antoszkiewicz, B.; Łukaszkiewicz, J.; Rosłon-Szeryńska, E.; Wysocki, C.; Wiśniewski, P. Invasive species and maintaining biodiversity in the natural Areas—Rural and urban—Subject to strong anthropogenic pressure. *J. Ecol. Eng.* **2018**, *19*, 14–23. [CrossRef]
46. Pickett, S.T.; Cadenasso, M.L.; Childers, D.L.; McDonnell, M.J.; Zhou, W. Evolution and future of urban ecological science: Ecology in, of, and for the city. *Ecosyst. Health Sustain.* **2016**, *2*, e01229. [CrossRef]
47. US Fish and Wildlife Service. Managing Invasive Plants Concept, Principles and Practices. 2021. Available online: https://www.fws.gov/invasives/stafftrainingmodule/methods/chemical/impacts.html (accessed on 2 March 2022).
48. Semeraro, T.; Scarano, A.; Buccolieri, R.; Santino, A.; Aarrevaara, E. Planning of Urban Green Spaces: An Ecological Perspective on Human Benefits. *Land* **2021**, *10*, 105. [CrossRef]
49. Klaus, V.H.; Kiehl, K. A conceptual framework for urban ecological restoration and rehabilitation. *Basic Appl. Ecol.* **2021**, *52*, 82–94. [CrossRef]

land

MDPI

Review

How Do Different Modes of Governance Support Ecosystem Services/Disservices in Small-Scale Urban Green Infrastructure? A Systematic Review

Sina Razzaghi Asl [1,*] and Hamil Pearsall [1]

Department of Geography and Urban Studies, Temple University, Philadelphia, PA 19122, USA
* Correspondence: tuk98809@temple.edu

Abstract: As cities are facing environmental and societal challenges, including climate change, rapid urbanization, and the COVID-19 pandemic, scholars and policymakers have recognized the potential of small-scale urban green infrastructures (UGI), such as rain gardens and street trees, to support important ecosystem services (ES) during periods of crisis and change. While there has been considerable research on the design, planning, engineering, and ecology of small-scale UGI, the governance modes of such spaces to support ES and manage ecosystem disservices (EDS) have received significantly less research attention. In this article, we provide a systematic review to evaluate how different modes of governance support different ES in small-scale green infrastructure. We evaluated governance in six types of small-scale green infrastructure: small parks, community gardens, vacant lands, rain gardens, green roofs, and street trees. Our review examines the different characteristics of four new governance approaches, including adaptive, network, mosaic, and transformative to understand their bottom-up nature and applicability in governing ES/disservices of small-scale UGI. Each governance mode can be effective for managing the ES of certain small-scale UGI, given their associations with principles such as resilience thinking, connectivity, and active citizenship. Our synthesis highlights knowledge gaps at the intersection between governance arrangements and ES in small-scale UGI. We conclude with a call for further research on the environmental and contextual factors that moderate the linkages between governance modes and ES/EDS in different types of UGI.

Keywords: governance; small-scale UGI; ES; ecosystem disservices; UGI principles; systematic review

1. Introduction

Urban land area is predicted to triple from 2000 to 2030 [1]. This rapid urbanization will negatively impact local and regional ecosystem functions and will exacerbate consequences of the climate change, and reduce the adaptive capacity of urban areas to cope with a changing climate [2,3]. Urban green infrastructure (UGI) such as green roofs, green facades, public parks, urban forests, urban wetlands, and unmanaged green sites [4], provide nature-based solutions (NBS) that offer a promising avenue for climate change adaptation in cities to reduce the negative environmental impacts of urbanization, such as the urban heat island effect and altered precipitation patterns. UGI supports a wide range of ES at different spatial levels including but not limited to provisioning (e.g., food, and freshwater), regulating (e.g., urban temperature regulations, noise reduction, air purification, pollination, runoff mitigation, and waste treatment), socio-cultural (tourism, recreation, cognitive development, social cohesion), and supporting (e.g., habitat for biodiversity diversity), with fewer documented health benefits (e.g., good health, mortality) [5–13]. Research has also documented EDS associated with UGI, including conflicts with grey infrastructure, air pollution, and green gentrification [14–16].

The management and governance of UGI provide an important mechanism for balancing the services and disservices of UGI in cities. Governance can be defined as a process of collective decision making that allows different stakeholders to include their needs and expectations [17]. Different governance approaches have been widely used in the field of natural resource management to protect and sustain resources [18–20]. There has been an increase in scholarship on UGI governance with growing recognition of the importance of UGI for addressing stormwater runoff, urban heating, and air pollution. Governance is a critical component of effective UGI implementation as cities experience significant changes such as extreme events, pandemics, and biodiversity loss [21]. The role of UGI may also undergo abrupt, surprising change [22,23]. For example, in response to government restrictions such as stay-at-home, social distancing, and quarantine policies during the recent COVID-19 pandemic, UGI became an important resource for socio-cultural and regulatory ES because green spaces provided physical and mental health benefits [24]. However, given urbanization trends, as well as environmental change in existing urban areas, there is a need for cities to develop suitable environmental governance approaches actively and intentionally to address pressing societal challenges [25]

We review governance approaches in the context of small-scale UGI to evaluate how different governance models address ES. UGI governance is defined here as the "processes, interactions, organizations, and decisions" related to greenspace provision and administration, as defined by Lawrence et al. [26]. Although there is no consensus to define UGI, one of the most recently published studies indicates UGI as a multiscalar concept, which can range from small-scale green infrastructure such as rain gardens, pocket parks, community gardens, and green lands, to large-scale facilities targeting the protection and preservation of the natural habitats [27]. Small-scale UGI offers a unique opportunity to enhance ES while minimizing disservices [28–31]. First, the effectiveness of UGI is largely dependent on interconnected social and ecological processes that need to be properly managed and planned at the local scale while also connected to broader scale policies. Because small-scale UGI is often decentralized and has very different governance processes from large-scale UGI [32], the management and governance of small-scale UGI can be more responsive to local social and ecological needs. Second, considering that large- and medium-scale green areas are usually covered with a large area of single-species allergenic species in cities, highly diverse small-scale green spaces can significantly reduce the risk of allergenicity from urban green spaces. [33–35]. Furthermore, several recent studies have suggested that large- and medium-scale green spaces are associated with gentrification outcomes, whereas small-scale green spaces may limit increases in property values [36–38]. Finally, despite a wide variety of literature on the design, planning, engineering, and ecology of small-scale UGI, the governance dimensions of such spaces to support ES have, to date, received significantly less research attention [39,40].

Over the last few decades, many countries have developed their governance practices to optimize ES in small-scale UGI to cope with growing challenges such as water scarcity, biodiversity loss, institutional shortcomings, citizen participation, fiscal austerity, shortcomings of top-down management, lack of environmental knowledge, lack of political stability, and mismatch between boundaries and the scale of ES [40–43]. However, there is little research that compares how different governance approaches address ES in small-scale green infrastructure. To address this gap, we synthesize literature on small-scale UGI management and governance, drawing on diverse geographic contexts to provide a better understanding of existing approaches' characteristics and exploring the conditions under which UGI governance approaches may emerge. We do not aim to investigate the suitability or adaptability of these approaches with different UGI types. Instead, we provide a synthesis of how different governance approaches address ES in the six studied types of small-scale UGI.

2. Literature Review

2.1. Urban Green Infrastructure and ES

Green infrastructure (GI) is a relatively new concept, and several studies have proposed different definitions for GI. The two most cited definitions are from Benedict [44] who defines GI as "an interconnected network of green space that conserves natural ecosystem values and functions and provides associated benefits to human populations", and the European Commission [45], which defines GI as "a strategically planned network of natural and semi-natural areas with other environmental features designed and managed to deliver a wide range of ES. It incorporates green spaces (or blue if aquatic ecosystems are concerned) and other physical features in terrestrial (including coastal) and marine areas. On land, GI is present in rural and urban settings". According to these definitions, key characteristics of GI, including multifunctionality, ES, ecological networks, connectivity, and multiscalar, serve as boundary concepts among various policymakers, planners, and researchers to guide UGI planning and designing [46].

Urban ecosystem services (UES) have multiple benefits for human health and well-being in the face of rapid urbanization, land-use transformation, and climate change crisis [47]. ES can be defined as "the benefits people obtain from ecosystems" [48]. UES is supported by a diverse green infrastructure type including but not limited to parks, urban forests, farmlands, vacant lots, and gardens. UES can be divided into four categories according to the Millennium ecosystem assessment [48]: provisioning services (materials obtained from ecosystems), regulating services (benefits obtained from the regulation by ecosystem process), habitat or supporting services (essentials to produce all ES) and cultural services (non-material benefits obtained from ecosystems). Research suggests that small-scale green infrastructure can moderate the negative environmental impacts of rapid urbanization and climate change by contributing to recreation, mitigating air pollution, cooling surface, and air temperatures, and retaining stormwater run-off [49]. For example, green roofs and walls may improve air quality and flood control management or street trees can reduce exposure to pollution in urban areas [14,50,51]. Moreover, community gardens in urban neighborhoods not only provide food but can also have health, social and aesthetic benefits for the local community [52–54]. Green spaces and urban trees can also mitigate air temperature through transpiration, evaporation, shading, and modifying wind-flow mechanisms [55]. A study by Peschardt et al. [56] indicates that small-scale green spaces have socializing benefits because they provide spaces for neighbors to interact, whereas other services such as noise reduction and carbon storage are less associated with small-scale green spaces compared with large-scale green infrastructure due to their lower compactness or density.

Table 1 summarizes the ES provided by six types of small-scale GI examined in this study. As can be seen, small-scale UGI provides a wide variety of benefits, albeit some UGI types, such as community gardens, may provide a larger range of services than others. A review of articles, in this case, shows that most studies have focused on green spaces' benefits and the impacts of some types of UGI, such as rain gardens or pocket parks on human health, microclimate regulation, and socio-cultural services. In contrast, few studies have examined how different modes of governance might shape ecosystem service provision in different types of UGI.

Table 1. Urban ES provided by six studied small-scale green infrastructures.

ES	Some Examples of UGI and Their Impacts in Literature
Provisioning	Community gardens can address food security in urban areas [57,58]
Supporting	Street trees offer key conservation opportunities for pollinators [59], they also reduce the negative effects of urbanization on birds [60]; green roofs can have ecological significance by attracting and supporting urban fauna [61]; vacant lands can support insects' habitats [62]
Regulating	Vacant lands have cooling effects in urbanized areas [63]; green roofs have large impact on the urban heat island effect, positive effect on street canyon air quality, and stormwater management [64–66]; rain gardens may provide considerable carbon potential, offsetting the whole carbon footprint [67]; street trees can reduce air quality depending on the aspect ratio as well as stormwater [68,69]; community gardens can reduce surface runoff [70]

Table 1. *Cont.*

ES	Some Examples of UGI and Their Impacts in Literature
Socio-Cultural	Small parks offer health benefits [62,71]; green roofs offer recreational and experimental benefits for residents [72]; community gardens as learning environments for sustainability [73]; vacant lots may provide social and cultural values for local communities [74]

2.2. *Urban Green Infrastructure and EDS*

While UGI has several benefits, it also sometimes produces EDS that are frequently overlooked [75]. The concept of EDS refers to the negative impacts that ecosystems can have on humans and their environs [76]. According to Lyytimaki and Sipila [76], EDS are "functions of ecosystems that are perceived as negative for human well-being" and can be brought on by natural or political occurrences such as floods, earthquakes, wildfires, or conflicts. For example, small-scale UGI such as street trees may provide allergies associated with grass pollen and damage to properties [77–80]. Some species release a significant amount of biogenic volatile organic compounds (VOCs), which, when combined with nitrogen oxides (NOx), can create particulate matter, secondary organic aerosol, and ozone, which exacerbate respiratory diseases such as asthma [81]. In addition, research shows that the risk of vegetables and soil contaminated by heavy metals and pollutants in community gardens and green roofs can be considered EDS [82]. There is no agreement on how to classify EDS in relation to ES, despite the fact that certain research has split it into various groups [73,83–86]. Better understanding of the conditions under which EDS arises will help policymakers, practitioners, and communities reduce these negative impacts. While urban areas depend on ES, understanding disservices are of paramount importance from a governance lens. Since EDS reduces public support for UGI, it is important to reduce these negative impacts to optimize UGI for sustainability. For example, in the Mediterranean region, the ornamental patterns of the urban areas imply significant pollen risk from woody species such as plane trees or cypresses, as the most allergenic ornamental species [87,88]. Some studies such as those conducted by Von Döhren and Haase [72] and Sousa-Silva et al. [88] have provided a reliable overview of the environmental and health issues produced by different types of urban trees. Table 2 summarizes some examples of EDS provided by six types of small-scale GI examined in this study.

Table 2. Urban ecosystem disservices are provided by six studied small-scale green infrastructures.

EDS	Some Examples of UGI and Their Impacts in Literature
EDS	Tall and leafy trees may block the views [89]; Vacant lands may be unsafe and ugly [89]; Some plant species may create allergenic pollen [90–92]; Tree roots may cause sidewalk pavement problems [93]; Community gardens may get contaminated by greywater irrigation from contaminated drainage channels or streams [94]; Increasing UGI results in an increase in hornet species [95]; Urban trees produce green waste resulting in public health issues [14]

2.3. *A Need for New Governance Approaches*

Enhancing urban resilience and sustainability in the face of "wicked problems" are key challenges for UGI governance [95]. According to Andersson et al. [96] and Jerome [97], small-scale UGI can contribute multiple co-benefits to support a wide variety of ES. However, there are still some barriers and uncertainties to governing and managing different types of GI worldwide.

One of the important challenges for governance in existing small-scale UGI, such as pocket parks or vacant land uses, is that they can be temporary or short-term land uses. For example, a study in Detroit, Texas found that ragweed populations are more common in vegetated vacant lots, making the transition management of these lots crucial to avoiding significant effects on allergenic pollen burdens [98]. Thus, if cities rely on the ES that these spaces provide, there is a need for governance mechanisms that either provide long-term security for these spaces or support a more adaptive, flexible, and dynamic governance approach to cope with the temporary negative consequences of these spaces [99,100]. Kabisch [101] states the major challenges for green infrastructure governance in Berlin as financial constraints, loss of expertise, and low awareness of such spaces' benefits at the local scale. Fox-Kamper et al. [102] found the major barriers to community garden governance include unsecured land tenure, community engagement, and lack of long-term governance support. A study by Guitart et al. [103] shows that the main challenge for community garden governance in the United States is land tenure where gardeners lack long-term access to land. Furthermore, some scholars have highlighted the issue of changing governance settings and GI data inconsistency as some of the most important challenges GI are facing [104]. Undoubtedly, one of the most important barriers to implementing GI, such as rain gardens, is their costs. These facilities can be expensive to install and maintain, which in turn reduces the willingness of planners and owners to shift toward them [25].

Moreover, urban governance is challenging given environmental justice (EJ) issues in terms of UGI equitable distribution, transparent procedures, and sufficient recognition of various actors' needs and perceptions [2]. Availability, accessibility, and attractiveness of small-scale UGI for different social groups and inhabitants are among the most important issues that EJ research has recently addressed [105,106]. For instance, Sanchez and Reames [107] address spatial equity in green roof distribution in Detroit, MI, and show that green roofs were concentrated in the wealthiest part of Detroit's urban core with a predominantly white population. Consequently, an emerging focus in environmental governance is how different governance approaches can broaden access and participation to diverse social groups, particularly marginalized or vulnerable groups. A potential opportunity for small-scale UGI to promote environmental justice lies in its need for local governance, which can place decision making in the hands of local communities and give them ownership over these spaces. In addition to promoting equitable governance, local ownership may reduce disservices, such as green gentrification, which has been identified as a concern by researchers and non-profit sectors in recent years [108]. In other words, an equitable distribution, experience, and understanding of UGI throughout the cities is an important goal of UGI governance.

Significant shifts have occurred within environmental decision making on UGI in the past 20 years. These shifts have yielded collaborative and bottom-up management approaches to guarantee future success in the face of rapid urbanization, climate change, and major societal disruptions, such as the COVID-19 pandemic. UGI government styles based on centralized decision making, public budgets, top-down, and bureaucratic arrangements have been replaced increasingly by new horizontal approaches of environmental decentralized governance focused on the fluidity between top-down and bottom-up approaches. This new emerging paradigm shift largely emphasizes the concepts of flexibility, collaboration, coordination, awareness, adaptation, inclusiveness, knowledge generation, and transparency [109–113]. As a result, a range of new democratic governance approaches is in use under conditions of uncertainty, complexity, instability, and unpredictability to include different stakeholders' voices in the UGI decision making process and problem-solving. The uncertainty and complexity of managing ES at the local scale is related to socio-political (e.g., population growth), economic pressure (e.g., shrinking budgets), and environmental changes (e.g., climate change). New UGI governance approaches are intended to better address multiple stressors of urbanization and climate change by utilizing ES and harnessing disservices [19,38,114–116]. Over the last several decades, a wide variety of governance arrangements have been proposed, including "state governance" of publicly owned vacant lands and community gardens, and "networked governance" of public-private partnerships for local parks to the "self-governance/market-based" approach of guerilla gardening. However, examining the applicability of different new governance approaches and policies to co-create and co-manage UGI is an important research direction.

3. Research Methods

3.1. Study Selection

This study conducted a systematic literature review on new governance approaches of small-scale UGI following the PRISMA procedure introduced by Moher et al. [117]. We conducted a systematic review to address the need for a review, critique, and possible reconceptualization of the diverse and interdisciplinary knowledge base on UGI, ES, and governance approaches [118]. The methodological approach used PRISMA key processes to construct the sampling frame, as shown in Figure 1: study planning and identification, screening and selection of publications, and content analysis of the selected documents.

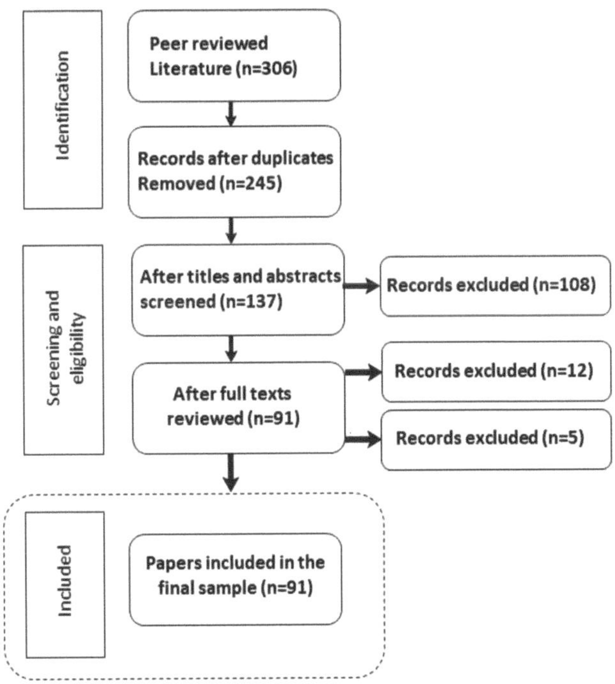

Figure 1. Flow diagram showing the methodology based on PRISMA procedure adapted from Moher et al. (2009).

First, a set of keywords in Google Scholar were used to identify studies on small-scale UGI and governance in urban settings. Six different small-scale UGI were selected, including small parks, community gardens, vacant lands, rain gardens, green roofs, and street trees in this article (Figure 2). All different combinations of six UGI were searched using Scopus and the Web of Science (WOS) databases as the search engine, with the search field set to 'keywords', and the document type set to 'article' or 'review'. To find relevant literature, five separate search queries (garden included both rain garden and community garden) were used, each with a different two-way combination of (keyword category-related) search phrases (Query 1: Park AND governance; Query 2: Garden AND governance; Query 3: Vacant land/lot AND governance; Query 4: Tree AND governance; Query 5: Green roof AND governance). To consolidate and deepen the review, the second round of searches was begun by setting the search field to a two-way combination of (green space/governance) and (nature-based solution/governance). The year of the publications was not filtered; hence, the sample collected from the databases contained all years of publication.

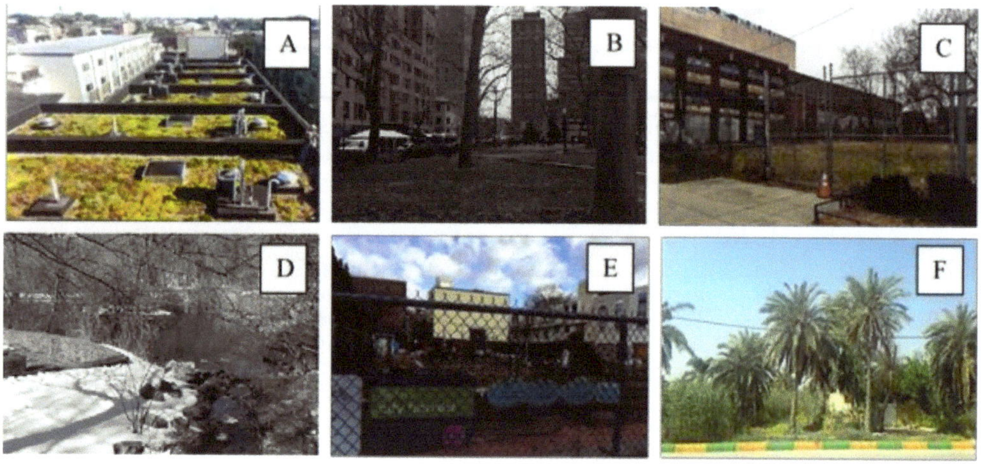

Figure 2. Diverse small-scale urban green infrastructure. (**A**) A small green roof on top of a residential building, USA. (**B**) A local public park, USA. (**C**) A privately owned piece of vacant land, USA. (**D**) Rain garden development, USA. (**E**) A community garden, USA. (**F**) Street trees, Iran. Photos taken by the authors.

3.2. Literature Synthesis

The search returned a total of 306 articles. A selection of 245 articles was found after duplicates were eliminated and they were ready for analysis. Since this review seeks to investigate only bottom-up governance approaches, traditional top-down, government-led arrangements were excluded from the final sample. In total, 108 articles were disqualified in the first round of screening (at the level of the title and abstract). The resulting sample's titles, abstracts, and full text were assessed for relevancy considering only six types of small-scale UGI and new governance modes. Since this study addressed the link between ES and UGI, only articles focused on different types of ES through governance approaches were included. Articles with a rural or regional focus and large green infrastructures such as urban parks or vast vacant lands were excluded. Two runs of full-text screening were conducted; the first resulted in the exclusion of 12, and the second pass resulted in the exclusion of an additional 5.

The systematic review was completed by analyzing 91 academic articles through qualitative content analysis [119]. Content analysis is a research approach for testing theoretical concerns and improving data comprehension in which a condensed number of concepts or categories characterizing a reality, a theory, or a study topic can be obtained [120]. Our review included publications with a focus on concepts and models for governance of ES in different types of small-scale green infrastructure. Since the objective of content analysis of sample documents is to highlight the applicability of new governance modes in the context of six small-scale UGI, the following questions guided the analysis: (1) What types of ES are associated with small-scale GSI? (2) What are the principles of new governance approaches associated with six small-scale UGI? (3) Which new governance modes were associated with different types of small-scale UGI? The authors were then able to determine the most prevalent governance modes and their core principles in the literature by carefully interpreting the articles and all the ensuing categories.

4. Results and Discussion

Based on our review of 91 articles, it was possible to identify four new, bottom-up governance approaches to small-scale UGI: adaptive, mosaic, network, and transformative. Despite some overlaps among these modes of governance, their nature and principles vary to some degree based on contextual, environmental, and social parameters. However, the main similarity among these modes of governance was their bottom-up and multi-agent characteristics, which help policy makers find the most suitable solutions to manage complex urban green areas. The results of this review demonstrate that although the ES and governance modes are not directly linked, their potential relationships may be discerned through the identification of several principles that are applicable in different UGI cases. Some studies have directly focused on these principles, whereas others indirectly described them. Table 2 shows a set of principles for each governance mode which were applied regarding ES in studied small-scale UGI.

4.1. Date and Type of Studies

A systematic review of 91 articles indicated that the governance aspect of UGI is a growing topic of interest in academic research. The earliest articles were published in 2000 in the *Journal of Health and Places* about community gardening development and management. Based on their methods, the articles were divided into four groups: experimental, observational, discussion, and review (Figure 3). Discussion studies were primarily theoretical without data collection, whereas experimental studies involved intervention and primary data collection, observational studies used secondary data to analyze a phenomenon without intervention, and review studies conducted a systematic or scoping literature review. The most common study type was experimental (n = 36), followed by discussion (n = 27), observational (n = 23), and review (n = 5). In general, surveys, including interviews and questionnaires from stakeholders, were the most prevalent kind of experimental study. In addition, we found that community gardens and street trees were the most common types of UGI discussed in the reviewed articles. The specific characteristics of community gardens including but not limited to a socially inclusive, location in central parts of the neighborhoods, and a need to make fluidity between up-down and down-top management can lead them to be one of the most debatable themes in the UGI governance literature. Street trees were also amongst those UGIs that were largely reviewed thanks to their public nature, largely known benefits, and probably the fact that they are often apparently co-managed by both residents and municipalities. However, the adaptability and transformability of the governance modes of these elements in the face of societal and environmental changes (e.g., recent pandemic) can be further analyzed in different contexts to help policymakers to plan more mental and physical resilience for the future [119,120].

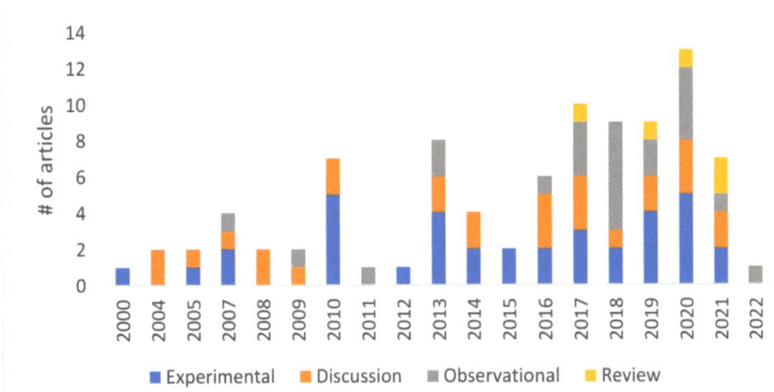

Figure 3. Publication dates and types for all articles (n = 91).

4.2. Governance Modes

Adaptive and co-management governance approaches were most reported in literature ($n = 48$), followed by other governance arrangements including mosaic ($n = 30$), network ($n = 10$), and transformative ($n = 3$). The three common principles used in literature belonged to adaptive governance including adaptability with the highest number ($n = 45$ of 91 adaptive groups), diversity of stakeholders ($n = 39$ of 91 articles), and flexibility ($n = 36$ of 91 articles), followed by self-governance as a core principle of mosaic governance ($n = 29$ of 91 articles). Moreover, the systematic review showed that the two principles of connectivity and diversity of stakeholders/actors emerged in two governance categories of adaptive/mosaic and adaptive/networked, respectively, making them more important in the case of decentralized and flexible UGI management.

The four governance modes and their principles are illustrated in Table 3.

Table 3. Four major governance approaches for six small-scale UGS and their principles in the literature.

Governance Models	Principles		Number of Studies
Adaptive governance	**Adaptability** is the capacity of actors to influence resilience.	45	48
	Diversity of stakeholders facilitates collaboration among institutions and jurisdictions.	39	
	Flexibility allows stakeholders to adapt their needs and expectations to new opportunities.	36	
	Social learning allows actors to share their values, experiences, and actions.	23	
	Connectivity facilitates negotiations and collaborations across horizontal (collaborative) and vertical (hierarchical) connections.	22	
	Resilience thinking is about how to learn to live with change and make use of it.	17	
Mosaic governance	**Self-governance** strengthens the autonomy of citizens to shape their own bottom-up initiatives and rules.	29	30
	Active citizen enhances the ability of people to organize themselves in a multiform manner.	26	
	Polycentricity allows multiple centers of governance to interact with each other across diverse scales and actors.	26	
	Connectivity fosters social and ecological resilience through linking actors.	18	
	Stewardship focuses on collaborative management activity.	14	
	Reflexivity allows to include the perspectives, values, and norms of a variety of actors.	11	
Networked governance	**Knowledge sharing** allows exchanging information between local stakeholders.	7	10
	Social networks facilitate social interactions between actors.	7	
	Diversity of actors allows the presence of various actors, often multi-level.	6	
	Decentralization transfers organization activities to several local actors.	4	
Transformative governance	**Social innovation** is the design of new solutions to imply transformative changes.	3	3
	Transition management accelerates the sustainability transition through the participatory process of visioning, learning, and experimenting.	3	
	Regime shifts are large, abrupt, persistent changes in the structure and function of ecosystems.	2	
	Long-termism allows improving the long-term future of ecosystems.	2	
	Panarchy means drastic transformative changes.	1	

4.3. Adaptive Governance

Adaptive governance (AG) as the most frequent mode mentioned in the literature, is defined as *"an outgrowth of the theoretical search for modes of managing uncertainty and complexity in socio-ecological systems"* [25]. AG appears in almost more than half of the papers examined (48). Allen and Gunderson [121] define it as *"the institutional framework that deals with social and political dimensions of resource management and that allow adaptive management to function"*. AG arrangements include a network of multiple public and private actors to cope with uncertainty and complexity in small-scale UGI. According to Webb et al. [122], involving a diverse group of stakeholders in multi-level governance will facilitate collaboration among institutions and jurisdictions, which is critical for addressing sustainability and resilience in complex urban systems. AG can respond to the uncertainty and complexity of complex socio-ecological systems and increase the capacity of UGI to tackle social and environmental circumstances such as the COVID-19 pandemic.

This approach mainly focuses on the concepts of flexibility, connectivity, and learning in form of policies, formal mechanisms, and regulatory standards which are required to maintain small-scale UGI such as stormwater management systems or rain gardens. Flexibility in adaptive governance arrangement allows stakeholders to adapt their needs and expectations to new opportunities and drive UGI regulations and policies [115]. This approach presents many opportunities for leveraging local UGI to support ES. For instance, converting vacant land into stormwater management systems requires flexible and responsible management structure tools, such as land banks to help municipalities identify suitable vacant parcels [123]. Without the flexibility of AG, it would be time-consuming and difficult for municipal agencies to respond to social and environmental disturbances [27].

Close connections among different actors facilitate negotiations and collaborations across horizontal (collaborative) and vertical (hierarchical) connections. In all types of small-scale UGI bottom-up and top-down connections are needed because city governments may support the establishment of community gardens, vacant lands, trees, rain gardens, stormwater management sites, and green roofs through the provision of land, funding, and regulations, whereas local communities may engage, utilize, and maintain them. Multiple actors are linked through various mechanisms such as collectively managed urban green spaces [38,124].

Social learning is part of the collaborative process of small-scale UGI adaptive governance in which actors can share their values, experiences, actions, and socio-ecological memories of ES through an active process of both formal and informal reflection [125–127]. For example, Lin and Egerer [128] emphasized the role of social learning as one of the characteristics of AG that supports food provision services in community gardens in which farmers learn through their experiences (e.g., light and water availability, soil properties, and ground cover management). Moreover, AG supports a cycle of monitoring and learning to better support ES and reduce disservices.

4.4. Mosaic Governance

The concept of mosaic governance (MG) in the context of UGI elevates the role of a diverse range of active citizens in UGI planning and citizen-led greening management and initiatives [129,130]. In this review, MG was mentioned in about 30 documents out of 91 analyzed. According to Buijs [131], active citizenship is defined as "*citizens' ability to organize themselves in a multiform manner, to mobilize resources and to act in the public to protect rights and take care of common goods*" (p.1). It may be separate from or connected to local authorities' arrangements [132]; however, local authorities usually can support active citizen practices in UGI. Active citizenship can leverage local UGI in the face of a financial crisis. For instance, during the recent pandemic and its consequent social and economic restrictions, active local citizens in community gardens might play a significant and creative role in managing, producing, and marketing local food for the neighborhoods [133]. Mattijssen et al. [134] propose a new term "green self-governance" to respond to the critical points including equal representation of non-active citizens and instrumentalization of citizens in the active citizenship approach. Green self-governance is defined as "a specific form of governance in which citizens play a major role in realizing, protecting, and/or managing green public space" [134]. In contrast to traditional centralized governance approaches for UGI, self-governance arrangements aim to strengthen the autonomy of citizens to shape their bottom-up initiatives and rules (e.g., citizen maintenance of local vacant lands).

MG approaches emphasize the connectivity and multifunctional nature of UGI and use reflexive notions of stewardship to facilitate the active citizenship process. Connectivity in mosaic-oriented governing of green infrastructure's ES involves two dimensions. First, linking UGI can create support and protect the functions and benefits of mosaic UGI that individual UGI cannot provide alone. Second, the connectivity between local authorities and residents is critical to the success of the UGI regime to foster social and ecological resilience. Gulsurd et al. [135] called this approach "reflexive co-governance", in which citizens play an active role in creating healthy ecosystems. In contrast to traditional governance approaches, in MG or reflexive arrangements, both local government and local citizens can contribute to the delivery and management of ES in small-scale UGI. In this approach, local authorities recognize the legitimacy and autonomy of individuals as active citizens [131]. Thus, this approach includes polycentric governance in which multiple centers of governance interact with each other across diverse scales and actors. It also emphasizes environmental stewardship at local levels, where a self-organized approach to responding to small-scale heterogeneity (e.g., size, quality, range of activities, etc.) is highly recommended within a collaborative management activity [136–142]. According to some studies [143–145], contemporary urban environmental stewardship activity allows residents to protect and manage small parks, trees, and community gardens that provide ES. For example, Langemeyer et al. [146] found that civic urban gardens as a new way of connected UGI can enhance stewardship action much better rather than traditional allotment gardens.

4.5. Networked Governance

Networks can facilitate the relationship between formal governance arrangements and the informal social processes that affect the local governance of UGI. Of the 10 papers where it was addressed, only 5 of them defined networked governance (NG). According to Nochta and Skelcher [147], networked governance is defined as "*systems of coordination that seek to guide and steer multi-actor interactions to solve complex public policy problems*". Network governance of small-scale UGI creates a space for cooperation between various actors to facilitate the co-production of UGI policies based on knowledge sharing, which in turn can increase social and ecological resilience [148]. A community-based management approach is one example of networked governance that includes collective action and self-governance of ES that relies on strong connections between stakeholders at local levels [149]. For example, the networks among community gardeners help them increase social resilience and capital through sharing knowledge about food productivity essentials such as soil contaminant risks. Although networks may create unevenness in the distribution of power and resources [150], mismatches with ecological scales [139], and formal or informal dialogues [151,152], they are powerful tools to build and maintain community garden development [153].

Some scholars propose the networked governance approach as a type of adaptive governance in which the characteristics of actors and the patterns of interactions between different actors' matter and are important for improving the performance of ecosystems [154,155]. Key to this approach is the concept of diversity, which means the presence of various actors, often multi-level, are involved in the process of local green infrastructure planning [156]. For instance, a networked governance approach to street tree planting and maintenance requires the involvement of different actors including local municipalities, non-profit organizations, community groups, and individual volunteers. Some of these actors (e.g., municipalities and NGOs) mainly play financial and supportive roles, whereas others (e.g., individuals and community groups) may participate in tree planting and stewardship activities. The majority of UGI governance networks, such as residential green roofs, are informal and differ from traditional hierarchical modes of government [152]. Additionally, the concept of "ecological networks" is central to understanding the linkages between network governance, ES, and UGI. Ecological networks are defined as "tools for improving biodiversity and ecological connectivity among habitats which are designed to consider different levels of nature protection" [157]. This definition is also aligned with the concept of green infrastructure as a network of green spaces that provide multiple benefits.

More recently, a concept of hybrid governance has also been applied to UGI specifically nature-based solutions (NBS) governance. Toxopeus et al. [158] define hybrid governance as a tool "to drive innovation and deliver co-benefits to multiple stakeholders, representing a demand-driven, cost-effective realization of sustainable urban infrastructure". Hybrid governance as a type of polycentric and multi-level governance runs parallel to the networked governance approach in local UGI due to its participatory and democratic nature. For example, over the last decade, public and non-public actors (e.g., businesses) in the U.S. context have collaborated to create the financial resource and technical synergies to install green roofs and solar PV. However, non-public actors have not had a role in the development of policies and regulation processes, which in turn has raised the issue of equity [159].

4.6. Transformative Governance

The literature regarding the application of transformative governance (TG) in the context of UGI is limited. Only three articles directly focused on the application of TG in the context of UGI. Transformative governance is defined as *"an approach to environmental governance that has the capacity to respond to, manage, and trigger regime shifts in coupled social-ecological systems (SESs) at multiple scales"*. [25]. While the adaptive governance approach emphasizes maintaining and adapting existing ecosystem regimes, transformative governance arrangements pursue regime shifts to create new systems to better support ecosystems and human health and well-being. Regime shifts, as direct consequences of cross-scale dynamics of socio-ecological systems, can result in drastic transformative changes in different aspects of ES that UGI can provide at a local level called "panarchy", which is central to the concept of transformation [25]. Panarchy theory stresses cross-scale links, in which processes at one scale influence operations at other scales, influencing the system's overall dynamics [160]. For example, the recent COVID-19 outbreak occurred quickly and became widespread, which subsequently created considerable changes in food security, nutrition, and food systems worldwide. As Mejia et al. [133] state, the community gardens have had a significant role during the pandemic by providing fresh food and supporting human well-being and social benefits. According to this concept, small-scale UGI is considered in some cases (e.g., rain gardens) as a transformative governance response to climate change issues at a large scale and as an example of an adaptive governance approach in other situations (e.g., street trees).

Given the complex nature of small-scale UGI, TG aims to alter the nature of UGI through innovative approaches, such as nature-based solutions (NBS). Frantzeskaki and Bush [161] indicate the role of intermediaries (systemic, regime-based, niche, process, and user) as facilitators of transformative governance in the case of urban trees in Australia. This approach leverages systematic changes in UGI systems to address social and ecological challenges such as climate change. In small-scale UGI, examples of transformative changes may include the greening of vacant lands or the shifts from gray to green roofs. For instance, Kabisch [162] explains how local participation as a key feature of new governance modes can transform an urban brownfield site into a multifunctional urban park in Leipzig, Germany.

Transformative governance is considered part of transition management. Transition management has been defined as "a specific form of multi-level governance. Whereby state and non-state actors are brought together to co-produce and coordinate policies iteratively and evolutionarily on different policy levels" [163]. Loorbach [164] describes the tenets of transition management as (1) long-term focus, (2) uncertainties and surprise, (3) networks and self-steering, and (4) innovation. Social innovation is a new term used for environmental governance and management in the transition arena. It refers to a set of new concepts, products, and processes that not only seek to meet basic social needs but also change routine flows and arrangements to better utilize the urban and social systems [165]. Green infrastructure can be considered an innovative response to complex and challenging urban systems and can support transformative governance [166].

Social innovation is a process of change that can facilitate the emergence of transformative governance in the context of UGI. According to Elmqvist et al. [167], social innovation can play an important role in urban transformations by proposing new alternatives and configurations for preserving the functionality of a system such as UGI. For example, community gardens serve as an example of social innovation because they enhance the value of provisioning and regulating ES such as food production and air quality improvements [142]. Social innovation at the local scale can create physical and community space for optimizing the usage of UGI for ES. For instance, transforming vacant lands into UGI can provide opportunities for provisioning, regulating, and socio-cultural ES [72,102]. Likewise, due to the unused nature of urban vacant lots and the greening opportunities, they are mostly documented under the transformative governance approach in the literature. Nevertheless, there is room to examine their applicability for implementing some temporary socio-cultural services, short-term activities or policies (e.g., land bank in the U.S.) which can be matched with the adaptive governance approach [27,102,123]. More recently, digital tools such as smart ecosystems and the internet of trees as digital networks are new concepts that can support the environmental transformative governance approach [168,169].

Our analysis showed that all four networks, mosaic, transformative, and adaptive governance approaches are applicable for different provisioning, regulating, cultural, and supporting ES through practicing principles in six small-scale UGI, as shown in Table 2; however, these occur to varying degrees, depending on their principles' applicability in UGI cases. In other words, each governance mode can be effective for managing the ES of certain small-scale UGI, given their relationships with principles. First, the principles of resilience thinking and connectivity are interconnected with food and water supply in UGI governance. For example, social resilience can be developed using connectivity concepts between different stakeholders and governments to produce and distribute food through community gardens. Connectivity among diverse actors can support mosaic and adaptive governance arrangements as well, which can build the resilience of ES through social networks, preventing disturbances, and maintenance of biodiversity [165]. Second, involving citizens and integrating citizen science concepts in small-scale UGI as part of mosaic governance can address cultural ES in some cases, such as developing a small new pocket park with volunteers [73]. The literature shows that active citizenship has a significant role in governing cultural ES such as environmental education, recreation, or social cohesion, in turn, can promote mental and physical health outcomes [132]. Third, the concept of socio-ecological networks, which are central to governance networks, can enhance both vertical and horizontal connectivity between multi-level government arrangements and habitats, respectively, to support biodiversity conservation practices. For example, actors (e.g., government, citizens, or NGOs) can balance supply and demand in green spaces by conserving native plants or using interventions such as "community-based management" which can create new supply nodes connected to demanders [149].

Our study has a few limitations. This review did not directly analyze the political and financial leadership and arrangement concepts in the management of governance of small-scale UGI. It is possible that the examples mentioned from different geographical areas represent the type of political system or governance and may not reflect the conditions of other existing examples well. Therefore, further analysis in terms of clarifying the potential interlinkages between government and political structures, and their capacities for governing local UGI is needed [170]. Furthermore, we did not identify the potential differences between the four modes of governance in terms of how they might address and resolve issues related to EDS. This approach needs to be further analyzed by highlighting synergies and tradeoffs between each UGI type, the applicable governance approach, and examining the crossovers and combined use of the four governance approaches in different ways [116]. It should be noted too that we could not provide more empirical evidence of analysis of the situation about small-scale UGI due to the word number limitations for publication. Finally, our review did not identify the best governance mode for each type of ES/EDS associated with UGI, as well as the characteristics of a successful governance structure, which can be investigated based on more empirical studies.

Figure 4 shows a Sankey diagram mapping the systematic review of the literature regarding the potential linkages between new governance modes and six types of UGI. As is shown, vacant lots and green roofs were among those small-scale UGI which were less documented and there is a need to address this gap by applying new governance arrangements to optimize their benefits for urban environments and residents. To advance this approach, more studies are needed to evaluate potential links between different governance modes and their associated ES in different contexts and types of UGI. Additionally, future studies can focus on the applicability and suitability of governance approaches to offer equity, resilience, and sustainability concepts in the face of climate change and societal disruptions such as the COVID-19 pandemic.

Figure 4. Sankey diagram showing the link of governance modes and UGI types. The thickness of each band corresponds to the number of studies involving the linkage.

5. Conclusions

Our review provides a first overview of the applicable governance approaches for six types of small-scale green infrastructure including small parks, community gardens, vacant lots, rain gardens, green roofs, and street trees. This review focused on the most common governance approaches, including adaptive, mosaic, networked, and transformative approaches for managing the ES and EDS of selected small-scale UGIs. This study provided some insights for policymakers, planners, researchers, and those who are interested in investigating the linkages between UGI governance approaches and ES. First, it demonstrated a novel attempt to categorize new governance approaches for managing ES in the context of small-scale UGI. This review showed that each governance approach had its and specific principles for framing ES in different small-scale UGI which are moderated by contextual factors conducting nuanced linkages between governance approaches and ESES. Second, as our review indicated that the specific characteristics of small-scale UGI such as location, ownership, the type of provided ES, and their potential for further development may determine the mode of governance. However, cultural, political, economic, and government structures in different contexts may affect this relationship to some degree. It is extremely important to examine different governance approaches in diverse contexts, as well as the potential of combined governance modes for other UGI types. In other words, the flexible adoption of different governance arrangements rather than only selecting a certain mode may become useful for managing multifunctional UGI types or adapting to environmental and societal change.

Finally, although there are multiple studies regarding different governance modes in natural resources management at national or state scales, this study highlighted the current status quo of knowledge and future potential research directions regarding the linkages between diverse new governance types and ES/EDS of small-scale UGI. The research gaps can help urban planners, green infrastructure planners, and urban ecologists pursue suitable governance approaches for better utilizing different types of services provided by small-scale GI in ES within cities.

Author Contributions: Conceptualization, methodology, and writing were carried out by S.R.A.; and H.P. reviewed and edited previous draft versions and provided supervision. All authors have read and agreed to the published version of the manuscript.

Funding: This research received no external funding.

Institutional Review Board Statement: Not applicable.

Informed Consent Statement: Not applicable.

Data Availability Statement: Data are available from the first author on request.

Acknowledgments: The authors acknowledge and thank the helpful comments provided by an anonymous reviewer.

Conflicts of Interest: The authors declare no conflict of interest in the manuscript, or in the decision to publish the results.

References

1. Seto, K.C.; Guneralp, B.; Hutyra, L.R. Global forecasts of urban expansion to 2030 and direct impacts on biodiversity and carbon pools. *Proc. Natl. Acad. Sci. USA* **2012**, *109*, 16083–16088. [CrossRef] [PubMed]
2. Venter, Z.S.; Hjertager Krog, N.; Barton, D.N. Linking green infrastructure to urban heat and human health risk mitigation in Oslo, Norway. *Sci. Total Environ.* **2020**, *709*, 136193. [CrossRef] [PubMed]
3. Zuniga-Teran, A.A.; Gerlak, A.K.; Mayer, B.; Evans, T.P.; Lansey, K.E. Urban resilience and green infrastructure systems: Towards a multidimensional evaluation. *Curr. Opin. Environ. Sustain.* **2020**, *44*, 42–47. [CrossRef]
4. Van Oijstaeijen, W.; Van Passel, S.; Cools, J. Urban green infrastructure: A review on valuation toolkits from an urban planning perspective. *J. Environ. Manag.* **2020**, *267*, 110603. [CrossRef]
5. Nieuwenhuijsen, M.J. Green Infrastructure and Health. *Annu. Rev. Public Health* **2021**, *42*, 317–328. [CrossRef]
6. European Environmental Agency [EEA]. Green Infrastructure and Territorial Cohesion: The Concept of Green Infrastructure and Its Integration into Policies Using Monitoring Systems (EEA Technical Report No. 18). 2011. Available online: http://www.upv.es/contenidos/CAMUNISO/info/U0632842.pdf (accessed on 10 February 2022).
7. Benedict, M.A.; McMahon, E.T. *Green Infrastructure: Linking Landscapes and Communities*; Island Press: Washington, DC, USA, 2012.
8. Eisenman, T.S. Frederick Law Olmsted, green infrastructure, and the evolving city. *J. Plan. Hist.* **2013**, *12*, 287–311. [CrossRef]
9. Taylor Lovell, S.; Taylor, J.R. Supplying urban ecosystem srvices through multifunctional green infrastructure in the United States. *Landsc. Ecol.* **2013**, *28*, 1447–1463. [CrossRef]
10. Hansen, R.; Pauleit, S. From multifunctionality to multiple ecosystem srvices? A conceptual framework for multifunctionality in green infrastructure planning for urban areas. *Ambio* **2014**, *43*, 516–529. [CrossRef]
11. Albert, C.; Haaren, C.V. Implications of applying the green infrastructure concept in landscape planning for ecosystem srvices in peri-urban areas: An expert survey and case study. *Plan. Pract. Res.* **2015**, *32*, 227–242. [CrossRef]
12. Bertram, C.; Rehdanz, K. Preferences for cultural urban ecosystem srvices: Comparing attitudes, perception, and use. *Ecosyst. Serv.* **2015**, *12*, 187–199. [CrossRef]
13. Andersson, E.; Tengo, M.; McPhearson, T.; Kremer, P. Cultural ecosystem srvices as a gateway for improving urban sustainability. *Ecosyst. Serv.* **2015**, *12*, 165–168. [CrossRef]

14. Escobedo, F.J.; Kroeger, T.; Wagner, J.E. Urban forests and pollution mitigation: Analyzing ES and disservices. *Environ. Pollut.* **2011**, *159*, 2078–2087. [CrossRef]
15. Von Dohren, P.; Haase, D. Ecosystem disservices research: A review of the state of the art with a focus on cities. *Ecol. Indic.* **2015**, *52*, 490–497. [CrossRef]
16. Maya-Manzano, J.M.; Fernández-Rodríguez, S.; Monroy-Colín, A.; Silva-Palacios, I.; Tormo-Molina, R.; Gonzalo-Garijo, A. Allergenic pollen of ornamental plane trees in a Mediterranean environment and urban planning as a prevention tool. *Urban For. Urban Green.* **2017**, *27*, 352–362. [CrossRef]
17. Lemos, M.C.; Agrawal, A. Environmental Governance. *Annu. Rev. Environ. Resour.* **2006**, *31*, 297–325. [CrossRef]
18. Rist, S.; Chidambaranatha, M.; Escobar, C.; Weismann, U.; Zimmermann, A. Moving from sustainable management to sustainable governance of natural resources: The role of social learning processes in rural India, Bolivia and Mali. *J. Rural. Stud.* **2007**, *23*, 23–37. [CrossRef]
19. Lockwood, M.; Davidson, J.; Curtis, A.; Stratford, E.; Griffith, R. Governance Principles for Natural Resource Management. *Soc. Nat. Resour.* **2010**, *23*, 986–1001. [CrossRef]
20. Georgescu, M.; Morefield, P.E.; Bierwagen, B.G.; Weaver, C.P. Urban adaptation can roll back warming of emerging megapolitan regions. *Proc. Natl. Acad. Sci. USA* **2014**, *111*, 2909–2914. [CrossRef]
21. Whitehead, M. *Environmental Transformations: A Geography of the Anthropocene*; Routledge: New York, NY, USA, 2014.
22. Gunderson, L.H.; Holling, C.S. (Eds.) *Panarchy: Understanding Transformations in Human and Natural Systems*; Island Press: Washington, DC, USA, 2002.
23. Derkzen, M.L.; Teeffelen, A.J.V.; Nagendra, H.; Verburg, P.H. Shifting roles of urban green space in the context of urban development and global change. *Curr. Opin. Environ. Sustain.* **2017**, *29*, 32–39. [CrossRef]
24. Heo, S.; Lim, C.; Bell, M. Relationships between Local Green Space and Human Mobility Patterns during COVID-19 for Maryland and California, USA. *Sustainability* **2020**, *12*, 9401. [CrossRef]
25. Chaffin, B.C.; Garmenstani, A.S.; Gunderson, L.H.; Benson, M.H.; Angeler, D.G.; Arnold, C.A.; Cosens, B.; Craig, R.K.; Ruhl, J.B.; Allen, C.R. Transformative Environmental Governance. *Annu. Rev. Environ. Resour.* **2016**, *41*, 399–423. [CrossRef]
26. Lawrence, A.; De Vreese, R.; Johnston, M.; van den Bosch, C.C.K.; Sanesi, G. Urban forest governance: Towards a framework for comparing approaches. *Urban For. Urban Green.* **2013**, *12*, 464–473. [CrossRef]
27. Arthur, N.; Hack, J. A multiple scale, function, and type approach to determine and improve Green Infrastructure of urban watersheds. *Urban For. Urban Green.* **2022**, *68*, 127459. [CrossRef]
28. Green, O.; Garmestani, A.S.; Albro, S.; Ban, N.C.; Berland, A.; Burkman, C.E.; Gardiner, M.M.; Gunderson, L.; Hopton, M.E.; Schoon, M.L.; et al. Adaptive governance to promote ecosystem srvices in urban green spaces. *Urban Ecosyst.* **2016**, *19*, 77–93. [CrossRef]
29. Caparrós Martínez, J.; Milán-García, J.; Rueda-López, N.; de Pablo-Valenciano, J. Green infrastructure and water: An analysis of global research. *Water* **2020**, *12*, 1760. [CrossRef]
30. Nordh, H.; Østby, K. Pocket parks for people—A study of park design and use. *Urban For. Urban Green.* **2013**, *12*, 12–17. [CrossRef]
31. Kerishnan, P.B.; Maruthaveeran, S. Factors contributing to the usage of pocket parks–A review of the evidence. *Urban For. Urban Green.* **2021**, *58*, 126985. [CrossRef]
32. Zhang, D.; Gersberg, R.M.; Ng, W.J.; Tan, S.K. Conventional and decentralized urban stormwater management: A comparison through case studies of Singapore and Berlin, Germany. *Urban Water J.* **2017**, *14*, 113–124. [CrossRef]
33. Cariñanos, P.; Casares-Porcel, M. Urban green zones and related pollen allergy: A review. Some guidelines for designing spaces with low allergy impact. *Landsc. Urban Plan.* **2011**, *101*, 205–214. [CrossRef]
34. Wolch, J.; Byrne, J.; Newell, J. Urban green space, public health, and environmental justice: The challenge of making cities 'just green enough'. *Landsc. Urban Plan.* **2014**, *125*, 234–244. [CrossRef]
35. Wong, G.K.L.; Jim, C.Y. Do vegetated rooftops attract more mosquitoes? Monitoring disease vector abundance on urban green roofs. *Sci. Total Environ.* **2016**, *573*, 222–232. [CrossRef] [PubMed]
36. Branas, C.C.; South, E.; Kondo, M.C.; Hohl, B.C.; Bourgois, P.; Wiebe, D.J.; MacDonald, J.M. Citywide cluster randomized trial to restore blighted vacant land and its effects on violence, crime, and fear. *Proc. Natl. Acad. Sci. USA* **2018**, *115*, 2946–2951. [CrossRef] [PubMed]
37. Rigolon, A.; Németh, J. Green gentrification or 'just green enough': Do Park location, size and function affect whether a place gentrifies or not? *Urban Stud.* **2020**, *57*, 402–420. [CrossRef]
38. Kim, S.K.; Wu, L. Do the characteristics of new green space contribute to gentrification? *Urban Stud.* **2021**, *59*, 360–380. [CrossRef]
39. Cruz, N.; Rode, P.; McQuarrie, M. New urban governance: A review of current themes and future priorities. *J. Urban Aff.* **2018**, *41*, 1–9. [CrossRef]
40. Moher, D.; Liberati, A.; Tetzlaff, J.; Altman, D.G.; The PRISMA Group. Preferred Reporting Items for Systematic Reviews and Meta-Analyses: The PRISMA Statement. *PLoS Med.* **2009**, *6*, e1000097. [CrossRef]
41. Torraco, R. Writing integrative literature reviews: Guidelines and examples. *Hum. Resour. Dev. Rev.* **2005**, *4*, 356–367. [CrossRef]
42. Mayring, P. Qualitative content analysis. *Forum Qual. Soc. Res.* **2000**, *1*, 1089. Available online: http://nbn-resolving.de/urn:nbn:de:0114-fqs0002204 (accessed on 25 February 2022).
43. Elo, S.; Kyngas, H. The qualitative content analysis process. *J. Adv. Nurs.* **2008**, *62*, 107–115. [CrossRef]
44. Benedict, M.; McMahon, E. Green infrastructure: Smart conservation for the 21st century. *Renew. Resour. J.* **2002**, *20*, 12–17.

45. European Commission. Communication from the Commission to the European Parliament, the Council, the European Economic and Social Committee and the Committee of the Regions. Green Infrastructure (GI)–Enhancing Europe's Natural Capital; Brussels. Available online: https://eur-lex.europa.eu/resource.html?uri=cellar:d41348f2-01d5-4abe-b817-4c73e6f1b2df.0014.03/DOC_1&format=PDF (accessed on 10 January 2022).
46. Schleyer, C.; Lux, A.; Mehring, M.; Görg, C. Ecosystem Services as a Boundary Concept: Arguments from Social Ecology. *Sustainability* **2017**, *9*, 1107. [CrossRef]
47. Gómez-Baggethun, E.; Gren, Å.; Barton, D.N.; Langemeyer, J.; McPhearson, T.; O'Farrell, P.; Andersson, E.; Hamstead, Z.; Kremer, P. Urban ecosystem srvices. In *Urbanization, Biodiversity and Ecosystem Srvices: Challenges and Opportunities*; Springer: Dordrecht, The Netherlands, 2013. [CrossRef]
48. Millennium Ecosystem Assessment. *Ecosystems and Human Well-Being: Synthesis*; Island Press: Washington, DC, USA, 2005.
49. Derkzen, M.; Van Teeffelen, A.; Verburg, P.H. Green infrastructure for urban climate adaptation: How does residents' views on climate impacts and green infrastructure shape adaptation preferences? *Landsc. Urban Plan.* **2017**, *157*, 106–130. [CrossRef]
50. Irga, P.; Burchett, M.D.; Torpy, F.R. Does urban forestry have a quantitative effect on ambient air quality in an urban environment? *Atmos. Environ.* **2015**, *120*, 173–181. [CrossRef]
51. Amorim, J.; Engardt, M.; Johansson, C.; Ribeiro, I.; Sannebro, M. Regulating and Cultural ecosystem srvices of Urban Green Infrastructure in the Nordic Countries: A Systematic Review. *Int. J. Environ. Res. Public Health* **2021**, *18*, 1219. [CrossRef]
52. Diekmann, L.O.; Gray, L.C.; Thai, C.L. More Than Food: The Social Benefits of Localized Urban Food Systems. *Front. Sustain. Food Syst.* **2020**, *4*, 534219. [CrossRef]
53. Breuste, J.; Qureshi, S.; Li, J. Scaling down the ecosystem services at a local level for urban parks of three megacities. *Hercynia-Okol. Umw. Mitteleur.* **2013**, *46*, 1–20.
54. Soga, M.; Gaston, K.; Yamaura, Y. Gardening is beneficial for health: A meta-analysis. *Prev. Med. Rep.* **2017**, *5*, 92–99. [CrossRef]
55. Gill, S.E.; Handley, J.F.; Pauleit, S. Adapting cities for climate change: The role of green infrastructure. *Built Environ.* **2007**, *33*, 115–133. [CrossRef]
56. Steenkamp, J.; Cilliers, E.J.; Cilliers, S.S.; Lategan, L. Food for Thought: Addressing Urban Food Security Risks through Urban Agriculture. *Sustainability* **2021**, *13*, 1267. [CrossRef]
57. Pena, J.C.; Martello, F.; Ribeiro, M.C.; Armitage, R.A.; Young, R.J.; Rodrigues, M. Street trees reduce the negative effects of urbanization on birds. *PLoS ONE* **2017**, *12*, e0174484. [CrossRef]
58. Gardiner, M.; Burkman, C.E.; Prajzner, S.P. The value of urban vacant land to support arthropod biodiversity and ES. *Environ. Entomol.* **2013**, *42*, 1123–1136. [CrossRef]
59. Xing, Q.; Hao, X.; Lin, Y.; Hang, T.; Ke, Y. Experimental investigation on the thermal performance of a vertical greening system with green roof in wet and cold climates during winter. *Energy Build.* **2019**, *183*, 105–117. [CrossRef]
60. Yang, J.; Yu, Q.; Gong, P. Quantifying air pollution removal by green roofs in Chicago. *Atmos. Environ.* **2008**, *42*, 7266–7273. [CrossRef]
61. Zheng, X.; Zou, Y.; Lounsbury, A.W.; Wang, C.; Wang, R. Green roofs for stormwater runoff retention: A global quantitative synthesis of the performance. *Resour. Conserv. Recycl.* **2021**, *170*, 105577. [CrossRef]
62. Kavehei, E.; Jenkins, G.A.; Adame, M.F.; Lemckert, C. Carbon sequestration potential for mitigating the carbon footprint of green stormwater infrastructure. *Renew. Sustain. Energy Rev.* **2018**, *94*, 1179–1191. [CrossRef]
63. Eisenman, S.T.; Churkin, G.; Jariwal, S.P.; Kumar, P.; Lovasi, G.S.; Pataki, D.E.; Weinberger, K.R.; Whitlow, T.H. Urban trees, air quality, and asthma: An interdisciplinary review. *Landsc. Urban Plan.* **2019**, *187*, 47–59. [CrossRef]
64. Berland, A.; Shiflett, S.; Shuster, W.D.; Garmestani, A.S.; Goddard, H.C.; Herrmann, D.; Hopton, M.E. The role of trees in urban stormwater management. *Landsc. Urban Plan.* **2017**, *162*, 167–177. [CrossRef]
65. Gittleman, M.; Farmer, C.J.Q.; Kremer, P.; McPhearson, T. Estimating stormwater runoff for community gardens in New York City. *Urban Ecosyst.* **2017**, *20*, 129–139. [CrossRef]
66. Cohen, D.A.; Marsh, T.; Williamson, S.; Han, B.; Derose, K.P.; Golinelli, D.; McKenzie, T. The Potential for Pocket Parks to Increase Physical Activity. *Am. J. Health Promot.* **2014**, *28*, S19–S26. [CrossRef]
67. Mesimäki, M.; Hauru, K.; Lehvävirta, S. Do small green roofs have the possibility to offer recreational and experiential benefits in a dense urban area? A case study in Helsinki, Finland. *Urban For. Urban Green.* **2019**, *40*, 114–124. [CrossRef]
68. Corkery, L. Community Gardens as a Platform for Education for Sustainability. *Aust. J. Environ. Educ.* **2004**, *20*, 69–75. [CrossRef]
69. Kim, G. The Public Value of Urban Vacant Land: Social Responses and Ecological Value. *Sustainability* **2016**, *8*, 486. [CrossRef]
70. Clarke, M.; Davidson, M.; Egerer, M.; Anderson, E.; Fouch, N. The underutilized role of community gardens in improving cities' adaptation to climate change: A review. *People Place Policy* **2019**, *12*, 241–251. [CrossRef]
71. Lyytimaki, J.; Sipil, M. Hopping on one leg—The challenge of ecosystem disservices for urban green management. *Urban For. Urban Green.* **2009**, *8*, 309–315. [CrossRef]
72. Von Dohren, P.; Haase, D. Risk assessment concerning urban ecosystem disservices: The example of street trees in Berlin, Germany. *Ecosyst. Serv.* **2019**, *40*, 101031. [CrossRef]
73. Peschardt, K.K.; Schipperijn, J.; Stigsdotter, U.K. Use of Small Public Urban Green Spaces (SPUGS). *Urban For. Urban Green.* **2012**, *11*, 235–244. [CrossRef]

74. Teixeira, F.Z.; Bachi, L.; Blanco, J.; Zimmermann, I.; Welle, I.; Carvalho-Ribeiro, S.M. Perceived ecosystem ssrvices (ES) and ecosystem disservices (EDS) from trees: Insights from three case studies in Brazil and France. *Landscape Ecol.* **2019**, *34*, 1583–1600. [CrossRef]
75. Armstrong, D. A Survey of Community Gardens in Upstate New York: Implications for Health Promotion and Community Development. *Health Place* **2000**, *6*, 319–327. [CrossRef]
76. Baldock, K.C.R.; Goddard, M.A.; Hicks, D.M.; Kunin, W.E.; Mitschunas, N.; Morse, H.; Osgathorpe, L.M.; Potts, S.G.; Robertson, K.M.; Scott, A.V.; et al. A systems approach reveals urban pollinator hotspots and conservation opportunities. *Nat. Ecol. Evol.* **2019**, *3*, 363–373. [CrossRef]
77. Roman, L.A.; Conway, T.M.; Eisenman, T.S.; Koeser, A.K.; Barona, C.O.; Locke, D.H.; Jenerette, G.D.; Östberg, J.; Vogt, J. Beyond 'trees are good': Disservices, management costs, and tradeoffs in urban forestry. *Ambio* **2021**, *50*, 615–630. [CrossRef]
78. Wooster, E.I.F.; Fleck, R.; Torpy, F.; Ramp, D.; Irga, P.J. Urban green roofs promote metropolitan biodiversity: A comparative case study. *Build. Environ.* **2022**, *207*, 108458. [CrossRef]
79. Russo, A.; Escobedo, F.J.; Girella, G.T.; Zerbe, S. Edible green infrastructure: An approach and review of provisioning ecosystem services and disservices in urban environments. *Agric. Ecosyst. Environ.* **2017**, *242*, 53–66. [CrossRef]
80. Pearsall, H. Staying cool in the compact city: Vacant land and urban heating in Philadelphia, Pennsylvania. *Appl. Geogr.* **2017**, *79*, 84–92. [CrossRef]
81. Curtis, A.J.; Helmig, D.; Baroch, C.; Daly, R.; Davis, S. Biogenic volatile organic compound emissions from nine tree species used in an urban tree-planting program. *Atmos. Environ.* **2014**, *95*, 634–643. [CrossRef]
82. Lyytimaki, J. Bad nature: Newspaper representation of ecosystem disservices. *Urban For. Urban Green.* **2014**, *13*, 418–424. [CrossRef]
83. Shackleton, C.M.; Ruwanza, S.; Sinasson Sanni, G.K.; Bennett, S.; Lacy, P.; Modipa, R.; Mtati, N.; Sachikonye, M.; Thondhlana, G. Unpacking Pandora's Box: Understanding and categorising ecosystem disservices for environmental management and human wellbeing. *Ecosystem* **2016**, *19*, 587–600. [CrossRef]
84. Wu, S.; Huang, J.; Li, S. Classifying ecosystem disservices and comparing their effects with ES in Beijing, China. *arXiv* **2020**, arXiv:2001.01605.
85. Campagne, C.S.; Roche, P.K.; Salles, J.-M. Looking into Pandora's Box: Ecosystem disservices assessment and correlations with ecosystem services. *Ecosyst. Serv.* **2018**, *30*, 126–136. [CrossRef]
86. Lara, B.; Rojo, J.; Fernández-González, F.; González-García-Saavedra, A.; Serrano-Bravo, M.D.; Pérez-Badia, R. Impact of Plane Tree Abundance on Temporal and Spatial Variations in Pollen Concentration. *Forests* **2020**, *11*, 817. [CrossRef]
87. Pecero-Casimiro, R.; Fernández-Rodríguez, S.; Tormo-Molina, R.; Silva-Palacios, I.; Gonzalo-Garijo, Á.; Monroy-Colín, A.; Coloma, J.F.; Maya-Manzano, J.M. Producing Urban Aerobiological Risk Map for Cupressaceae Family in the SW Iberian Peninsula from LiDAR Technology. *Remote Sens.* **2020**, *12*, 1562. [CrossRef]
88. Sousa-Silva, R.; Smargiassi, A.; Kneeshaw, D.; Dupras, J.; Zinszer, K.; Paquette, A. Strong variations in urban allergenicity riskscapes due to poor knowledge of tree pollen allergenic potential. *Sci. Rep.* **2021**, *11*, 10196. [CrossRef] [PubMed]
89. Lyytimäki, j.; Kjerulf Petersen, L.; Normander, B.; Bezák, P. Nature as a nuisance? Ecosystem services and disservices to urban lifestyle. *Environ. Sci.* **2008**, *5*, 161–172. [CrossRef]
90. D'Amato, G.; Cecchi, L.; Bonini, S.; Nunes, C.; Annesi-Maesano, I.; Behrendt, H.; Liccardi, G.; Popov, T.; van Cauwenberge, P. Allergenic pollen and pollen allergy in Europe. *Allergy* **2007**, *62*, 976–990. [CrossRef] [PubMed]
91. Cariñanos, P.; Grilo, F.; Pinho, P.; Casares-Porcel, M.; Branquinho, C.; Acil, N.; Andreucci, M.B.; Anjos, A.; Bianco, P.M.; Brini, S.; et al. Estimation of the Allergenic Potential of Urban Trees and Urban Parks: Towards the Healthy Design of Urban Green Spaces of the Future. *Int. J. Environ. Res. Public Health* **2019**, *16*, 1357. [CrossRef]
92. Hamilton, W.D. Sidewalk/curb-breaking tree roots. 1. Why tree roots cause pavement problems. *Arboric. J.* **1984**, *8*, 37–44. [CrossRef]
93. Wu, C.; Li, X.; Tian, Y.; Deng, Z.; Yu, X.; Wu, S.; Shu, D.; Peng, Y.; Sheng, F.; Gan, D. Chinese Residents' Perceived ES and Disservices Impacts Behavioral Intention for Urban Community Garden: An Extension of the Theory of Planned Behavior. *Agronomy* **2022**, *12*, 193. [CrossRef]
94. Azmy, M.M.; Hosaka, T.; Numata, S. Responses of four hornet species to levels of urban greenness in Nagoya city, Japan: Implications for ecosystem disservices of urban green spaces. *Urban For. Urban Green.* **2016**, *18*, 117–125. [CrossRef]
95. Hagemann, F.; Randrup, T.B.; Ode Sang, A. Challenges to implementing the urban ecosystem service concept in green infrastructure planning: A view from practitioners in Swedish municipalities. *Socio-Ecol. Pract. Res.* **2020**, *2*, 283–296. [CrossRef]
96. Andersson, E.; Enqvist, T.; Tengo, M. Stewardship in urban landscapes. In *Science and Practice of Landscape Stewardship*; Bieling, C., Plieninger, T., Eds.; Cambridge University Press: Cambridge, UK, 2017.
97. Jerome, G. Defining community-scale green infrastructure. *Landsc. Res.* **2017**, *42*, 223–229. [CrossRef]
98. Katz, D.S.W.; Connor Barrie, B.T.; Cary, T.S. Urban ragweed populations in vacant lots: An ecological perspective on management. *Urban For. Urban Green.* **2014**, *13*, 756–760. [CrossRef]
99. Nemeth, J.; Langhorst, J. Rethinking urban transformation: Temporary uses for vacant land. *Cities* **2013**, *40*, 143–150. [CrossRef]
100. Dennis, M.; Armitage, R.P.; James, P. Social-ecological innovation: Adaptive responses to urban environmental conditions. *Urban Ecosyst.* **2016**, *19*, 1063–1082. [CrossRef]

101. Kabisch, N. Ecosystem service implementation and governance challenges in urban green space planning—The case of Berlin, Germany. *Land Use Policy* **2015**, *42*, 557–567. [CrossRef]
102. Fox-Kamper, R.; Wesener, A.; Munderlein, D.; Sondermann, M.; McWilliam, W.; Kirk, N. Urban community gardens: An evaluation of governance approaches and related enablers and barriers at different development stages. *Landsc. Urban Plan.* **2017**, *170*, 59–68. [CrossRef]
103. Guitart, D.; Pickering, C.; Byrne, J. Past results and future directions in urban community gardens research. *Urban For. Urban Green.* **2012**, *11*, 364–373. [CrossRef]
104. Feltynowski, M.; Bergier, T.; Kabisch, N.; Laszkiewicz, E.; Strohbach, M.; Kronenberg, J. Challenges of urban green space management in the face of using inadequate data. *Urban For. Urban Green.* **2017**, *31*, 56–66. [CrossRef]
105. Kronenberg, J.; Haase, A.; Laszkiewicz, E.; Antal, A.; Baravikova, A.; Biernacka, M.; Dushkova, D.; Filcak, R.; Haase, D.; Ignatieva, M.; et al. Environmental justice in the context of urban green space availability, accessibility, and attractiveness in post socialist cities. *Cities* **2020**, *106*, 102862. [CrossRef]
106. Silva, C.; Viegas, I.; Panagopoulos, T.; Bell, S. Environmental Justice in Accessibility to Green Infrastructure in Two European Cities. *Land* **2018**, *7*, 134. [CrossRef]
107. Sanchez, L.; Reames, T. Cooling Detroit: A socio-spatial analysis of equity in green roofs as an urban heat island mitigation strategy. *Urban For. Urban Green.* **2019**, *44*, 126331. [CrossRef]
108. Anguelovski, I.; Connolly, J.J.T.; Masip, L.; Pearsall, H. Assessing green gentrification in historically disenfranchised neighborhoods: A longitudinal and spatial analysis of Barcelona. *Urban Geogr.* **2017**, *39*, 458–491. [CrossRef]
109. MacKenzie, A.; Pearson, L.J.; Pearson, C.J. A framework for governance of public green spaces in cities. *Landsc. Res.* **2018**, *44*, 444–457. [CrossRef]
110. Gunningham, N.; Holley, C. Next-generation environmental regulation: Law, regulation and governance. *Annu. Rev. Law Soc. Sci.* **2016**, *12*, 273–293. [CrossRef]
111. Lo, C. Going from government to governance. In *Global Encyclopedia of Public Administration, Public Policy and Governance*; Farazmand, A., Ed.; Springer: Cham, Switzerland, 2018.
112. Harrington, E.; Hsu, D. Roles for government and other sectors in the governance of green infrastructure in the U.S. *Environ. Sci. Policy* **2018**, *88*, 104–115. [CrossRef]
113. Depietri, Y. Planning for urban green infrastructure: Addressing tradeoffs and synergies. *Curr. Opin. Environ. Sustain.* **2022**, *54*, 101148. [CrossRef]
114. Armitage, D. Adaptive Capacity and Community-Based Natural Resource Management. *Environ. Manag.* **2005**, *35*, 703–715. [CrossRef]
115. Newig, J.; Fritsch, O. Environmental governance: Participatory, multi-level—And effective? *Environ. Policy Gov.* **2009**, *19*, 197–214. [CrossRef]
116. Spotswood, E.N.; Benjamin, M.; Stoneburner, L.; Wheeler, M.M.; Beller, E.E.; Balk, D.; McPhearson, T.; Kuo, M.; McDonald, R.I. Nature inequity and higher COVID-19 case rates in less-green neighbourhoods in the United States. *Nat. Sustain.* **2021**, *4*, 1092–1098. [CrossRef]
117. Dennis, M.; James, P. Site-specific factors in the production of local urban ecosystem srvices: A case study of community-managed green space. *Ecosyst. Serv.* **2016**, *17*, 208–216. [CrossRef]
118. Breen, A.; Giannotti, E.; Molina, M.F.; Vásquez, A. From "Government to Governance"? A Systematic Literature Review of Research for Urban Green Infrastructure Management in Latin America. *Front. Sustain. Cities* **2020**, *2*, 572360. [CrossRef]
119. Amundsen, H.; Berglund, F.; Westskog, H. Overcoming barriers to climate change adaptation—a question of multilevel governance? *Environ. Plan. C Govern. Policy* **2010**, *28*, 276–289. [CrossRef]
120. Aronson, M.; Lepczyk, C.A.; Evans, K.L.; Goddard, M.A.; Lerman, S.B.; MacIvor, J.S.; Nilon, C.H.; Vargo, T. Biodiversity in the city: Key challenges for urban green space management. *Front Ecol. Environ.* **2017**, *15*, 189–196. [CrossRef]
121. Allen, C.; Gunderson, L. Pathology and failure in the design and implementation of adaptive management. *J. Environ. Manag.* **2011**, *92*, 1379–1384. [CrossRef] [PubMed]
122. Webb, R.; Bai, X.; Stafford Smith, M.; Costanza, R.; Griggs, D.; Moglia, M.; Neuman, M.; Newman, P.; Newton, P.; Norman, B.; et al. Sustainable urban systems: Co-design and framing for transformation. *Ambio* **2018**, *47*, 57–77. [CrossRef] [PubMed]
123. Kim, G. An integrated system of urban green infrastructure on different types of vacant land to provide multiple benefits for local communities. *Sustain. Cities Soc.* **2017**, *36*, 116–130. [CrossRef]
124. Colding, J.; Barthel, S. The potential of 'Urban Green Commons' in the resilience building of cities. *Ecol. Econ.* **2013**, *86*, 156–166. [CrossRef]
125. Plummer, R.; FitzGibbon, J.E. Connecting adaptive co-management, social learning and social capital through theory and practicein. In *Adaptive Co-Management: Collaboration, Learning and Multi-Level Governance*; Armitage, D., Berkes, F., Doubleday, N., Eds.; University of British Columbia Press: Vancouver, BC, Canada, 2007; pp. 38–61.
126. Krasny, M.; Tidball, K.G. Community Gardens as Contexts for Science, Stewardship, and Civic Action Learning. *Cities Environ.* **2009**, *2*, 8. [CrossRef]
127. Barthel, S.; Folke, C.; Colding, J. Social–ecological memory in urban gardens—Retaining the capacity for management of ecosystem services. *Glob. Environ. Chang.* **2010**, *20*, 255–265. [CrossRef]

128. Lin, B.; Egerer, M.H. Global social and environmental change drives the management and delivery of ecosystem services from urban gardens: A case study from Central Coast, California. *Glob. Environ. Chang.* **2020**, *60*, 102006. [CrossRef]
129. Coffey, B.; Bush, J.; Mumaw, L.; de Kleyn, L.; Furlong, C.; Cretney, R. Towards good governance of urban greening: Insights from four initiatives in Melbourne, Australia. *Aust. Geographer* **2020**, *51*, 189–204. [CrossRef]
130. Ordonez, C.; Threlfall, C.; Livesley, S.; Kendal, D.; Fuller, R.; Davern, M.; der Ree, R.; Hochuli, D.F. Decision-making of municipal urban forest managers through the lens of governance. *Environ. Sci. Policy* **2020**, *104*, 136–147. [CrossRef]
131. Buijs, A.E.; Mattijssen, T.J.M.; van der Jagt, A.P.N.; Ambrose-Oji, B.; Andersson, E.; Elands, B.H.M.; Møller, M.S. Active citizenship for urban green infrastructure: Fostering the diversity and dynamics of citizen contributions through mosaic governance. *Curr. Opin. Environ. Sustain.* **2017**, *22*, 1–6. [CrossRef]
132. Jansson, M.; Vogel, N.; Fors, H.; Randrup, T. The governance of landscape management: New approaches to urban open space development. *Landsc. Res.* **2018**, *44*, 952–965. [CrossRef]
133. Mejia, A.; Bhattacharya, M.; Nigon-Crowley, A.; Kirkpatrick, K.; Katoch, C. Community gardening during times of crisis: Recommendations for community-engaged dialogue, research, and praxis. *J. Agric. Food Syst. Community Dev.* **2020**, *10*, 13–19. [CrossRef]
134. Mattijssen, J.M.T.; Buijs, A.A.E.; Elands, B.H.M.; Arts, B.J.M.; Van Dam, R.I.; Donders, J.L.M. The Transformative Potential of Active Citizenship: Understanding Changes in Local Governance Practices. *Sustainability* **2019**, *11*, 5781. [CrossRef]
135. Gulsrud, N.; Hertzog, K.; Shears, I. Innovative urban forestry governance in Melbourne? Investigating "green placemaking" as a nature-based solution. *Environ. Res.* **2018**, *161*, 158–167. [CrossRef]
136. Walker, B.; Holling, C.S.; Carpenter, S.R.; Kinzig, A. Resilience, Adaptability and Transformability in Social–ecological Systems. *Ecol. Soc.* **2004**, *9*, 5. [CrossRef]
137. Jerome, G.; Mell, I.; Shaw, D. Re-defining the characteristics of environmental volunteering: Creating a typology of community-scale green infrastructure. *Environ. Res.* **2017**, *158*, 399–408. [CrossRef]
138. Feindt, P.H.; Weiland, S. Reflexive governance: Exploring the concept and assessing its critical potential for sustainable development. Introduction to the special issue. *J. Environ. Policy Plan.* **2018**, *20*, 661–674. [CrossRef]
139. Ernstson, H.; Barthel, S.; Andersson, E.; Borgstrom, S.T. Scale-crossing brokers and network governance of urban ecosystem services: The case of Stockholm. *Ecol. Soc.* **2010**, *15*, 28. [CrossRef]
140. Thomas, K.; Littlewood, S. From Green Belts to Green Infrastructure? The Evolution of a New Concept in the Emerging Soft Governance of Spatial Strategies. *Plan. Pract. Res.* **2010**, *25*, 203–222. [CrossRef]
141. Trogrlic, R.S.; Rijke, J.; Dolman, N.; Zevenbergen, C. Rebuild by Design in Hoboken: A Design Competition as a Means for Achieving Flood Resilience of Urban Areas through the Implementation of Green Infrastructure. *Water* **2018**, *10*, 553. [CrossRef]
142. Rusciano, V.; Civero, C.; Scapato, D. Social and Ecological High Influential Factors in Community Gardens Innovation: An Empirical Survey in Italy. *Sustainability* **2020**, *12*, 4651. [CrossRef]
143. Ng, H. Recognizing the edible urban commons: Cultivating latent capacities for transformative governance in Singapore. *Urban Stud.* **2020**, *57*, 1417–1433. [CrossRef]
144. Mattijssen, T.; Buijs, A.; Elands, B.; Arts, B. The 'Green' and 'Self 'in Green Self-Governance—A Study of 264 Green Space initiatives by Citizens. *J. Environ. Policy Plan.* **2018**, *20*, 96–113. [CrossRef]
145. Connolly, J.; Svendsen, E.S.; Fisher, D.R.; Campbell, L.K. Organizing urban ecosystem services through environmental stewardship governance in New York City. *Landsc. Urban Plan.* **2013**, *109*, 76–84. [CrossRef]
146. Langemeyer, J.; Camps-Calvet, M.; Calvet-Mir, L.; Barthel, S.; Gomez-Baggethun, E. Stewardship of urban ecosystem services: Understanding the value(s) of urban gardens in Barcelona. *Landsc. Urban Plan.* **2017**, *170*, 79–89. [CrossRef]
147. Notcha, T.; Skelcher, C. Network governance in low-carbon energy transitions in European cities: A comparative analysis. *Energy Policy* **2020**, *138*, 111298.
148. Bixler, P.; Lieberknecht, K.; Atshan, S.; Zutz, C.P.; Richter, S.M.; Belaire, J.A. Reframing urban governance for resilience implementation: The role of T network closure and other insights from a network approach. *Cities* **2020**, *103*, 102726. [CrossRef]
149. Metzger, J.P.; Fidelman, P.; Sattler, C.; Schroter, B.; Maron, M.; Eigenbrod, F.; Fortin, M.; Hohlenwerger, C.; Rhodes, J. Connecting governance interventions to ecosystem services provision: A socio-ecological network approach. *People Nat.* **2020**, *3*, 266–280. [CrossRef]
150. Ghose, R. The complexities of citizen participation through collaborative governance. *Space Polity* **2005**, *9*, 61–75. [CrossRef]
151. Nyseth, T. Network Governance in Contested Urban Landscapes. *Plan. Theory Pract.* **2008**, *9*, 497–514. [CrossRef]
152. Chaffin, B.C.; Floyd, T.M.; Albro, S.L. Leadership in informal stormwater governance networks. *PLoS ONE* **2019**, *14*, e0222434. [CrossRef] [PubMed]
153. Ghose, R.; Pettygrove, M. Urban Community Gardens as Spaces of Citizenship. *Antipode* **2014**, *46*, 1092–1112. [CrossRef]
154. Carlsson, L.; Sandstrom, A. Network governance of the commons. *Int. J. Commons* **2008**, *2*, 33–54. [CrossRef]
155. Sandstrom, A.; Rova, C. Adaptive Co-management Networks: A Comparative Analysis of Two Fishery Conservation Areas in Sweden. *Ecol. Soc.* **2010**, *15*, 14. [CrossRef]
156. Keast, R. Network governance. In *Handbook on Theories of Governance*; Ansell, C., Torfing, J., Eds.; Edward Elgar Publishing: Cheltenham, UK, 2016; pp. 442–453.
157. Magaudda, S.; Ascanio, R.; Muccitelli, S.; Palazzo, A.L. 'Greening' green infrastructure. Good Italian practices for enhancing green infrastructure through the common agricultural policy. *Sustainability* **2020**, *12*, 2301. [CrossRef]

158. Toxopeus, H.; Kotsila, P.; Conde, M.; Katona, A.; van der Jagt, A.; Polzin, F. How 'just' is hybrid governance of urban nature-based solutions? *Cities* **2020**, *105*, 102839. [CrossRef]
159. Reames, T. Distributional disparities in residential rooftop solar potential and penetration in four cities in the United States. *Energy Res. Soc. Sci.* **2020**, *69*, 101612. [CrossRef]
160. Allen, C.; Angeler, D.G.; Garmestani, A.; Gunderson, L.H.; Holling, C.S. Panarchy: Theory and application. *Ecosystems* **2014**, *17*, 578–589. [CrossRef]
161. Frantzeskaki, N.; Bush, J. Governance of nature-based solutions through intermediaries for urban transitions—A case study from Melbourne, Australia. *Urban For. Urban Green.* **2021**, *64*, 127262. [CrossRef]
162. Kabisch, N. Transformation of urban brownfields through co-creation: The multi-functional Lene-Voigt Park in Leipzig as a case in point. *Urban Transform.* **2019**, *1*, 2. [CrossRef]
163. Kemp, R.; Loorbach, D.; Rotmans, J. Transition management as a model for managing processes of co-evolution towards sustainable development. *Int. J. Sustain. Dev. World Ecol.* **2007**, *14*, 78–91. [CrossRef]
164. Loorbach, D. Transition Management for Sustainable Development: A Prescriptive, Complexity-Based Governance Framework. *Gov. Int. J. Policy Adm. Inst.* **2010**, *23*, 161–183. [CrossRef]
165. Biggs, R.; Schlüter, M.; Schoon, M.L. (Eds.) *Principles for Building Resilience: Sustaining ES in Social–Ecological Systems*; Cambridge University Press: Cambridge, UK, 2014.
166. Schaffler, A.; Swilling, M. Valuing green infrastructure in an urban environment under pressure—The Johannesburg case. *Ecol. Econ.* **2013**, *86*, 246–257. [CrossRef]
167. Elmqvist, T.; Andersson, E.; Frantzeskaki, N.; McPhearson, T.; Olsson, P.; Gaffney, O.; Takeuchi, K.; Folke, C. Sustainability and resilience for transformation in the urban century. *Nat. Sustain.* **2019**, *2*, 267–273. [CrossRef]
168. Gabrys, J. Smart forests and data practices: From the Internet of Trees to planetary governance. *Big Data Soc.* **2020**, *7*, 2053951720904871. [CrossRef]
169. Møller, M.S.; Olafsson, A.S.; Vierikko, K.; Sehested, K.; Elands, B.; Buijs, A.; van den Bosch, C.K. Participation through place-based e-tools: A valuable resource for urban green infrastructure governance? *Urban For. Urban Green.* **2018**, *40*, 245–253. [CrossRef]
170. Boulton, C.; Dedekorkut-Howes, A.; Byrne, J. Governance Factors Shaping Greenspace Provision: From Theory to Practice. *Plan. Theory Pract.* **2021**, *22*, 27–50. [CrossRef]

land

MDPI

Article

The Embeddedness of Nature-Based Solutions in the Recovery and Resilience Plans as Multifunctional Approaches to Foster the Climate Transition: The Cases of Italy and Portugal

Elena Di Pirro [1,*,†], Rúben Mendes [2,†], Teresa Fidélis [2], Lorenzo Sallustio [1], Peter Roebeling [3,4], Marco Marchetti [1] and Bruno Lasserre [1]

1 DiBT: Dipartimento di Bioscienze e Territorio, Università degli Studi del Molise, Contrada Fonte Lappone, 86090 Pesche, Italy
2 GOVCOPP: Research Unit on Governance, Competitiveness and Public Policies, Department of Environment and Planning, Campus Universitario de Santiago, University of Aveiro, 3810-193 Aveiro, Portugal
3 CESAM: Centre for Environmental and Marine Studies, Department of Environment and Planning, Campus Universitario de Santiago, University of Aveiro, 3810-193 Aveiro, Portugal
4 Wageningen Economic Research, Wageningen University and Research, Droevendaalsesteeg 4, 6708 PB Wageningen, The Netherlands
* Correspondence: elena.dipirro@unimol.it
† These authors contributed equally to this work.

Abstract: European countries recently prepared recovery and resilience plans (RRPs) to recover from the pandemic crisis and reach climate neutrality. Nature-Based Solutions (NBS) are recognized as crucial drivers to fostering climate transition while addressing other challenges. Accordingly, RRPs offer the opportunity to promote the adoption of NBS. This article assesses the NBS embeddedness in the policy discourse of Italian and Portuguese RRPs and how they are considered to meet climate–and related environmental–targets. We conducted a discourse analysis based on two steps, (*i*) a quantitative analysis to classify different nature-related terms into four categories—biophysical elements, general environmental concepts, threats and challenges, and NBS—and estimate their frequency in the text; (*ii*) a qualitative analysis to understand the relationship between the categories of challenges and NBS as well as the dedicated investments. The results show that NBS are barely mentioned, with a frequency in the texts for the NBS category of 0.04% and 0.01%, respectively, in Italian and Portuguese RRPs. Narratives are mainly built around general concepts such as resilience and sustainability with nature scarcely considered as an ex novo solution to meet challenges. Notwithstanding, Italy invests 330 M in the implementation of urban forests, while in Portugal, no specific NBS interventions have been considered so far. To date, both countries are primarily orienting the climate transition toward reducing emissions instead of combining these measures with multifunctional NBS to address environmental and socio-economic challenges.

Keywords: discourse analysis; environmental policies; green deal; NextGenerationEU; bio-based economy; climate change; urban forests

1. Introduction

The European Union is addressing the recovery from the pandemic crisis by investing in a stimulus package worth EUR 2.018 trillion at current prices. It consists of a combination of the EU's long-term budget for 2021 to 2027 and the NextGenerationEU [1]. The latter is a temporary instrument to stimulate the recovery with scope and ambition without precedent, including investments and reforms to accelerate the ecological and digital transition, support education, and achieve greater gender, territorial and generational equality. To access NGEU funds, each member state had to prepare a national recovery and resilience plan (RRP) for the period 2021–2026, according to the criteria established by Article 18 of Regulation no. 2021/241/EU. One-third of the overall EU budget aims to finance the European Green Deal, under which Europe aims to become the first climate-neutral continent by 2050, producing no more greenhouse gases than the ecosystems can naturally absorb. To reach this target, all the member states pledged to reduce net greenhouse gas emissions by at least 55% in 2030 as compared to their 1990 levels. Further ambitious environmental goals were set for all member states, such as zero soil sealing by 2050 [2] and a vast tree planting campaign (i.e., 3 billion trees by 2030; [3]). Achieving these goals relies on the transformation of all sectors of the EU's economy, requiring a paradigm shift for a transition to a circular, nature-positive, carbon-neutral, bio-based and equitable economy [4–7]. Therefore, the focus should not be only on the transformation of energy and transport systems, but also on measures across the economy to harness the potential of nature to contribute to both mitigating climate change and enhancing our resilience to its impacts [8]. As one of the main environmental challenges, climate change is already affecting Europe's ecosystems and human health, and it is expected to pose further threats to the ecosystem and socio-economic system shortly [9–11]. The RRP is thus an opportunity for all the member states to include and invest in a nature-based recovery, addressing the effects of climate change via adaptation and mitigation measures [4]. The latest IPCC report states that all scenarios that limit climate change to 1.5 °C rely on decreasing emissions, decarbonizing the economy as well as adopting land-use change mitigation strategies [12,13]. Accordingly, coupling the decrease in emission sources with the increase in carbon sinks through terrestrial ecosystems is one of the most reliable strategies to fight climate change [9,14,15]. Particularly, all nature-based approaches have emerged as a key instrument to face different challenges across sectors of society and business, also offering multiple cost-effective benefits to ecosystems and human wellbeing [12,16]. However, adopting a nature-based economic perspective means the explicit recognition of nature as both providing inputs and generating outputs for our economy [4,17]. Although it could be still difficult to assess the monetary value and the economic benefits of the outputs [18,19], it is recognized worldwide that there is a need to overcome the "business-as-usual" model based on resource exploitation, biodiversity loss, and carbon emission growth through investing in nature and fostering the transition to sustainable development [6,20].

In recent years, several nature-based approaches have become a key topic of contemporary research on sustainable development of urban and rural areas [21] such as ecological restoration, ecological engineering, green and blue infrastructure, ecosystem services, urban forestry, ecosystem-based management and adaptation, and eco-disaster risk reduction [22–24]. Since 2015 [25], the concept of Nature-Based Solutions (NBS) enclosed each of them under its 'umbrella' including all the approaches, with different terminology, that work with and enhance nature to help address multiple challenges [26]. Several studies have indeed shown how NBS are very efficient in facing extreme events related to climate change, by adaptation and mitigation actions (e.g., reducing flood risk, and storing CO_2) [10,27,28], contemporarily able to preserve human health [27,28], psychosocial well-being [29,30], improve air quality [31–33], and increase landscape connectivity [34,35].

Thanks to their capacity and multifunctionality, NBS are gaining momentum in the emerging policy discourse, and multiple initiatives raised to mainstream the NBS, encouraging their development for more sustainable and just communities [36]. Therefore, NBS are expected to further shape the policy narrative in global environmental decision-making [37]. Accordingly, at a global policy level, 66% of all signatories to the Paris Agreement included NBS for climate change mitigation and adaptation in their nationally determined contributions [38,39]. Furthermore, the EU claims to be a world leader in NBS through supporting numerous projects in the Research and Innovation Agenda [23,25]. These projects are proving to be a catalyst for research–practice partnerships [40], gathering insights regarding NBS performance, impacts assessments, and cost-effectiveness [8,41]. Consequently, NBS are tested in front-runner cities, demonstration sites, and urban living labs, and the EU is using their outcomes to upscale these initiatives to a broader public [10,23] and to facilitate their operationalization from experts to decision makers and stakeholders [42]. However, the contexts of urban living labs, as well as frontrunner cities and regions, are designed to provide flexible governance conditions, supportive decision makers, and policy instruments [43], hence scarcely representing the complicated real-life contexts of practice [39]. Furthermore, working with limited and scattered case studies, and often only at the local and municipal level, increases the difficulties to spread the gained knowledge to other contexts and scales, rising the issues related to planning silos [44,45]. This overlaps with the fact that the policy instruments for NBS implementation are mainly restricted to the municipal level (i.e., focusing on urban planning [46]) and not to the landscape, country or higher levels [47]. So far, the processes to mainstream and institutionalize NBS into national policy are still not clear, and this concept, with its huge potential, is suffering multiple incorporation difficulties as already observed for other environmental concepts (e.g., ecosystem services) [45].

Given the planned investments to reach global and European targets (e.g., RRP) as well as the high NBS capacity to help in this path, there is still a need to enhance knowledge regarding the NBS inclusion at national and regional policy levels [16,48]. It is necessary to urgently strengthen policy frameworks at the national level [49] to enhance NBS multifunctionality in favor of climate mitigation and adaptation, biodiversity conservation and human well-being as a whole [12,44,48,50]. Therefore, as economic instruments are usually recognized as enablers for a successful NBS uptake [51], taking benefit from the investments related to the recovery from the pandemic crisis is probably the once-in-a-lifetime opportunity to systematically introduce NBS in the member states policy framework.

In this work, we explored if member states have seized this opportunity by analyzing how the role of nature is embedded in the narrative of RRP documents, and how NBS are framed as an investment to foster the climate transition. Narrative and discourse analysis have been applied to other environmental policies, processes, or plans [52] to assess the embeddedness of particular topics since different narrative approaches can influence decision making and knowledge production [37]. Particularly, we focused on two case studies, Italy and Portugal. Both countries are heavily impacted by climate change and are studied by several projects focusing on NBS and related approaches (e.g., H2020, LIFE). Firstly, we conducted a discourse analysis based on two stages. A quantitative analysis to collect different nature-related words included in both the RRPs, classifying them into four different categories of terms. After that, we conducted a qualitative analysis to understand the way NBS are included in the text and how they are translated into actions, interventions as well as investments. Finally, we presented a comparative analysis between the two member states, highlighting the current state, pros, cons, possible ways forward, and future challenges.

2. Background

Description of the Recovery and Resilience Plans in Italy and Portugal

The recovery and resilience plans (RRPs) aim to mitigate the economic and social impact of the coronavirus pandemic and to enhance EU sustainability, resilience, as well as its ability to face climatic and digital transitions' challenges. The EU regulation sets six major areas of intervention (pillars) on which all RRPs have to focus: green transition; digital transformation; economic cohesion, productivity, and competitiveness; social and territorial cohesion; health, economic, social, and institutional resilience; and policies for the next generation. The green transition pillar derives directly from the Green Deal and thus shares the dual goal of achieving a reduction in greenhouse gas emissions of 55% compared to the 1990 scenario by 2030 and, in turn, to achieve climate neutrality by 2050. The regulation of the NGEU stipulates that (*i*) at least 37% of planned investment and reform should support climate goals, and (*ii*) all the investments and reforms must respect the principle of "do no significant harm" to the environment [1].

The RRP is, in each member state, a reform plan mainly based on fostering economic growth and increasing job opportunities. The guidelines for the development of RRP identify under the name of "Components" the areas where aggregate investments and the respective reforms to reach specific objectives. In accordance, the investment lines need to be matched to a reform strategy aimed at improving the regulatory and legal conditions of the context and to steadily increase the country's equity, efficiency, and competitiveness. Each Component reflects reforms and investment priorities in the area of intervention to address specific challenges by building a coherent package of complementary measures.

The expected economic growth in terms of gross domestic product is similar between the two countries analyzed. Both countries start with a growth in the gross domestic product of 1.5%, expected to lift to 2.5% in Italy and to 2.4% in Portugal, by 2026, employing economic resources about 12 times higher in Italy than Portugal. The expected economic growth is up to 240,000 and 50,000 new jobs, respectively.

The Italian RRP is organized into 16 Components, in turn comprising 63 reforms and 163 investments, financing a total of EUR 191.5 billion. The Components and the respective reforms and investments are grouped into six missions: digitalization (40.3 billion) ecological transition (59.5 billion), sustainable mobility (25.4 billion), research and education (30.9 billion), social cohesion and inclusion (19.9 billion), and health (15.6 billion). The ecological transition takes the highest percentage of the total funding program with respect to the other missions [53].

The Portuguese RRP is organized into 20 Components that, in turn, comprise 37 reforms and 83 investments, financing a total of EUR 16.6 billion. The Components and the respective reforms and investments are grouped in three structuring dimensions: resilience (11.1 billion), climate transition (3 billion), and digital transition (2.5 billion). Both the transitions—climate and digital—represent 33% of the total funding program, while the remaining resources are dedicated to the resilience dimension, which encompasses the aspect of social vulnerabilities, economic and territorial resilience [54].

3. Materials and Methods

The methodology used in this work is based on a discourse analysis [52] conducted in both the original RRPs. In our understanding of discourse analysis, discourses are defined as "socio-cultural meaning structures identified through general characteristics of text, speech or the symbolic aspect of actions" [52]. Narratives are instead adopted by different stakeholders (e.g., policymakers, NGOs, and research institutes) to frame and legitimize their work associated with or adapted to a certain discourse [55]. In our work, we divided the discourse analysis into two different steps, quantitative and qualitative. Firstly, we conducted a content analysis of the RRPs considering different nature-related terms and we grouped them into four categories, Biophysical elements (I), General environmental concepts (II), Threats and challenges (III), and NBS (IV). Grouping the terms into categories was instrumental, as nature can be framed in the narrative of policies according to different aspects and functions, which is reinforced by the growing use of discourse analysis to study environmental challenges in policy topics [52]. As visible in Table 1, in category I, we considered the most common biophysical elements (e.g., tree). In category II, we considered the concepts that are usually included in the policy narratives (e.g., the environment in a broader meaning) but not associated with physical elements or established solutions. Particularly, some of the concepts that have become hegemonic in the policy discourse (e.g., resilience and sustainability) by functioning as a linguistic political mechanism, despite their frequent decoupling from objectives, indicators, and outcomes in policy achievement, from environmental conservation to social equity [56]. In category III, we considered threats and environmental challenges (e.g., climate change, biodiversity loss) that can be potentially addressed by NBS or related approaches (e.g., green infrastructure, urban forests), as already proposed in the literature [57]. Lastly, in category IV, we considered approaches and methods that conceptualize nature as a solution to face multiple challenges (e.g., urban forestry); thus, we considered them under the umbrella of NBS. In accordance with this classification, quantitative analysis is performed as an instrumental step to help understand the overall term frequency patterns shown in both documents and then employed to orientate the qualitative step by focusing on the relationship between specific groups of terms. Accordingly, in the qualitative step, we aim to investigate if nature is framed as a solution to meet the socio-economic and environmental challenges mentioned in the text and if traditional approaches (e.g., brown infrastructure [58], grey interventions [59]) are envisaged to address the challenges. In line with the aim of this work, we thus focused the qualitative analysis only on the last two categories, i.e., threats and challenges, and NBS. Through a coding process, we investigated the relationship between terms included in the threats and challenges category with terms included in the NBS category, selected in the quantitative step. To perform both the steps of the analysis, we used the Atlas.Ti (version 8.3), a software used in social science research that assists in both the qualitative and quantitative steps of the research (for a detailed review of the software, please see [60]).

Table 1. Coded terms and associated words, aggregated into four categories, "biophysical elements"—category I, "general environmental concepts" related to environment and ecological transition (hereafter "general environmental concepts")—category II, "threats and challenges" potentially addressed by NBS (hereafter "threats and challenges")—category III, and "different ecosystem-based approaches" (hereafter "NBS")—category IV. The use of the "*" at the end of the word accounted for both the plural and all the related words.

	Terms
Category I Biophysical elements	Tree* Air Territor* ǀ Land Water ǀ Wetland* ǀ Irrigation* Soil Ecosystem* Biodiversity ǀ Habitat* ǀ Specie* Riparia* Forest* Agroecologic* Sea ǀ Marin* ǀ Coastal
Category II General environmental concepts	Natural Capital Circular Economy Green* Climate Transition ǀ Green transition Ecological Transition Unsustainab* Resilie* Sustain* Resist* Natur* Ecologi*
Category III Threats and challenges	Climate change* Land take ǀ Soil sealing ǀ Urbaniz* Pollut* Biodiversity loss ǀ Ecosystem fragmentation ǀ Habitat fragmentation Hydrological risk ǀ Landslide risk ǀ Floods Drought Heat Island ǀ Thermic stress ǀ Heat wave Desertifi* Energy efficiency
Category IV Nature-based solutions	Ecological Network* ǀ Ecological Connect* Natural Park* ǀ National Park* ǀ Protected/Natural area* ǀ Marine area* Nature/Ecosystems/Biodiversity/Landscape conservation Ecological/Natural/Environmental restoration Ecosystem based approach* Ecosystem service* Renatur* Nature based solution* Blue Infrastructure* Green/Ecological Corridor* ǀ Green Infrastructure Natural engineering solution* ǀ Bioclimatic architecture solution* Permeab* Urban Forest* Forestation* ǀ Reforestation ǀ Afforestation ǀ Forestry Green area* ǀ Green space* ǀ Garden*

3.1. *Quantitative Step*

First, we conducted a preliminary analysis of both RRPs in the original languages (i.e., Italian and Portuguese) to identify and collect all nature-related terms enclosed in the documents (please see Table S1 in the Supplementary Materials; the terms are selected in both original languages). In addition, through a grey literature review, we identified challenges and threats related to climate change that can be addressed by NBS [41] and collected them along the texts. In total, we found 46 different nature-related terms and grouped them into four categories able to explain the different roles and relationships with nature, namely "biophysical elements", "general environmental concepts" related to environment and ecological transition (hereafter "general environmental concepts"), "threats and challenges" potentially addressed by NBS (hereafter "threats and challenges"), and "different ecosystem-based approaches" (hereafter "NBS"). Specifically, Table 1 shows the 46 entries derived from the content analysis and classified according to the four categories. When necessary, for some terms, we also considered other associated words, i.e., both plural and singular (expressed in the table with the *) as well as the synonymous or close meanings, e.g., Heat Island* | Thermal stress | Heatwave* (Table 1). All the words identified (i.e., singular, plural, synonymous) were associated and coded as a single term (i.e., each of the 46 entries in Table 1) through the Autocoding tool included in the software Atlas.Ti (version 8.3) [60]. The search was conducted using the same principles of searching as in the scientific databases; the use of the "*" at the end of the search accounted for all the related words, e.g., "ecologi*" accounted for "ecological", "ecologically", etc. Hence, all the words selected for each entry of the table were counted as references and assigned to the respective term. In this way, we built a database with the number of references per term shown throughout the document. During the Autocoding process, we excluded the words that were not related to the meaning in the search, e.g., when the word "nature" is presented as "the nature of the problem", the word "nature" was excluded from the counting as it is not relevant with the meaning of our interest. The number of references for each term was then (*i*) summed up under each category to analyze the relative percentage of the category out of the total words of the document and (*ii*) analyzed as the relative percentage out of the total references counted in the document. These metrics allow us to discuss and compare the different RRPs in both absolute and relative terms.

3.2. *Qualitative Step*

In the qualitative step, we focused our attention on the terms included in the category of threats and challenges (III) and NBS (IV) identified in the quantitative step. We considered NBS, according to Eggermont et al. [61], as follows: Type 1 NBS, no or minimal intervention in ecosystems for maintaining ecosystem services supply (e.g., protected areas and conservation measures); Type 2, management approaches for improving the ecosystem services supply compared to what would be obtained with a more conventional intervention (e.g., multifunctional agricultural and forests management); and Type 3, creating new ecosystems (e.g., green roofs). For each of the terms included in the threats and challenges category, we thus investigated when they are addressed by NBS (i.e., Type 1, 2, 3) or by a traditional or grey approach (i.e., absence of NBS). We used an open coding approach to assess how the categories of threats and challenges (III) and NBS (IV) were framed in the policy discourse and then used an axial coding approach to relate the two categories and understand if and how NBS are being considered to address the challenges (for the different coding approaches please see [62]). Exploring the relationship between the terms in these two categories we proposed a critical reflection, inspired by critical discourse analysis and eco-linguistic, regarding the capacity of the government to seize the opportunity to include NBS to foster the climate transition and meet the challenges presented in the two policy documents (for further applications of critical discourse analysis in environmental and policy discourse see [63,64]).

4. Results

4.1. Quantitative Analysis

Of the 46 terms analyzed in both the RRPs, in the Italian RRP, we found 1410 references out of 111,178 total words of the whole document (1.27%), while in the Portuguese RRP, we found 1505 references out of the 127,171 words of the whole document (1.18%). All the categories show a frequency below 1% in both documents, with a similar pattern in both countries (Figure 1). Italy shows a higher relative frequency than Portugal in categories II, III, and IV (i.e., general environmental concepts, threats and challenges, and NBS, respectively) while only in category I (i.e., biophysical elements) was this ratio is reversed. The most frequent category is general environmental concepts (II), with, respectively, 0.67% and 0.58%, in Italy and Portugal, followed by the biophysical elements with, respectively, 0.45% and 0.49%, the threats and challenges with, respectively, 0.11% and 0.10%, and lastly, the NBS with values lower than, respectively, 0.04% and 0.01%.

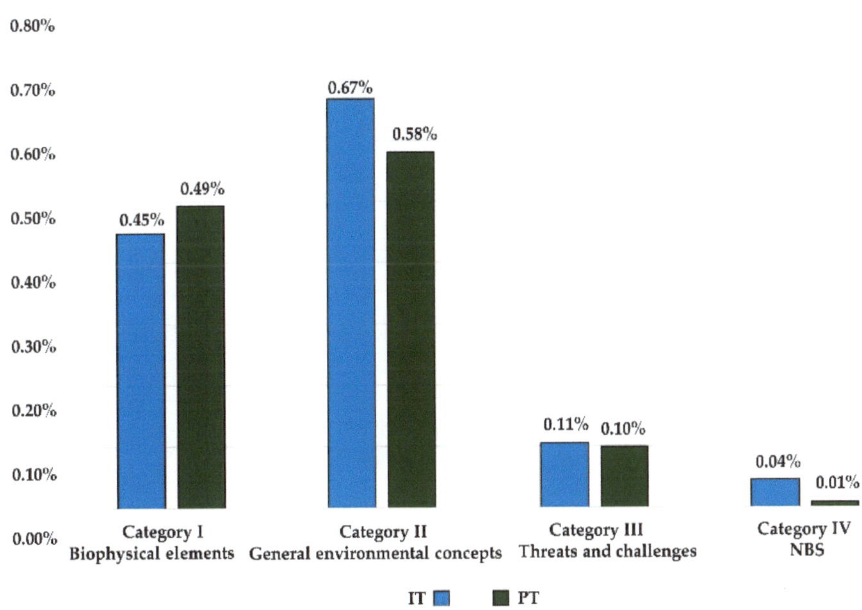

Figure 1. Frequency (%) of each category in the text of both resilience and recovery plans presented as the percentage of terms by the total amount of words in each document.

In Italy, the 46 coded terms show 1410 references throughout the document, with clear differences in their frequency among the four considered categories (Figure 2). Particularly, in category I (total 500 references), all the terms show at least one reference in the text. For ease of reading and exposure, we present the results referring to the first word for each entry (Table 1). The two most used terms are territory (i.e., 236 references) and water (i.e., 131), followed by biodiversity (i.e., 43) and ecosystem* (i.e., 34). In category II (total 733 references), the most frequent term is resilience (i.e., 337), followed by sustainability (i.e., 154) and green (i.e., 93), and ecology (i.e., 52). Climate transition, unsustainability, and resistance are terms completely absent in the Italian RRP. In category III (tot 117 references), all the terms show at least one reference, and the most frequent is energy efficiency (i.e., 46), followed by pollution (i.e., 25) and climate change (i.e., 23). In category IV, (total 50 references), the most frequent terms are gardens and green areas (i.e., 14), protected areas (i.e., 10), followed by restoration, nature conservation, renaturalization, reforestation, urban forest and ecosystem services. References to nature-based solutions, green infrastructure, blue infrastructure, naturalistic engineering, and permeability are absent.

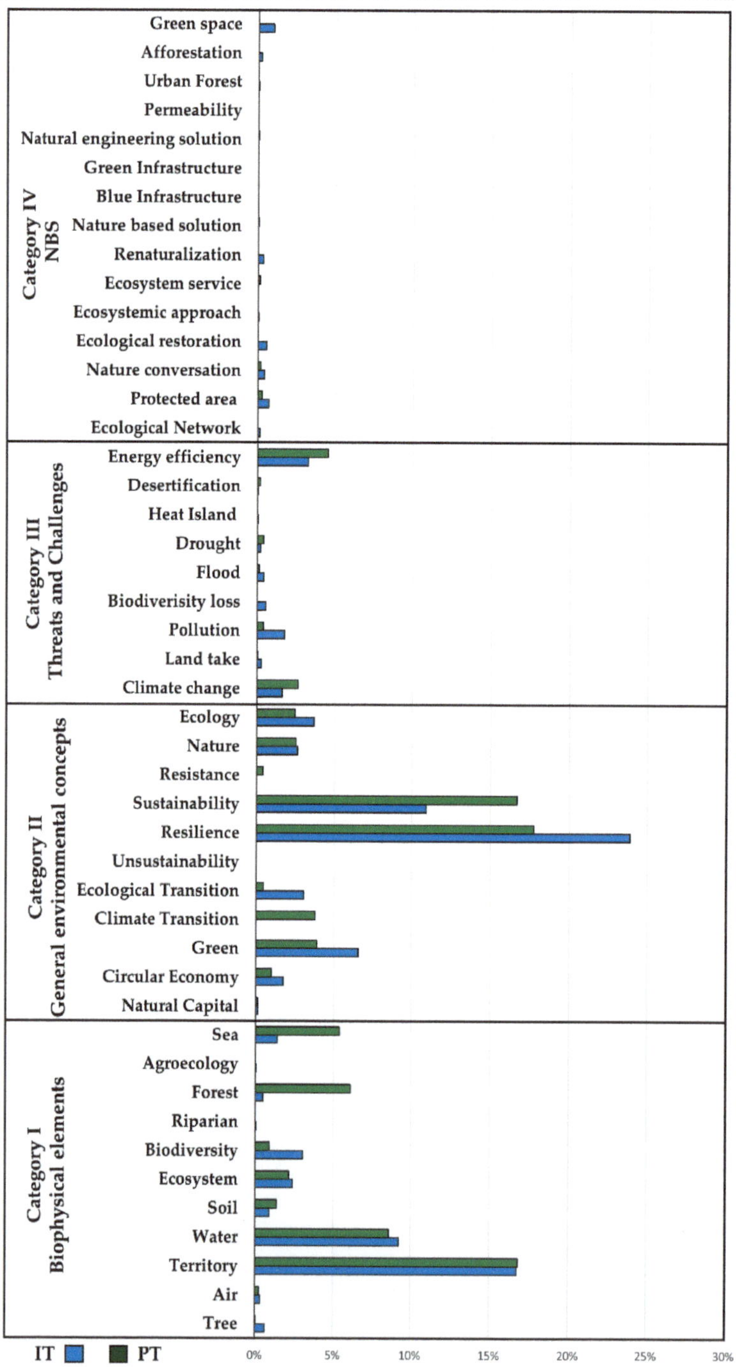

Figure 2. Relative frequency (%) of terms in the text of both national resilience and recovery plans, presented as the percentage of terms by the total amount of coded terms. We reported only the first word of the coded terms, for the complete list please refer to Table 1.

In Portugal, the 46 coded terms show 1505 references throughout the document, with a similar relative distribution frequency in four categories compared to the Italian results. In this country, particularly, in category I (total 629 references), territory and water are confirmed as the two most frequent terms (253 and 130, respectively), followed by forest (i.e., 92) and sea (i.e., 81), while riparian and agro-ecology are absent. In category II (total 739 references), resilience and sustainability are confirmed as the two most frequent of the category (267 and 251, respectively), followed by green and climate transition (59 and 57, respectively). Analogously to the Italian RRP, the concept of unsustainability is absent. Category III (total 126 references) is composed mainly of energy efficiency and climate change (68 and 40, respectively), while the other terms coded as challenges appear in the text less than six times, with habitat fragmentation and heatwaves absent. Considering category IV (total 11 references), the terms protected areas, natural conservation, ecosystem service, NBS, and ecological engineering appear less than five times.

4.2. *Qualitative Analysis: The Role of Nature in the Narratives and Investments Envisaged to Meet the Challenges*

Following the quantitative analyses, and considering the previous results, we explored across the documents how the threats and challenges presented in both plans (117 references for Italy and 126 for Portugal) were envisaged to be addressed and the respective mitigation role of the NBS mentioned (50 references for Italy and 11 for Portugal). We found that the inclusion of nature takes different meanings and roles in the policy narrative (e.g., nature as a resource, as hazard, etc.) and in the investments to tackle the challenges. Particularly, both plans included conservation and protection approaches (NBS Type 1) as well as management action (NBS Type 2) to face threats and challenges, while only the Italian plan includes and invests in NBS Type 3 (i.e., building new ecosystems). However, we found that the policy responses to the threats and challenges considered in the work are still mainly oriented toward conventional approaches or grey infrastructures (i.e., the absence of specific references to NBS in the texts). This is also confirmed by the fact that the investments to foster the ecological and climate transitions (Mission 2 in Italy and Dimension 2 in Portugal) are largely dedicated to mitigation measures (i.e., decreasing the emissions of industries and transports, decarbonization strategies), while poor emphasis is paid to adaptation measures in both policy narratives. Indeed, in the area of climate reforms and investments, Italy's major challenges include strengthening the energy efficiency of buildings (about EUR 15 billion), improving the management of waste and water resources (about EUR 12 billion), as well sustainable mobility (about EUR 35 billion). Similarly, Portugal's challenges include strengthening the energy efficiency of buildings (EUR 610 million) and sustainable mobility (EUR 967 million), as well as diversifying energy sources, hydrogen and decarbonization of the industries (EUR 1 billion summing up the two investments).

Accordingly, energy efficiency stands in both plans as the most referenced challenge to be addressed to foster the ecological and climate transition and fight climate change. Italy planned to increase the energy efficiency of buildings and facilities, particularly on farms and agricultural enterprises (Mission 2 Component 1; investment 2.2; hereafter, all the Italian investments are reported in the following short form, M2C1; i2.2), school and judicial buildings, as well as private buildings (M2C3; i1.1, i1.2, i2.1), reaching a potential surface of intervention of about 40 million m^2, and investing in total about EUR 15.3 billion. There is no mention to NBS or related approaches, despite the document proposed specifically structural works (e.g., thermal insulation, solar or photovoltaic panels). Similarly, Portugal also proponed structural interventions aimed at reducing emissions and energy expenditure with an overall investment of EUR 610 million (Component 13; hereafter, all the Portuguese investments are reported in the following short form, C13), explicitly referring to the possibility of NBS inclusion such as green roofs or, more generally, bioclimatic architectural solutions without envisaging any investment.

Among the other challenges here considered, large investments were dedicated to water management. In both RRPs, they are articulated to both face flood vulnerability (referenced six and two times, Italy and Portugal, respectively) and water scarcity, i.e., drought (three and six times, respectively). Nature takes the double meaning of hazard and biophysical resource to be preserved, being scarcely considered as a solution to actively address water management in an explicit manner. In the Italian document, the conservation, monitoring, and requalification of the territory are framed as possible strategies to mitigate the flood and hydrogeological vulnerability, providing investment up to EUR 2.5 billion (M2C4; i2.1). On the other hand, water scarcity is mainly addressed through investments in new grey infrastructure and traditional interventions, investing in Italy EUR 2 billion (M2C4; 4.1) and in Portugal EUR 390 million (C9).

Similarly, also in the investments dedicated to the sea and coastal areas, nature is conceptualized as a resource to be preserved and restored, but the narrative of the two RRPs shows different objectives. On one hand, the Italian plan recognizes the importance of the challenges related to sea-level rise as a cause of marine and biodiversity loss. Therefore, the protection and sustainable management of marine natural capital and restoration of coastal areas are mentioned in two different investments with a total of EUR 670 million (M2C4; i3.5 with EUR 400 million plus M3C2; i1.1 with EUR 270 million). On the other hand, in the Portuguese plan, the sea and coastal areas are mainly framed as an economic asset, and the narrative is embedded in terms such as "sea economy" or "sea potentialities", thus dedicating most of the investment to enhancing the sea and coastal-related economy (C10—EUR 252 million).

Both countries identified the integrated management of croplands and forests as crucial to preserve cultural and natural heritage and enhance job opportunities, but even in this case, the narrative of the documents is oriented to emphasize different objectives and aspects between countries. In the Italian RRP, the investment is oriented to foster the sustainable use of environmental resources (e.g., timber production), encourage "slow tourism" [65], and the energy autonomy of mountain and rural communities (Green Communities project, M2C1; i3.2 with EUR 140 million). Particularly, in Italy, these territories are referred as "inner areas" [66], and they are already subject to specific national policies and investments aiming to reduce the socio-economic gap with cities through enhancing a more sustainable and bio-based economy [67,68]. In the Portuguese RRP, an entire Component and related investments (C8—EUR 615 million) are dedicated to forests. However, the investment objective is mainly oriented to forests' management to increase the resistance and resilience to wildfire, framed as the main threat to Portuguese forests. Accordingly, the investments and reforms are mainly focused on the importance of the risk prevention for the population and biodiversity. Furthermore, the document refers to silviculture actions as a way to enlarge the portion of managed areas, increasing productivity and economic opportunities. Nonetheless, the Portuguese document explicitly recognizes the role of forest management to improve the potential of forests as a carbon sink, emphasizing their mitigation potential, also including the conservation and enhancement of biodiversity and natural capital to ensure the ecosystem services supply.

Analogously, the protection and enhancement of natural and cultural capital are identified in the Italian plan as an opportunity to foster culture and tourism without increasing threats related to land take and urbanization. In this regard, an intervention is planned to restore and requalify 5000 Italian historical parks and gardens in urban and peri-urban contexts (14 references) (M1C3; i2.3 with EUR 300 million allocated). The narrative related to this investment thus recognizes not only the cultural and social value of gardens, but also their importance in increasing ecosystem services supply can, in turn, improve human health and well-being.

The narrative of the Italian document builds an even more specific and explicit language, recognizing the value of restoring vulnerable ecosystems (e.g., riparian) and strengthening the ecological connectivity with new ecosystems to mitigate pollution, reducing hydrogeological risk, and fighting habitat fragmentation, pollution, and degradation. The investment of EUR

360 million (M2C4; i3.3) thus includes the ecological restoration of one of the most degraded, fragmented, and polluted areas in Italy (i.e., Po 'valley), providing for widespread renaturation interventions along all the ecological corridor (i.e., 1500 ha). Similarly, the investment related to the enhancement of urban green areas (M2C4; i3.1) provides EUR 330 million for urban forestry interventions, specifically planting 6.6 million new trees in the 14 Italian metropolitan cities for mitigating pollution in densely inhabited areas.

As opposed to the Italian RRP, in the Portuguese one, we could not find any measures clearly referring to the construction of new ecosystems (i.e., Type 3). Therefore, any measure is comparable to the renaturation or to the implementation of urban forests, as envisaged in the Italian RRP. Except for the unique reference to the possible NBS implementation to promote energy efficiency in residential areas (i.e., green roofs), the inclusion of nature in urban contexts is absent from the Portuguese RRP. The absence of terms encountered in the quantitative step, e.g., green spaces, green and blue infrastructures, confirms that nature is conceptualized in the Portuguese policy narrative mainly as elements belonging to the rural areas and not framed as a solution to tackle the urban challenges. Accordingly, urbanization, habitat fragmentation, and heat islands, challenges usually related to urban contexts, are absent in the Portuguese RRP. Furthermore, the threat of pollution is scarcely mentioned (six references), focusing only on a reduction in sources of pollutants. Besides the unique reference to prevent pollution in the sea, we could not find any other relation between nature and pollution nor investments that use nature as a way to deal with pollution issues.

5. Discussion

The RRP aims to promote a robust recovery of the economies achieving climate neutrality by 2050 and reducing greenhouse gas emissions by 55% compared to the 1990 scenario by 2030. The regulation of the NextGenerationEU required at least 37% of planned investment and reform to reach climate goals. Hepburn et al. [69] analyzed 300 global rescue and recovery policies from COVID-19 highlighting that the packages seeking synergies between economic and climate goals have better prospects for increasing national wealth, by enhancing productive, social, physical, and natural capital [69]. Following this logic, the former statement suggests that all EU member states might be considered on the right track, regarding both the expected economic growth and climate goals the EU set.

The environmental threats considered in this paper are all directly or indirectly correlated with the challenges of climate change and could be faced with NBS. We thus excluded other challenges that may be addressed in the literature by NBS, such as public health and social cohesion [28,70,71], as they cannot be limitedly associated with the causes and effects of climate change. As a consequence of our analysis, which focused on both policy discourse and allocated investments, we can state that NBS do not represent the main policy narrative in RRPs to respond to the environmental threats and challenges associated with climate change. Indeed, both plans identified the improvement of energy efficiency and renewable energy, and the decarbonization of industry and transport as the most relevant levers for reaching the climate goals, mainly financing interventions for reducing greenhouse gas emissions. However, even if measures to limit the temperature increase to 1.5 °C will be successful, some impacts will continue to increase due to climate system feedback and inertia (e.g., sea-level rise) [13,38]. According to Hepburn et al. [69], "natural capital investments for ecosystem resilience and regeneration including restoration of carbon-rich habitats and climate-friendly agriculture" stands as one of five policies with the highest potential on both economic multiplier and climate impact metrics. Notwithstanding, the NBS implementation to foster climate adaptation remains a neglected measure in both documents as well as their use to foster mitigation is scarcely mentioned and funded, even though NBS proved highly efficient for both measures [25,72] in the context of different initiatives and projects [42]. Italy and Portugal currently stand among the countries showing more literature related to NBS [45]. Although this research effort, the scientific outcomes have probably struggled to be translated into the policy narrative of the RRP, especially in Portugal. However, within RRP framework, the Italian Government has funded two new research Centers specifically aiming to increase sustainability in urban contexts—even establishing and upscaling NBS—namely the Sustainable Mobility Center and the National Biodiversity Future Center, with a total amount of about EUR 640 million [73]. This confirms the awareness of policymakers on the pivotal role of research in this sector.

5.1. Comparative Analysis of the Discourses

Both member states analyzed in this work show in their discourses a strong focus on ecological transition and green revolution (Italy) and climate transition (Portugal), with a dedicated section in the RRP (Mission 2 and Dimension 2, respectively). In the Italian RRP, the term climatic transition is excluded from the discourse. In the Portuguese RRP, three concepts of transitions appear related to environmental issues, namely climate transition, ecological transition, and green transition, which often appear linked, thus hindering the possibility to assert different meanings to each. As already shown in the literature, a variety of approaches to conceptualize transition appear in policy that often overlap, while being also distinct and divergent in their approaches and scopes [74]. We thus highlight how the absence of a clear definition of these concepts increases their mixed-use, and ambiguous meaning, complicating the policy discourse and the attribution of specific targets to foster the transition. Furthermore, these two missions alone do not reach the 37% of budget required by the EU for climate objectives (31% for Italy, 18% for Portugal). In both plans, other contributions are diffused and spread in other missions and dimensions to accomplish the climate targets, further increasing the confusion about terms, objectives and investments. Nevertheless, in the Italian document, Mission 2 (M2—ecological transition and green revolution) covers the largest portion of investments (59.5 billion euros out of 191.5 billion invested in the plan) which considers decarbonization, and nature protection and management as complementary aspects for fostering the ecologic transition and the green revolution. Portugal, on the other hand, adopts a different strategy differentiating the management of the territory into the resilience dimension (C8 and C9—Forests and Hydric management) from the dimension dedicated to climate transition (D2), which steers the investments exclusively to the reduction in greenhouse gases emissions, the increase in renewable energy sources, and the reduction in primary energy use. This division is probably due to the conceptualization of forest management mainly oriented to fight wildfires and, similarly, water management to fight water scarcity, thus neglecting the inclusion of NBS in fostering climate transition.

The lack of clarity and specific targets can be also confirmed by the results of the quantitative analysis. The narrative of both documents is framed around the terms included in general environmental concepts, showing the highest frequency (0.67%, 0.58%, in Italy and Portugal, respectively) with respect to the other categories, biophysical elements (0.45%, 0.49%), threats and challenges (0.11%, 0.10%), and lastly, NBS (0.04%, 0.01%). In addition to the concepts of climate and ecological transitions previously mentioned, among the general environmental concepts coded we found the terms with the highest relative frequency out of the coded terms, i.e., resilience, sustainability and green. These terms display their functions as a linguistic and ideological political mechanism often disconnected from specific objectives and outcomes. As illustrated by Tahvilzadeh et al. [56], "Sustainability discourse did not make any effective climate or environmental protection policies possible, nor did it have clout enough to combat rampant social inequalities". Analogously, the narrative of "green" can raise some contradictory interpretations [75]. In the PRRs, most of the references to the term "green" are not related to NBS, such as green infrastructure, green space or green area, but instead are referred to as an eco-friendly behavior or approach, e.g., "green economy", "green transition" "green communities" or "green islands". Given the heterogeneity and multiple interpretations, all these narratives, on the one hand, can serve multiple discourses (e.g., sustainable development and de-growth; [55]), but on the other hand, can overshadow ecological safeguarding and social equity concerns [56].

As shown in Figure 2, the terms classified in the NBS category are the least referenced in the text. Among these, both countries show higher references for NBS Type 1 and 2, i.e., nature conservation and management as well as different typologies of parks and protected areas. In Italy, these actions are mentioned within different investments across the document, recognizing the value of nature as a resource to be protected or restored (further details in the results section). Portugal dedicates a Component to increasing the management of forests and in turn the resilience to wildfires. Despite biodiversity conservation and enhancement of natural capital are considered as an objective of this component, no mention to the maintenance of native forest is made. The absence of this reference can be significant given that the Portuguese's forests ecosystems are strongly threatened by alien species, e.g., eucalyptus [76], and appropriate silvicultural measures applied to native forest can help in improving their resistance to alien species invasions [77]. Furthermore, forests are mainly conceptualized in the document as an element of the rural areas, thus neglecting the urban dimension.

Considering NBS Type 3, Portugal fails in the allocation of specific investments and interventions, listing only green roofs and NBS among the possible approaches to improve the energy efficiency of buildings. Although there is no reference in the text regarding nature-based solutions, green and blue infrastructures, and ecologic engineering, Italy foresees two important forestry interventions such as the plantation of 6.6 million trees in the 14 metropolitan cities and the ecological restoration of riparian ecosystems of the Po' valley. In the Italian history up to the mid-1970s, numerous reforestation interventions in mountainous and rural areas have already been experimented, with laws, funding, and large-scale implementations aimed to regulate runoff, preventing soil erosion and landslides (for further information see [78]). The two investments envisaged in RRP together are close to EUR 700 million, representing one of the largest structural investments ever allocated in terms of NBS implementation in Italy in recent decades. However, we highlight that, despite the huge investment in absolute terms, this represents in relative terms approximately 0.36% of the total investments envisaged in the Italian RRP (EUR 191.5 billion), and that an extra budget dedicated to other Italian cities could have helped to mitigate other environmental challenges and extend their effects to critical areas out of the major cities [49]. The different approach to urban forests between the two countries might be explained by the differences in research interest between Mediterranean countries found and described by Krajter Ostoić et al. [79]. Accordingly, Italy stands as the leading scientific force in the thematic of urban forests implementation, especially for the air pollution mitigation [79]. In addition, Italy already experimented the inclusion of urban forests in the political context (i.e., Decree on Climate, 2019) allocating EUR 30 million for their implementation [49,80].

5.2. *Missed and Potential Opportunities to Include Nature-Based Solutions*

In both documents, we identified a series of investments that explicitly mention the value of nature as a resource to be preserved or restored, but do not yet include NBS. We believe that in these investments there may be room for possible implementations of NBS, referring to the currently available scientific literature. Among these, we certainly include water management (both flood and water scarcity) [19,59,81,82], soil restoration and water quality improvement [83,84], industrial land regeneration [2,85], wastewater management [86], coastal protection [87], biomass crops for sustainable biofuel production [88], and energy efficiency [89,90].

Under the Green Deal, the EU already invited all member states to reach specific environmental targets across different action plans (e.g., circular economy and zero pollution), strategies (e.g., Forestry and Biodiversity Strategy to 2030) and laws (e.g., European climate law), to improve the quality of ecosystems and human life in the next decade [2]. Considering the link between the Green Deal and the RRP, we found that these environmental targets were not fully included in the investments and reforms in the analyzed documents. It could be argued that the RRP is not the proper document to include considerations and actions related to the protection, management and/or implementation of nature, given that RRPs are reform plans primarily providing investments to recover and increase the economic growth of the member states. However, the term 'nature-positive economy' has recently emerged in the context of sustainable business and finance [55], and the vital role of NBS in this economic shift has been presented in a recent EU report [4]. The latter profiles "some of the economic activities where nature-based enterprises are engaged in the delivery of NBS—generating new jobs, innovations, skills, and wider economic impacts, achieved through a nature-based approach respecting the needs of the environment and communities". As a consequence, we argue that the RRP could have been the ideal arena to bring NBS and nature-based enterprises [91] systematically and methodically into policy and reform, not binding them only to a strategic level or constraining them into an "eco-friendly" narrative.

6. Conclusions

This article assessed the embeddedness of NBS in the recovery and resilience plans of Italy and Portugal. In the narrative of both plans, we observed the dominance of generic concepts such as resilience, sustainability and green, supported by different typologies of "transitions" to reach the climate goals set out in the Green Deal. Ecological, green and climate transitions are used within the individual document and among the documents as synonymous. Furthermore, we observed that the category of NBS and related approaches is the least frequent in both plans, and we found indicative the total lack of specific terms such as, nature-based solutions, green and blue infrastructures that are instead well-established in literature as well as in EU reports and financing initiatives. This happens although the recent EU effort to become a leader in NBS, investing in practical projects and research, providing for assessments and evaluation of NBS as well as involvement of stakeholders. Despite the several existent best practices, their outcomes are still scarcely considered and included in both RRPs.

The central aim of the documents is reducing emissions stated as mitigation measures, while adaptation measures are not central in the RRP. Italy shows two large investments, planting 6.6 million trees in the 14 metropolitan cities and the restoration of riparian ecosystems in the Po' Valley. These two investments, besides helping to fight climate change, will help in the path to reach other two important EU goals, the pledge to plant 3 billion trees by 2030, and the zero net land take by 2050. The case of Portugal is instead emblematic because it does not consider any of the other EU goals and continues to mostly limit the role of nature to its "use value" (e.g., sea economy), instead of working with it or imitating it to tackle the national challenges, according to the NBS definition [25] and in line with a "people and nature" perspective [92].

We are aware that the selection of terms and the division into categories might be considered dubious regarding its robustness, due to the lack of a rigorous approach in their definition and classification. However, we tried to reduce possible software limitations and researchers' bias for the counting and coding of terms, (*i*) working in the original language texts, (*ii*) incorporating a wide number of terms, and (*iii*) promoting complete transparency to readers by reporting (Table S1) the selection of terms in the original languages and not limiting them to their translation in Table 1.

With the results of this article, we aimed to provide a critical reflection on the missed opportunities to use nature to address multiple global challenges, not only from a protection and conservation perspective but also by directly promoting the use of NBS through the construction of new ecosystems, new enterprises and new jobs and, in turn, the promotion of a more bio-based and sustainable economy. Accordingly, we identified multiple investments in both plans that use vague language in explaining the approach planned to address some of the threats and challenges considered (e.g., restoring contaminated sites in Italy and fighting water scarcity in Portugal). In these cases, we are confident that the future research centers and open calls will face the challenges in a more specific and unambiguous way, drawing on the scientific literature to implement NBS and develop a more nature-positive path. This is of utmost importance as the RRPs aim to be the main source of reforms and funding opportunities for the next decades, thus potentially playing a crucial role in positively contributing to a transformative society and economy towards real sustainability.

Supplementary Materials: The following supporting information can be downloaded at: https://www.mdpi.com/article/10.3390/land11081254/s1, Table S1: Coded terms and associated words in the original languages.

Author Contributions: Conceptualization, E.D.P. and R.M.; methodology, R.M. and T.F.; software, R.M.; formal analysis, E.D.P. and R.M.; investigation and data curation, E.D.P.; validation, E.D.P., R.M. and T.F.; writing—original draft preparation, E.D.P. and R.M.; writing—review and editing, E.D.P., R.M., T.F., L.S., P.R., M.M. and B.L.; visualization, T.F., L.S. and P.R.; supervision, T.F. and L.S.; project administration, resources, M.M. and B.L.; funding acquisition, B.L. All authors have read and agreed to the published version of the manuscript.

Funding: This work has received funding from the research project "Establishing Urban FORest based solutions In Changing Cities" (EUFORICC), cod 20173RRN2S, funded by the PRIN 2017 program of the Italian Ministry of University and Research (project coordinator: C. Calfapietra).

Institutional Review Board Statement: Not applicable.

Informed Consent Statement: Not applicable.

Data Availability Statement: Not applicable.

Acknowledgments: Thanks for the financial support through the Research Unit on Governance, Competitiveness and Public Policies (GOVCOPP) (POCI-01-0145-FEDER-008540), funded by European Regional Development fund (FEDER) through COMPETE 2020, and the Foundation for Science and Technology (FCT)—Research Centre for Risks and Sustainability in Construction (RISCO), University of Aveiro, Portugal [FCT/UIDB/ECI/04450/2020]. Thanks are also due to FCT/MCTES for the financial support to CESAM (UIDP/50017/2020 + UIDB/50017/2020), through national funds and co-funding by European funds when applicable. R. Mendes also acknowledges the support from FCT through grant SFRH/BD/147883/2019.

Conflicts of Interest: The authors declare no conflict of interest.

References

1. European Union. *The EU's 2021–2027 Long-Term Budget and NextGenerationEU. Facts and Figures*; Publication office of the Europian Union: Luxembourg, 2021.
2. Sessa, M.R.; Russo, A.; Sica, F. Opinion Paper on Green Deal for the Urban Regeneration of Industrial Brownfield Land in Europe. *Land Use Policy* **2022**, *119*, 106198. [CrossRef]
3. European Commission. The 3 Billion Tree Planting Pledge For 2030. Available online: https://knowledge4policy.ec.europa.eu/publication/commission-staff-working-document-swd2021651-3-billion-tree-planting-pledge-2030_en (accessed on 10 May 2022).
4. European Union. *The Vital Role of Nature-Based Solutions in a Nature Positive Economy*; Publication office of the Uropian Union: Luxembourg, 2022.
5. Palahí, M.; Pantsar, M.; Costanza, R.; Kubiszewski, I.; Potočnik, J.; Stuchtey, M.; Nasi, R.; Lovins, H.; Giovannini, E.; Fioramonti, L.; et al. Investing in Nature as the True Engine of Our Economy: A 10-Point Action Plan for a Circular Bioeconomy of Wellbeing; 2020; ISBN 978-952-5980-92-9. Available online: https://efi.int/sites/default/files/files/publication-bank/2020/EFI_K2A_02_2020.pdf (accessed on 10 May 2022).
6. European Union. *Carbon Economy—Studies on Support to Research and Innovation Policy in the Area of Bio-Based Products and Services*; Publications Office of the European Union: Luxembourg, 2021.
7. Hetemäki, L.; Palahí, M.; Nasi, R. Seeing the Wood in the Forests. *Knowlege Action* **2020**, 1–18. [CrossRef]
8. Wild, T.; Bulkeley, H.; Naumann, S.; Vojinovic, Z.; Calfapietra, C.; Whiteoak, K. Nature-Based Solutions: State of the Art in EU-Funded Projects. *Luxemb. Publ. Off. Eur. Union* **2020**, 308. [CrossRef]
9. Kabisch, N.; Korn, H.; Stadler, J.; Bonn, A. *Nature-Based Solutions to Climate Change Adaptation in Urban Areas*; Kabisch, N., Korn, H., Stadler, J., Bonn, A., Eds.; Theory and Practice of Urban Sustainability Transitions; Springer International Publishing: Cham, Switzerland, 2017; ISBN 978-3-319-53750-4.
10. Veerkamp, C.; Ramieri, E., Romanovska, L.; Zandersen, M.; Förster, J.; Rogger, M.; Martinsen, L. *Assessment Frameworks of Nature-Based Solutions for Climate Change Adaptation and Disaster Risk Reduction*; European Topic Centre on Climate Change impacts, Vulnerability and Adaptation (ETC/CCA) Technical Paper 2021/3; European Environmental Agency: København, Denmark, 2021. [CrossRef]

11. Fischer, E.M.; Schär, C. Consistent Geographical Patterns of Changes in High-Impact European Heatwaves. *Nat. Geosci.* **2010**, *3*, 398–403. [CrossRef]
12. Seddon, N.; Chausson, A.; Berry, P.; Girardin, C.A.J.; Smith, A.; Turner, B. Understanding the Value and Limits of Nature-Based Solutions to Climate Change and Other Global Challenges. *Philos. Trans. R. Soc. B Biol. Sci.* **2020**, *375*, 20190120. [CrossRef]
13. Allen, M.R.; Dube, O.P.; Solecki, W.; Aragón-Durand, F.; Cramer, W.; Humphreys, S.; Kainuma, M.; Kala, J.; Mahowald, N.; Mulugetta, Y.; et al. Framing and Context. In *Global Warming of 1.5 °C*; An IPCC Special Report on the Impacts of Global Warming of 1.5 °C above Pre-Industrial Levels and Related Global Greenhouse Gas Emission Pathways, in the Context of Strengthening the Global Response to the Threat of Climate Change, Intergovernmental Panel on Climate Change (IPCC); Masson-Delmotte, V., Zhai, P., Pörtner, H.-O., Roberts, D., Skea, J., Shukla, P.R., Pirani, A., Moufouma-Okia, W., Péan, C., Pidcock, R., et al., Eds.; Cambridge University Press: Geneva, Switzerland, 2022; pp. 49–92. ISBN 9781009157940.
14. Marchetti, M.; Vizzarri, M.; Lasserre, B.; Sallustio, L.; Tavone, A. Natural Capital and Bioeconomy: Challenges and Opportunities for Forestry. *Ann. Silvic. Res.* **2015**, *38*, 62–73. [CrossRef]
15. Burrascano, S.; Chytrý, M.; Kuemmerle, T.; Giarrizzo, E.; Luyssaert, S.; Sabatini, F.M.; Blasi, C. Current European Policies Are Unlikely to Jointly Foster Carbon Sequestration and Protect Biodiversity. *Biol. Conserv.* **2016**, *201*, 370–376. [CrossRef]
16. Naumann, S.; Anzaldua, G.; Gerdes, H.; Frelih-Larsen, A.; Davis, M.; Berry, P.; Burch, S.; Sanders, M. Assessment of the Potential of Ecosystem-Based Approaches to Climate Change Adaptation and Mitigation in Europe. 2011. Available online: https://climate-adapt.eea.europa.eu/metadata/publications/assessment-of-the-potential-of-ecosystem-based-approaches-to-climate-change-adaptation-and-mitigation-in-europe (accessed on 10 May 2022).
17. McDonald, R.; Aljabar, L.; Aubuchon, C.; Birnbaum, H.; Chandler, C.; Toomey, B.; Daley, J.; Jimenez, W.; Trieschman, E.; Paque, J.; et al. *Funding Trees for Health. An Analysis of Finance and Policy Actions to Enable Tree Planting for Public Health*; The Nature Conservancy: Arlington County, VA, USA, 2017; Volume 36.
18. Vallecillo, S.; La Notte, A.; Zulian, G.; Ferrini, S.; Maes, J. Ecosystem Services Accounts: Valuing the Actual Flow of Nature-Based Recreation from Ecosystems to People. *Ecol. Modell.* **2019**, *392*, 196–211. [CrossRef]
19. Wild, T.C.; Henneberry, J.; Gill, L. Comprehending the Multiple 'Values' of Green Infrastructure – Valuing Nature-Based Solutions for Urban Water Management from Multiple Perspectives. *Environ. Res.* **2017**, *158*, 179–187. [CrossRef]
20. Sangha, K.K.; Gordon, I.J.; Costanza, R. Ecosystem Services and Human Wellbeing-Based Approaches Can Help Transform Our Economies. *Front. Ecol. Evol.* **2022**, *10*, 14. [CrossRef]
21. Albert, C.; Schröter, B.; Haase, D.; Brillinger, M.; Henze, J.; Herrmann, S.; Gottwald, S.; Guerrero, P.; Nicolas, C.; Matzdorf, B. Addressing Societal Challenges through Nature-Based Solutions: How Can Landscape Planning and Governance Research Contribute? *Landsc. Urban Plan.* **2019**, *182*, 12–21. [CrossRef]
22. Escobedo, F.J.; Giannico, V.; Jim, C.Y.; Sanesi, G.; Lafortezza, R. Urban Forests, Ecosystem Services, Green Infrastructure and Nature-Based Solutions: Nexus or Evolving Metaphors? *Urban For. Urban Green.* **2019**, *37*, 3–12. [CrossRef]
23. Faivre, N.; Fritz, M.; Freitas, T.; de Boissezon, B.; Vandewoestijne, S. Nature-Based Solutions in the EU: Innovating with Nature to Address Social, Economic and Environmental Challenges. *Environ. Res.* **2017**, *159*, 509–518. [CrossRef]
24. Cohen-Shacham, E.; Andrade, A.; Dalton, J.; Dudley, N.; Jones, M.; Kumar, C.; Maginnis, S.; Maynard, S.; Nelson, C.R.; Renaud, F.G.; et al. Core Principles for Successfully Implementing and Upscaling Nature-Based Solutions. *Environ. Sci. Policy* **2019**, *98*, 20–29. [CrossRef]
25. European Commission. *Towards an EU Research and Innovation Policy Agenda for Nature-Based Solutions & Re-Naturing Cities. Final Report of the Horizon 2020 Expert Group on "Nature-Based Solutions and Re-Naturing Cities" (Full Version)*; European Commission: Brussels, Belgium, 2015; ISBN 978-92-79-46051-7.
26. Castellar, J.A.C.; Popartan, L.A.; Pueyo-Ros, J.; Atanasova, N.; Langergraber, G.; Säumel, I.; Corominas, L.; Comas, J.; Acuña, V. Nature-Based Solutions in the Urban Context: Terminology, Classification and Scoring for Urban Challenges and Ecosystem Services. *Sci. Total Environ.* **2021**, *779*, 146237. [CrossRef] [PubMed]
27. Barboza, E.P.; Cirach, M.; Khomenko, S.; Iungman, T.; Mueller, N.; Barrera-Gómez, J.; Rojas-Rueda, D.; Kondo, M.; Nieuwenhuijsen, M. Green Space and Mortality in European Cities: A Health Impact Assessment Study. *Lancet Planet. Heal.* **2021**, *5*, e718–e730. [CrossRef]
28. van den Bosch, M.; Ode Sang, Å. Urban Natural Environments as Nature-Based Solutions for Improved Public Health—A Systematic Review of Reviews. *Environ. Res.* **2017**, *158*, 373–384. [CrossRef]
29. Spano, G.; D'este, M.; Giannico, V.; Carrus, G.; Elia, M.; Lafortezza, R.; Panno, A.; Sanesi, G. Are Community Gardening and Horticultural Interventions Beneficial for Psychosocial Well-Being? A Meta-Analysis. *Int. J. Environ. Res. Public Health* **2020**, *17*, 3584. [CrossRef]
30. Bratman, G.N.; Anderson, C.B.; Berman, M.G.; Cochran, B.; de Vries, S.; Flanders, J.; Folke, C.; Frumkin, H.; Gross, J.J.; Hartig, T.; et al. Nature and Mental Health: An Ecosystem Service Perspective. *Sci. Adv.* **2019**, *5*, eaax0903. [CrossRef]
31. Tallis, M.; Taylor, G.; Sinnett, D.; Freer-Smith, P. Estimating the Removal of Atmospheric Particulate Pollution by the Urban Tree Canopy of London, under Current and Future Environments. *Landsc. Urban Plan.* **2011**, *103*, 129–138. [CrossRef]
32. Sgrigna, G.; Baldacchini, C.; Dreveck, S.; Cheng, Z.; Calfapietra, C. Relationships between Air Particulate Matter Capture Efficiency and Leaf Traits in Twelve Tree Species from an Italian Urban-Industrial Environment. *Sci. Total Environ.* **2020**, *718*, 137310. [CrossRef] [PubMed]

33. Fusaro, L.; Marando, F.; Sebastiani, A.; Capotorti, G.; Blasi, C.; Copiz, R.; Congedo, L.; Munafò, M.; Ciancarella, L.; Manes, F. Mapping and Assessment of PM10 and O3 Removal by Woody Vegetation at Urban and Regional Level. *Remote Sens.* **2017**, *9*, 791. [CrossRef]
34. Valeri, S.; Zavattero, L.; Capotorti, G. Ecological Connectivity in Agricultural Green Infrastructure: Suggested Criteria for Fine Scale Assessment and Planning. *Land* **2021**, *10*, 807. [CrossRef]
35. Staccione, A.; Candiago, S.; Mysiak, J. Mapping a Green Infrastructure Network: A Framework for Spatial Connectivity Applied in Northern Italy. *Environ. Sci. Policy* **2022**, *131*, 57–67. [CrossRef]
36. van der Jagt, A.P.N.; Kiss, B.; Hirose, S.; Takahashi, W. Nature-Based Solutions or Debacles? The Politics of Reflexive Governance for Sustainable and Just Cities. *Front. Sustain. Cities* **2021**, *2*, 1–19. [CrossRef]
37. Melanidis, M.S.; Hagerman, S. Competing Narratives of Nature-Based Solutions: Leveraging the Power of Nature or Dangerous Distraction? *Environ. Sci. Policy* **2022**, *132*, 273–281. [CrossRef]
38. Chausson, A.; Turner, B.; Seddon, D.; Chabaneix, N.; Girardin, C.A.J.; Kapos, V.; Key, I.; Roe, D.; Smith, A.; Woroniecki, S.; et al. Mapping the Effectiveness of Nature-based Solutions for Climate Change Adaptation. *Glob. Chang. Biol.* **2020**, *26*, 6134–6155. [CrossRef]
39. Schröter, B.; Hack, J.; Hüesker, F.; Kuhlicke, C.; Albert, C. Beyond Demonstrators—Tackling Fundamental Problems in Amplifying Nature-Based Solutions for the Post-COVID-19 World. *Npj Urban Sustain.* **2022**, *2*, 4. [CrossRef]
40. Frantzeskaki, N.; McPhearson, T.; Collier, M.J.; Kendal, D.; Bulkeley, H.; Dumitru, A.; Walsh, C.; Noble, K.; Van Wyk, E.; Ordóñez, C.; et al. Nature-Based Solutions for Urban Climate Change Adaptation: Linking Science, Policy, and Practice Communities for Evidence-Based Decision-Making. *Bioscience* **2019**, *69*, 455–466. [CrossRef]
41. Raymond, C.M.; Pam, B.; Breil, M.; Nita, M.R.; Kabisch, N.; de Bel, M.; Enzi, V.; Frantzeskaki, N.; Geneletti, D.; Cardinaletti, M.; et al. *An Impact Evaluation Framework to Support Planning and Evaluation of Nature-Based Solutions Projects*; Centre for Ecology & Hydrology: Wallingford, UK, 2017.
42. Dumitru, A.; Wendling, L. *Evaluating the Impact of Nature-Based Solutions—A Handbook for Practitioners*; Publication office of the Uropian Union: Luxembourg, 2021; pp. 1–373. [CrossRef]
43. Sarabi, S.; Han, Q.; Romme, A.G.L.; de Vries, B.; Valkenburg, R.; Den Ouden, E.; Zalokar, S.; Wendling, L. Barriers to the Adoption of Urban Living Labs for Nbs Implementation: A Systemic Perspective. *Sustainability* **2021**, *13*, 13276. [CrossRef]
44. Di Pirro, E.; Sallustio, L.; Castellar, J.A.C.; Sgrigna, G.; Marchetti, M.; Lasserre, B. Facing Multiple Environmental Challenges through Maximizing the Co-Benefits of Nature-Based Solutions at a National Scale in Italy. *Forests* **2022**, *13*, 548. [CrossRef]
45. Mendes, R.; Fidélis, T.; Roebeling, P.; Teles, F. The Institutionalization of Nature-Based Solutions-a Discourse Analysis of Emergent Literature. *Resources* **2020**, *9*, 6. [CrossRef]
46. Cortinovis, C.; Olsson, P.; Boke-Olén, N.; Hedlund, K. Scaling up Nature-Based Solutions for Climate-Change Adaptation: Potential and Benefits in Three European Cities. *Urban For. Urban Green.* **2022**, *67*, 127450. [CrossRef]
47. Mendonça, R.; Roebeling, P.; Fidélis, T.; Saraiva, M. Policy Instruments to Encourage the Adoption of Nature-Based Solutions in Urban Landscapes. *Resources* **2021**, *10*, 81. [CrossRef]
48. UNEP Smart, Sustainable and Resilient Cities: The Power of Nature-Based Solutions. Available online: https://www.unep.org/resources/report/smart-sustainable-and-resilient-cities-power-nature-based-solutions (accessed on 10 May 2022).
49. Di Pirro, E.; Sallustio, L.; Sgrigna, G.; Marchetti, M.; Lasserre, B. Strengthening the Implementation of National Policy Agenda in Urban Areas to Face Multiple Environmental Stressors: Italy as a Case Study. *Environ. Sci. Policy* **2022**, *129*, 1–11. [CrossRef]
50. Davis, M.; Abhold, K.; Mederake, L.; Knoblauch, D. NBS in European and National Policy Framework. *Naturvation* **2018**, *50*, 1–52.
51. Sarabi, S.E.; Han, Q.; Romme, A.G.L.; de Vries, B.; Wendling, L. Key Enablers of and Barriers to the Uptake and Implementation of Nature-Based Solutions in Urban Settings: A Review. *Resources* **2019**, *8*, 121. [CrossRef]
52. Leipold, S.; Feindt, P.H.; Winkel, G.; Keller, R. Discourse Analysis of Environmental Policy Revisited: Traditions, Trends, Perspectives. *J. Environ. Policy Plan.* **2019**, *21*, 445–463. [CrossRef]
53. Presidenza del Consiglio dei Ministri Piano Nazionale Di Ripresa e Resilienza. Available online: https://italiadomani.gov.it/it/home.html (accessed on 10 May 2022).
54. MINISTÉRIO DO PLANEAMENTO PRR—Recuperar Portugal Construindo O Futuro. Available online: https://recuperarportugal.gov.pt/ (accessed on 10 May 2022).
55. D'Amato, D. Sustainability Narratives as Transformative Solution Pathways: Zooming in on the Circular Economy. *Circ. Econ. Sustain.* **2021**, *1*, 231–242. [CrossRef]
56. Tahvilzadeh, N.; Montin, S.; Cullberg, M. Functions of Sustainability: Exploring What Urban Sustainability Policy Discourse "Does" in the Gothenburg Metropolitan Area. *Local Environ.* **2017**, *22*, 66–85. [CrossRef]
57. Raymond, C.M.; Frantzeskaki, N.; Kabisch, N.; Berry, P.; Breil, M.; Nita, M.R.; Geneletti, D.; Calfapietra, C. A Framework for Assessing and Implementing the Co-Benefits of Nature-Based Solutions in Urban Areas. *Environ. Sci. Policy* **2017**, *77*, 15–24. [CrossRef]
58. Mander, Ü.; Kull, A.; Uuemaa, E.; Mõisja, K.; Külvik, M.; Kikas, T.; Raet, J.; Tournebize, J.; Sepp, K. Green and Brown Infrastructures Support a Landscape-Level Implementation of Ecological Engineering. *Ecol. Eng.* **2018**, *120*, 23–35. [CrossRef]
59. Vörösmarty, C.J.; Rodríguez Osuna, V.; Cak, A.D.; Bhaduri, A.; Bunn, S.E.; Corsi, F.; Gastelumendi, J.; Green, P.; Harrison, I.; Lawford, R.; et al. Ecosystem-Based Water Security and the Sustainable Development Goals (SDGs). *Ecohydrol. Hydrobiol.* **2018**, *18*, 317–333. [CrossRef]

60. Hwang, S. Utilizing Qualitative Data Analysis Software: A Review of Atlas.Ti. *Soc. Sci. Comput. Rev.* **2008**, *26*, 519–527. [CrossRef]
61. Eggermont, H.; Balian, E.; Azevedo, J.M.N.; Beumer, V.; Brodin, T.; Claudet, J.; Fady, B.; Grube, M.; Keune, H.; Lamarque, P.; et al. Nature-Based Solutions: New Influence for Environmental Management and Research in Europe. *Gaia* **2015**, *24*, 243–248. [CrossRef]
62. Corbin, J.; Strauss, A. *Basics of Qualitative Research: Techniques and Procedures for Developing Grounded Theory*; IV; SAGE Publication, Inc.: Thousand Oaks, CA, USA, 2015; ISBN 9781412997461.
63. Sumares, D.; Fidélis, T. Natura 2000 and the Narrative Nature of Nature: A Case for Critical Discourse Analysis. *J. Integr. Environ. Sci.* **2011**, *8*, 53–68. [CrossRef]
64. Yrjänä, L.; Rashidfarokhi, A.; Toivonen, S.; Viitanen, K. Looking at Retail Planning Policy through a Sustainability Lens: Evidence from Policy Discourse in Finland. *Land Use Policy* **2018**, *79*, 190–198. [CrossRef]
65. Lowry, L.L.; Lee, M. CittaSlow, Slow Cities, Slow Food: Searching for a Model for the Development of Slow Tourism. *Travel Tour. Res. Assoc. Adv. Tour. Res. Glob.* **2016**, *40*. Available online: https://www.semanticscholar.org/paper/CittaSlow%2C-Slow-Cities%2C-Slow-Food%3A-Searching-for-a-Lowry-Lee/fa6d9d7bfdc4f58207d3ae4dc7681f68c6409c3e (accessed on 10 May 2022).
66. Lucatelli, S.; Carlucci, C.; Guerrizio, M.A. A Strategy for "Inner Areas" in Italy. In *Education, Local Economy and Job Opportunities in Rural Area, Proceedings of the 2nd EURUFU Scientific Conference, Milan, Italy, 14–16 October 2015*; Gather, M., Luttmerding, A., Berding, J., Eds.; Transport and Spatial Planning Institute, University of Applied Sciences Erfurt: Erfurt, Germany, 2013; pp. 69–127.
67. De Toni, A.; Vizzarri, M.; Di Febbraro, M.; Lasserre, B.; Noguera, J.; Di Martino, P. Aligning Inner Peripheries with Rural Development in Italy: Territorial Evidence to Support Policy Contextualization. *Land Use Policy* **2021**, *100*, 104899. [CrossRef]
68. Marchetti, M.; Palahí, M. Perspectives in Bioeconomy: Strategies, Green Deal and Covid19. *For. Riv. di Selvic. ed Ecol. For.* **2020**, *17*, 52–55. [CrossRef]
69. Hepburn, C.; O'Callaghan, B.; Stern, N.; Stiglitz, J.; Zenghelis, D. Will COVID-19 Fiscal Recovery Packages Accelerate or Retard Progress on Climate Change? *Oxford Rev. Econ. Policy* **2020**, *36*, S359–S381. [CrossRef]
70. Liu, Y.; Wang, R.; Lu, Y.; Li, Z.; Chen, H.; Cao, M.; Zhang, Y.; Song, Y. Natural Outdoor Environment, Neighbourhood Social Cohesion and Mental Health: Using Multilevel Structural Equation Modelling, Streetscape and Remote-Sensing Metrics. *Urban For. Urban Green.* **2020**, *48*, 126576. [CrossRef]
71. Wolch, J.R.; Byrne, J.; Newell, J.P. Urban Green Space, Public Health, and Environmental Justice: The Challenge of Making Cities "Just Green Enough". *Landsc. Urban Plan.* **2014**, *125*, 234–244. [CrossRef]
72. IUCN. *IUCN Global Standard for Nature-Based Solutions: A User-Friendly Framework for the Verification, Design and Scaling up of NbS: First Edition*; IUCN, International Union for Conservation of Nature: Gland, Switzerland, 2020.
73. MUR PNRR: Nascono i 5 Centri Nazionali Di Ricerca. Available online: https://www.mur.gov.it/it/news/mercoledi-15062022/pnrr-nascono-i-5-centri-nazionali-la-ricerca (accessed on 29 June 2022).
74. Patterson, J.; Schulz, K.; Vervoort, J.; van der Hel, S.; Widerberg, O.; Adler, C.; Hurlbert, M.; Anderton, K.; Sethi, M.; Barau, A. Exploring the Governance and Politics of Transformations towards Sustainability. *Environ. Innov. Soc. Transitions* **2017**, *24*, 1–16. [CrossRef]
75. Scoones, I.; Leach, M.; Newell, P. *The Politics of Green Transformations*; Scoones, I., Leach, M., Newell, P., Eds.; Routledge: London, UK, 2015; ISBN 9781317601128.
76. Nunes, L.J.R.; Meireles, C.I.R.; Pinto Gomes, C.J.; Almeida Ribeiro, N.M.C. Historical Development of the Portuguese Forest: The Introduction of Invasive Species. *Forests* **2019**, *10*, 974. [CrossRef]
77. Sitzia, T.; Campagnaro, T.; Kowarik, I.; Trentanovi, G. Using Forest Management to Control Invasive Alien Species: Helping Implement the New European Regulation on Invasive Alien Species. *Biol. Invasions* **2016**, *18*, 1–7. [CrossRef]
78. Cantiani, P.; Di Salvatore, U.; Romano, R. Silvicultural Aspects of Artificial Black Pine Plantations: Analysis of Italian Regional Laws. *For. Riv. di Selvic. ed Ecol. For.* **2018**, *15*, 99–111. [CrossRef]
79. Krajter Ostoić, S.; Salbitano, F.; Borelli, S.; Verlič, A. Urban Forest Research in the Mediterranean: A Systematic Review. *Urban For. Urban Green.* **2018**, *31*, 185–196. [CrossRef]
80. Salbitano, F.; Sanesi, G. Decree on Climate 2019. What Resources Support Forests and Silviculture in Our Cities? *For. Riv. di Selvic. ed Ecol. For.* **2019**, *16*, 74–76. [CrossRef]
81. Lallemant, D.; Hamel, P.; Balbi, M.; Lim, T.N.; Schmitt, R.; Win, S. Nature-Based Solutions for Flood Risk Reduction: A Probabilistic Modeling Framework. *One Earth* **2021**, *4*, 1310–1321. [CrossRef]
82. World Bank. *Implementing Nature-Based Flood Protection: Principles and Implementation Guidance*; World Bank: Washington, DC, USA, 2017.
83. Keesstra, S.; Nunes, J.; Novara, A.; Finger, D.; Avelar, D.; Kalantari, Z.; Cerdà, A. The Superior Effect of Nature Based Solutions in Land Management for Enhancing Ecosystem Services. *Sci. Total Environ.* **2018**, *610–611*, 997–1009. [CrossRef] [PubMed]
84. Roebeling, P.C.; Van Grieken, M.E.; Webster, A.J.; Biggs, J.; Thorburn, P. Cost-Effective Water Quality Improvement in Linked Terrestrial and Marine Ecosystems: A Spatial Environmentaleconomic Modelling Approach. *Mar. Freshw. Res.* **2009**, *60*, 1150–1158. [CrossRef]
85. Cortinovis, C.; Geneletti, D. Mapping and Assessing Ecosystem Services to Support Urban Planning: A Case Study on Brownfield Regeneration in Trento, Italy. *One Ecosyst.* **2018**, *3*, e25477. [CrossRef]

86. Langergraber, G.; Castellar, J.A.C.; Pucher, B.; Baganz, G.F.M.; Milosevic, D.; Andreucci, M.B.; Kearney, K.; Pineda-Martos, R.; Atanasova, N. A Framework for Addressing Circularity Challenges in Cities with Nature-Based Solutions. *Water* **2021**, *13*, 2355. [CrossRef]
87. Kumar, P.; Debele, S.E.; Sahani, J.; Rawat, N.; Marti-Cardona, B.; Alfieri, S.M.; Basu, B.; Basu, A.S.; Bowyer, P.; Charizopoulos, N.; et al. Nature-Based Solutions Efficiency Evaluation against Natural Hazards: Modelling Methods, Advantages and Limitations. *Sci. Total Environ.* **2021**, *784*, 147058. [CrossRef] [PubMed]
88. Sallustio, L.; Harfouche, A.L.; Salvati, L.; Marchetti, M.; Corona, P. Evaluating the Potential of Marginal Lands Available for Sustainable Cellulosic Biofuel Production in Italy. *Socioecon. Plann. Sci.* **2022**, *82*, 101309. [CrossRef]
89. Kolokotsa, D.; Santamouris, M.; Zerefos, S.C. Green and Cool Roofs' Urban Heat Island Mitigation Potential in European Climates for Office Buildings under Free Floating Conditions. *Sol. Energy* **2013**, *95*, 118–130. [CrossRef]
90. Blanco, I.; Schettini, E.; Vox, G. Effects of Vertical Green Technology on Building Surface Temperature. *Int. J. Des. Nat. Ecodyn.* **2018**, *13*, 384–394. [CrossRef]
91. Kooijman, E.D.; McQuaid, S.; Rhodes, M.L.; Collier, M.J.; Pilla, F. Innovating with Nature: From Nature-Based Solutions to Nature-Based Enterprises. *Sustainability* **2021**, *13*, 1263. [CrossRef]
92. Mace, G.M. Whose Conservation? *Science* **2014**, *345*, 1558–1560. [CrossRef]

land

Article

Seasonal Variations in the Particulate Matter Accumulation and Leaf Traits of 24 Plant Species in Urban Green Space

Huong-Thi Bui [1,2,†], Uuriintuya Odsuren [1,†], Sang-Yong Kim [3] and Bong-Ju Park [1,2,*]

1 Division of Animal, Horticultural, and Food Sciences, Chungbuk National University, Cheongju 28644, Korea
2 Brain Korea 21 Center for Bio-Health Industry, Chungbuk National University, Cheongju 28644, Korea
3 Department of Plant Resources, Korea National Arboretum, Pocheon 11186, Korea
* Correspondence: bjpak@cbnu.ac.kr
† These authors contributed equally to this work.

Abstract: Particulate matter (PM), an extremely serious type of air pollution, leads to numerous human diseases. Mitigating PM in the urban city, where resident density has been increasing, has been a major challenge. The increase in residents leads to increasing traffic, the primary source of PM in urban areas. Plants play an important role in reducing PM and maintaining an ecological balance. For some Asian countries, such as Korea, with differing seasons and environmental conditions, PM accumulation and plant survival are greatly impacted by environmental conditions. In this study, we analyzed the amount of PM accumulation on the leaf surfaces and wax layers of 24 plant species during four seasons (spring, summer, autumn, and winter) to determine the PM accumulation in plants under different environmental conditions. The leaf traits of plant chlorophyll a (Chl a), chlorophyll b (Chl b), total chlorophyll (TChl), relative water content (RWC), leaf extract pH (pH), and leaf specific area (SLA) were analyzed to determine the influence of PM on plants and the relationship between PM and leaf traits. In this study, we found that the amount of PM accumulation differed among plants and seasons. Among the 24 plant species, plants *Pinus strobus*, *P. parviflora*, *P. densiflora*, *Euonymus japonicus*, and *Acer palmatum* were most adept at PM accumulation. Leaf structure, environmental conditions, such as PM concentration, and rainfall may be the main factors that impact the ability of plant leaves to accumulate PM. The plant leaf traits differed among the four seasons. PM accumulation on the leaf was negatively correlated with SLA (in all four seasons) and pH (in spring, summer, and autumn). PM was negatively correlated with Chl a, Chl b, and TChl in summer.

Keywords: air pollution; environmental conditions; large PM; coarse PM; wax layer

1. Introduction

Air pollution has increased over the past 50 years and is considered the world's largest environmental health problem [1]. Air pollution causes numerous diseases ranging from asthma to cancer, pulmonary illnesses, and heart disease that kill an estimated 7 million people worldwide every year, of which 4.2 million die from stroke, heart disease, lung cancer, and acute and chronic respiratory disease [2]. Particulate matter (PM) is the most dangerous type of air pollution and is a primary concern worldwide, particularly in developing countries. PM with a small diameter that can penetrate the lung alveoli negatively affects the respiratory system [3]. Particulate matter can originate from either anthropogenic or natural sources, such as agriculture or industrial activities, volcanic eruptions, soil erosion, sea salt, or desert sand. According to the aerodynamic diameter, PM can be classified as large PM (>10 μm), coarse (2.5–10 μm), fine (0.1–2.5 μm), and ultrafine ($\leq$0.1 μm), as reported by Sæbø et al. [4].

Plants play a vital role in mitigating urban pollution by accumulating PM on their leaves [5–9]. Plants that grow in urban environments improve air quality and act as a natural filter for PM. In Beijing, China, plants in the city center removed 774 tons of PM10 in one year [10]. However, species differ in their ability to accumulate PM. Plants accumulate PM on their leaf surfaces and epicuticular waxes, so the amount of PM accumulation depends on the macrostructure (plant height, petiole length) and microstructure of the leaf (leaf hairs, thickness wax layer, and stomatal density) [4,11]. Trees with large total leaf areas were considered the most effective type of plants for reducing PM [12]. Other studies have shown that plants growing low to the ground, such as shrubs, can accumulate more PM on the leaf because of exposed soil splash [13]. Several studies have demonstrated that the PM accumulation of needleleaf species is greater than that of broadleaf trees [14,15]. Additionally, the thickness of the wax layer is a factor that can influence the amount of PM accumulated on leaves [16]. Plants with thick wax accumulated more PM than those with thin wax layers. The long petiole length could increase the amount of PM removal from leaves via rain or wind. In addition, the plant's capacity to capture PM is influenced by several other factors, including environmental variables, such as wind and rain [17]. Wind and rain can wash PM from leaves, decreasing leaf PM accumulation [18]. However, the amount of PM washed from the leaf depends on the leaf structures, density, and area. Plants with a high density of leaves are less influenced by rain [19,20]. Furthermore, the concentration of PM in the atmosphere can directly influence the amount of PM accumulation on the leaf [21]. Under a high PM concentration, plants can accumulate more PM than plants grown in a lower PM concentration environment [22]. However, PM adversely affects plant life [23]. PM impacts plants' physical and biochemical characteristics [24]. PM influences the structural components of leaves, such as leaf area, leaf thickness, and wax amounts [19,25]. In addition, PM also impacts photosynthesis, leaf extract pH (pH), specific leaf area (SLA), and plant relative water content (RWC) [26]. Popek et al. [27] showed that the effectiveness of photosynthesis was reduced because of PM accumulation on the leaf surface. PM accumulation on the leaf leads to prevention of light absorption or blocking of stomata by reducing the total chlorophyll (TChl) of plants [28]. Therefore, PM also impacts the growth and productivity of plants. However, the impact of PM on plants depends greatly on plant responses [29]. Under different environmental conditions, the ability of PM to accumulate and the influence of PM on individual plant species can differ. The ability of PM accumulation and the tolerance to air pollution are important in selecting suitable plant species for improving urban air quality. However, a few researchers have addressed the correlation between PM accumulation on leaf and leaf traits in common plant species in Korea. Therefore, analyzing PM accumulation on plant leaves and the correlation between PM and leaf trait conditions can inform plant selections designed to improve air quality. Analyzing the complex correlation during different seasons helps to comprehensively determine the correlations among PM, plants, and the environment, which is a premise for selecting plant species to optimize the benefit of the plants in improving air quality in Korea.

2. Materials and Methods

2.1. Study Area and Sample Collection

The study site was located in Chungbuk National University in Cheong-Ju, South Korea, which is located at 36.6290° N, 127.4563° E (Figure 1). Twenty-four plant species that are commonly used for urban greening in Korea were selected for leaf sample collection. Leaf samples were collected from spring 2020 to winter 2021 in four time periods: at the beginning of June (late spring), at the beginning of August (summer), in October (autumn), and in February (late winter). These consisted of 11 evergreen species and 13 deciduous tree species in good condition (healthy and free from disease, insects, and pests) (Table 1). In winter, only the evergreen tree samples were collected because the leaves had fallen from the deciduous trees. For each plant species, we put a tag on the selected plant to ensure that the samples were collected from the same plant. Following each collection, the leaf samples were stored in paper bags and immediately transferred to the laboratory for analysis. The samples were collected after two weeks without heavy rain and between 8 am and 12 noon.

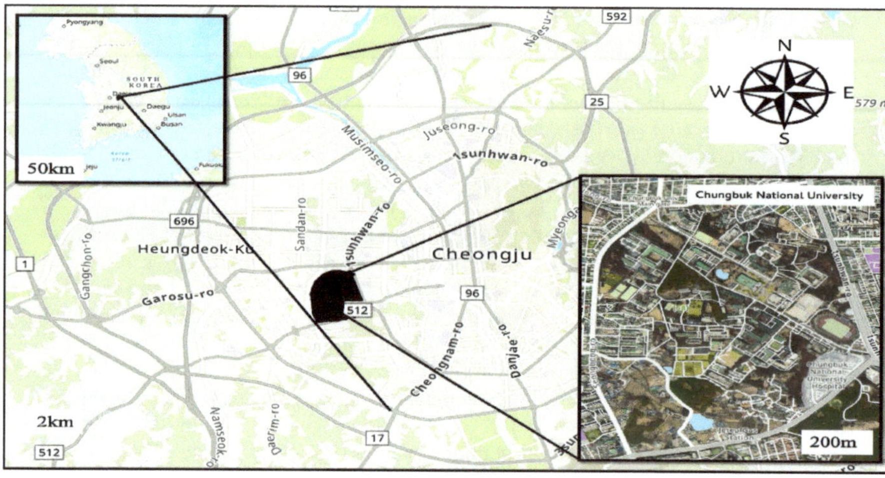

Figure 1. The location map of the sampling sites. Chungbuk National University, Cheongju City, Korea.

Table 1. List of landscaping plants analyzed in this study.

Species	Family	Habit	Type
Juniperus chinensis L.	Cupressaceae	Tree	Evergreen
Juniperus chinensis var. kaizuka Hort.	Cupressaceae	Tree	Evergreen
Pinus parviflora Siebold & Zucc.	Pinaceae	Tree	Evergreen
Pinus densiflora Siebold & Zucc.	Pinaceae	Tree	Evergreen
Chamaecyparis pisifera Siebold & Zucc.	Cupressaceae	Tree	Evergreen
Taxus cuspidata Siebold & Zucc.	Taxaceae	Tree	Evergreen
Abies holophylla Maxim.	Pinaceae	Tree	Evergreen
Picea abies (L.) H.Karst.	Pinaceae	Tree	Evergreen
Pinus strobus L.	Pinaceae	Tree	Evergreen
Platycladus orientalis (L.) Franco	Cupressaceae	Tree	Evergreen
Euonymus japonica Thunb.	Celastraceae	Shrub	Evergreen
Magnolia denudata Desr.	Magnoliaceae	Tree	Deciduous
Aesculus turbinata Blume	Hippocastanaceae	Tree	Deciduous
Rhododendron yedoense Maxim	Ericaceae	Shrub	Deciduous
Hibiscus syriacus L.	Malvaceae	Shrub	Deciduous
Acer palmatum Thunb.	Aceraceae	Tree	Deciduous
Cercis chinensis Bunge	Fabaceae	Shrub	Deciduous
Cornus officinalis Siebold&Zucc.	Cornaceae	Tree	Deciduous
Acer triflorum Kom.	Aceraceae	Tree	Deciduous
Zelkova serrata (Thunb.) Makino	Ulmaceae	Tree	Deciduous
Ginkgo biloba L.	Ginkgoaceae	Tree	Deciduous
Ligustrum obtusifolium Siebold & Zucc.	Oleaceae	Shrub	Deciduous
Prunus x *yedoensis* Matsum.	Rosaceae	Tree	Deciduous
Viburnum dilatatum Thunb.	Adoxaceae	Shrub	Deciduous

2.2. *Leaf Surface PM (sPM) and Epicuticular Wax (wPM) Analysis*

According to the method developed by Dzierzanowski et al. [30], leaf samples were washed with distilled water to collect PM on the leaf surface (sPM) and washed with chloroform to collect PM in wax (wPM). Approximately 300 cm^2 leaves of each plant species were put in a glass beaker and washed with 250 mL distilled water for 60 s. The glass beaker was placed on an ultrasonic cleaner (WUC-A22H, Daihan Scientific, Wonju,

Korea) for 6 min to ensure that all the PM was washed from the leaf surface. Then, the collected solution was passed through a metal sieve (diameter 100 μm mesh) to remove all particles with a diameter over 100 μm. After that, the solution water was filtered with two types of paper filters, type 91 and type 42, with pore sizes of 10 μm and 2.5 μm, respectively. Before filtering, paper filters were placed in a desiccator (DH. DeBG1K, Daihan Scientific, Wonju, Korea) for 48 h to control humidity, and then the filter papers were weighed. Following filtering, the PM was divided into two types: large PM (100–10 μm) and coarse PM (10–2.5 μm). The same method was used to evaluate the weight of the paper filter after filtration. After washing with water, the leaf area was measured by using a leaf area meter (LI-3100C, LI-COR Biosciences, Lincoln, NE, USA). Then, the leaf sample was washed with chloroform and filtered with filter paper by the same filtration methods to determine the amount of wPM. The amount of PM accumulation on the leaves of plants was evaluated using Equation (1):

$$PM = W_2 - W_1/A \quad (1)$$

where W_2 is the weight after filtration (g), W_1 is the weight before filtration (g), and A is the leaf area (cm^2).

2.3. *Analysis of Leaf Traits*

2.3.1. Chlorophyll (Chl a, Chl b, Total Chlorophyll (Tchl))

The concentration of chlorophyll (Chl a, Chl b, Tchl) was measured according to Lichtenthaler [31]. For each plant species, 0.05 g of fresh weight was placed on the different mortars. Liquid nitrogen was added and crushed. Approximately 10 mL of 100% acetone was added to the mortar, and the sample liquid was collected. A centrifuge (Cef-6, Daihan Scientific, Wonju, Korea) was used to homogenize the samples for 10 m at 4900 rpm. Ten milliliters of supernatant were collected, and the absorbance was analyzed at wavelengths of 470 nm, 616.6 nm, and 644.8 nm by using a spectrophotometer (UV-1800, Shimadzu, Kyoto, Japan). Chl a, Chl b, and TChl were calculated using Equation (2):

$$\begin{aligned} \text{Chlorophyll a} &= (11.24 \times A_{616.6}) - (2.04 \times A_{644.8}) \\ \text{Chlorophyll b} &= (20.13 \times A\ _{644.8}) - (4.19 \times A_{616.6}) \\ \text{Chlorophyll a + b} &= (7.05 \times A_{616.6}) + (18.09 \times A_{644.8}) \end{aligned} \quad (2)$$

where, $A_{616.6}$, $A_{644.8}$, and A_{470} are absorbance values at corresponding wavelengths.

2.3.2. Specific Leaf Area (SLA)

The SLA, which denotes the area per dry mass of the leaf, was measured following Ref. [20]. The leaf area was divided by leaf weight (indirectly indicating leaf thickness) and measured by an LI-3100C (LI-COR Biosciences, Lincoln, NE, USA). Then, the selected leaves were dried using a dry oven (HB-502S, Hanbaek Scientific, Bucheon, Korea) at 70 °C for 48 h to estimate their dry weights. SLA was determined using Equation (3):

$$\text{SLA } (cm^{-2} \cdot g^{-1}) = \text{Leaf area/dry weight} \quad (3)$$

2.3.3. Leaf Extract pH (pH)

The pH was measured using the method developed by Singh et al. [32]. For each plant species, 1.0 g of fresh leaves was placed in a test tube with 10 mL distilled water and homogenized at 2500 rpm for 3 min. Then, the pH value was measured by using a pH meter (HI 8424, Hana Instruments, Woonsocket, RI, USA).

2.3.4. Relative Water Content (RWC)

The RWC was determined according to the method developed by Turner [33]. The leaves were cut to a similar size (5 × 5 cm), and their fresh weight (FW) was immediately measured. After floating them in distilled water at 4 °C for 24 h, their turgid weight (TW) was determined. Finally, they were dried in an oven at 80 °C for 48 h and weighed to measure dry weight (DW). The RWC value was determined using Equation (4):

$$RWC\ (\%) = [(FW - DW)/(TW - DW)] \times 100 \tag{4}$$

where FW = fresh weight, TW = turgid weight, and DW = dry weight.

3. Statistical Analysis

All data were analyzed using SAS software version 9.4 (SAS Institute, Cary, NC, USA) for analysis of variance (ANOVA) with Duncan's multiple range test (DMRT). The significance level was set at 5%. Variations in the accumulated PM on different plant species and leaf traits in the four seasons were determined using a two-way ANOVA. The relationships between the amount of PM accumulation on the leaf and leaf traits were identified by using Pearson's correlation analysis. The presented data are given as the means with standard error (±SE).

4. Results and Discussion

4.1. PM Accumulation of Plant Species

In this study, we found that the amount of PM accumulation on leaf surfaces and wax layers differed among various plant species and sampling seasons. The amount of total PM accumulation on leaf surfaces of 24 plant species in the four seasons ranged from 8.32 to 148.43 $\mu g{\cdot}cm^{-2}$ in spring, 7.65 to 137.02 $\mu g{\cdot}cm^{-2}$ in summer, 3.65 to 122.84 $\mu g{\cdot}cm^{-2}$ in autumn, and 21.02 to 264.44 $\mu g{\cdot}cm^{-2}$ in winter (only on the evergreen plants). The amount of total PM accumulation on the wax layer of 24 plant species in spring, summer, autumn, and winter (only on the evergreen plants) ranged from 5.74 to 103.6 $\mu g{\cdot}cm^{-2}$, 4.12 to 116.67 $\mu g{\cdot}cm^{-2}$, 3.81 to 77.22 $\mu g{\cdot}cm^{-2}$, and 17.74 to 236.46 $\mu g{\cdot}cm^{-2}$, respectively. When comparing the amount of PM accumulation on leaves in the four seasons, we found that PM was highest in winter (needleleaf) and spring (broadleaf); conversely, the amount of PM was lowest in autumn. When comparing PM accumulation between needleleaf and broadleaf, the average PM accumulation on leaf surfaces and wax layers was higher on needleleaf than on broadleaf. In spring, the amount of PM accumulation on the leaf surfaces and the wax layers of 10 needleleaf ranged from 148.43 to 14.53 $\mu g{\cdot}cm^{-2}$ and from 103.63 to 16.62 $\mu g{\cdot}cm^{-2}$, respectively. Among the ten needleleaf plants, *P. strobus* showed the highest PM accumulation, followed by *P. parviflora* and *P. densiflora*. In contrast, the amount of PM accumulation in *J. chinensis* was the lowest. The amount of PM accumulation on the leaf surfaces and the wax layers of 14 broadleaf ranged from 23.75 to 8.32 $\mu g{\cdot}cm^{-2}$ and from 22.89 to 5.74 $\mu g{\cdot}cm^{-2}$, respectively. The plant species that showed the highest PM accumulation on leaf surfaces were *A. palmatum* followed by *E. japonicus* and *C. chinensis*, while *G. biloba* showed the lowest PM accumulation. The highest PM accumulations on the wax layers were the species *E. japonicus*, *P.* × *yedoensis*, and *C. officinalis*. Additionally, the plant species with the lowest PM accumulation was *V. dilatatum*. The amount of PM accumulation in needleleaf was higher than that in all the broadleaf plants except *J. chinensis* and *P. orientalis*. In summer, the amount of PM accumulation on the leaf surface and the wax layer of *P. strobus* was still the highest among needleleaf plants, with PM accumulations of 137.02 and 116.67 $\mu g{\cdot}cm^{-2}$ on the leaf surface and wax layer, respectively. Among the ten needleleaf, *P. orientalis* showed the lowest PM accumulation on both the leaf surfaces and the wax layer. For broadleaf plants, the amount of PM accumulation on the leaf surfaces ranged from 7.65 to 23.22 $\mu g{\cdot}cm^{-2}$, while the PM accumulation on the wax layer ranged from 4.25 to 13.83 $\mu g{\cdot}cm^{-2}$. Among 13 broadleaf, *A. palmatum* showed the highest PM accumulation on the leaf surface, and *P.* × *yedoensis* showed the highest accumulation on the wax layer. Conversely, *G. biloba* and *R. yedoense* showed the lowest PM accumulation on leaf surfaces and wax layers, respectively. In autumn, the amount of PM accumulation on the leaves of 10 needleleaf plants ranged from 8.86 to 122.84 $\mu g{\cdot}cm^{-2}$ on the leaf surfaces and from 8.52 to 77.22 $\mu g{\cdot}cm^{-2}$ on the wax layers, while, in summer, the highest and lowest PM accumulation on both the leaf surfaces and the wax layers were observed for *P. strobus* and *P. orientalis*, respectively. These plant species had higher PM accumulation than the other needleleaf plants, followed by *P. strobus*, *P. parviflora*, and *P. densiflora*. For broadleaf, the PM accumulation on the leaf surfaces ranged from 3.65 to 13.62 μg·cm^{-2} and on the wax layers ranged from 3.81 to 13.39 μg·cm^{-2}. *H. syriacus* had the highest PM accumulation on the leaf surface, while *P. × yedoensis* had the highest PM accumulation on the wax layer, followed by *E. japonicus* and *C. officinalis*. Conversely, *G. biloba* and *R. yedoense* had the lowest PM accumulation on the leaf surface and the wax layer, respectively. In winter, leaf samples of eleven evergreens (ten needleleaf and one broadleaf) were collected. The amount of PM accumulation on the leaf surfaces ranged from 21.02 to 264.44 μg·cm^{-2} and on the wax layers ranged from 17.74 to 236.46 μg·cm^{-2}. Among the 11 plant species, *P. strobus* showed the highest PM accumulation on the leaf surface, and *T. cuspidata* and *E. japonicus* showed the lowest PM accumulation on the leaf surface and the wax layer, respectively. Among the 24 plant species, *P. strobus* showed the highest PM accumulation, followed by *P. parviflora* and *P. densiflora*. The average PM accumulation on needleleaf was higher than the average PM accumulation on broadleaf by approximately three to over four times. Among broadleaf, *A. palmatum* and *E. japonicus* accumulated PM more effectively than other broadleaf (Table 2, Figure 2).

Table 2. The total PM accumulation on the leaf surface and the wax layer of 24 plant species in four seasons (spring, summer, autumn, and winter).

Species	Total sPM (μg·cm^{-2})				Total wPM (μg·cm^{-2})			
	Spring	Summer	Autumn	Winter	Spring	Summer	Autumn	Winter
J. chinensis	14.5 ± 4.1	12.5 ± 2.1	12.6 ± 1.7	26.5 ± 1.1	16.6 ± 0.9	13.6 ± 3.4	12.2 ± 4.7	30.3 ± 1.2
J. chinensis var. *kaizuka*	26.5 ± 1.6	12.8 ± 1.1	10.9 ± 1.5	31.3 ± 0.2	28.7 ± 1.8	13.3 ± 2.8	19.4 ± 6.8	38.1 ± 1.3
P. parviflora	113.9 ± 26.4	65.6 ± 7.1	54.1 ± 11.8	138.8 ± 7.5	40.4 ± 7.5	29.3 ± 6.6	56.8 ± 8.7	71.8 ± 8.4
P. densiflora	72.9 ± 2.7	75.1 ± 11.4	56.8 ± 6.0	72.7 ± 6.8	52.1 ± 5.6	47.8 ± 1.7	51.0 ± 10.2	81.3 ± 3.2
C. pisifera	24.8 ± 2.7	31.1 ± 7.5	11.5 ± 0.6	31.5 ± 0.9	23.5 ± 1.3	11.3 ± 0.6	16.5 ± 9.5	24.8 ± 3.2
T. cuspidata	25.7 ± 4.5	40.1 ± 5.0	9.0 ± 0.5	21.0 ± 0.7	24.3 ± 2.3	12.8 ± 1.3	10.4 ± 0.2	20.7 ± 0.5
A. holophylla	43.8 ± 3.6	17.5 ± 3.4	15.5 ± 0.3	67.5 ± 10.8	34.7 ± 11.1	35.3 ± 3.8	16.6 ± 2.5	46.0 ± 3.4
P. abies	30.7 ± 1.2	20.1 ± 0.5	22.7 ± 2.2	55.8 ± 4.9	28.4 ± 4.3	20.8 ± 0.7	23.3 ± 1.8	32.7 ± 3.3
P. strobus	148.4 ± 23.0	137.0 ± 31.8	122.8 ± 25.8	264.4 ± 31.1	103.6 ± 22.2	116.7 ± 5.4	77.2 ± 27.5	236.5 ± 23.1
P. orientalis	22.0 ± 2.3	11.6 ± 1.1	8.9 ± 0.9	28.7 ± 2.1	25.3 ± 1.3	9.3 ± 1.4	8.5 ± 0.8	28.8 ± 1.7
E. japonicus	23.7 ± 3.4	17.9 ± 0.3	12.9 ± 1.0	27.7 ± 3.2	22.9 ± 5.2	8.7 ± 0.5	10.7 ± 1.4	17.7 ± 1.6
M. denudata	10.9 ± 1.0	11.5 ± 0.8	9.0 ± 0.4	-	9.8 ± 0.7	6.7 ± 0.4	6.2 ± 0.8	-
A. turbinata	12.9 ± 0.8	9.9 ± 0.7	6.8 ± 1.1	-	9.2 ± 1.0	9.0 ± 0.5	4.5 ± 0.5	-
R. yedoense	16.0 ± 2.3	9.8 ± 0.2	9.8 ± 0.7	-	11.0 ± 0.2	4.1 ± 0.9	3.8 ± 0.7	-
H. syriacus	16.9 ± 0.4	16.6 ± 1.2	13.6 ± 1.9	-	12.0 ± 0.1	6.0 ± 0.3	6.3 ± 1.9	-
A. palmatum	23.8 ± 2.4	23.2 ± 9.9	12.6 ± 0.7	-	9.2 ± 0.6	6.6 ± 1.3	8.1 ± 1.3	-
C. chinensis	17.6 ± 2.1	10.8 ± 0.0	4.8 ± 0.7	-	10.3 ± 0.6	5.1 ± 0.1	7.3 ± 2.2	-
C. officinalis	16.2 ± 0.6	17.6 ± 0.9	12.7 ± 0.4	-	15.8 ± 1.3	5.2 ± 0.4	12.2 ± 1.2	-
A. triflorum	15.3 ± 0.8	13.5 ± 1.3	10.0 ± 0.1	-	13.5 ± 6.3	8.0 ± 2.1	6.3 ± 1.1	-
Z. serrata	10.0 ± 0.5	9.6 ± 0.6	3.7 ± 0.2	-	8.9 ± 06	4.3 ± 1.0	5.5 ± 0.0	-
G. biloba	8.3 ± 1.1	7.7 ± 2.4	7.2 ± 0.3	-	9.5 ± 1.5	6.3 ± 0.8	5.8 ± 0.5	-
L. obtusifolium	9.4 ± 0.9	16.2 ± 2.8	4.0 ± 0.3	-	11.1 ± 2.0	7.2 ± 0.5	7.7 ± 0.5	-
P. × yedoensis	15.4 ± 0.6	13.0 ± 1.0	5.7 ± 1.6	-	17.5 ± 2.9	13.8 ± 1.8	13.4 ± 1.4	-
V. dilatatum	10.6 ± 1.3	13.3 ± 1.5	7.0 ± 1.3	-	5.7 ± 1.1	4.9 ± 0.6	8.1 ± 1.1	-

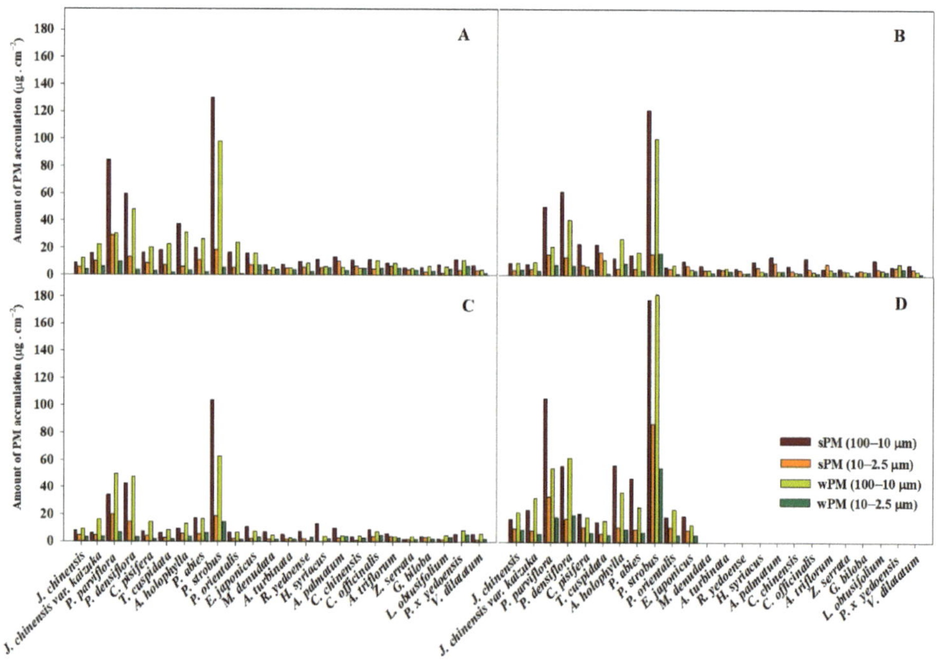

Figure 2. The amount of PM accumulation on the leaf surface and wax of 24 plant species. (**A**): Spring, (**B**): summer, (**C**): autumn, (**D**): winter.

The amount of wax on the leaves of the 24 plant species differed among both plant species and seasons. The amount of wax on the ten needleleaf during four seasons ranged from 81.20 to 755.04 μg·cm^{-2}, 55.95 to 1066.92 μg·cm^{-2}, 51.55 to 910.55 μg·cm^{-2}, and 106.74 to 3793.19 μg·cm^{-2} in spring, summer, autumn, and winter, respectively. In all the seasons, the amount of wax on *P. strobus* was the highest, followed by *P. densiflora*, while *T. cuspidata* showed the lowest amount of wax among the ten needleleaf. The wax on 13 broadleaf ranged from 15.68 to 95.41 μg·cm^{-2}, 15.76 to 191.50 μg·cm^{-2}, and 14.91 to 149.12 μg·cm^{-2} in spring, summer, and autumn, respectively. The amount of wax on *E. japonicus* was highest in spring, and the amount of wax on *V. dilatatum* was highest in summer and autumn. In winter, the amount of wax on *E. japonicus* was 160.86 μg·cm^{-2}. When comparing the amount of wax on needleleaf and broadleaf, we found that the amount of wax on all the needleleaf plants in spring, autumn, and summer was higher than that on broadleaf except for *T. cuspidata*. Among the 24 plant species, the amount of wax on *P. strobus* was higher than that of the other plant species (Figure 3).

Total sPM: (sPM (10–100 μm) + sPM2.5 (2.5–10 μm)), total wPM: (wPM (10–100 μm) + wPM2.5 (2.5–10 μm)).

Plant accumulation of PM occurs directly on leaves, so the leaf structure contributes to the ability of plants to accumulate PM. In this study, we found that the amount of PM accumulation on the leaves of 24 plant species was different among plant species and the four seasons. The amount of PM accumulation on the leaves of plants depends on leaf structures and environmental conditions [34]. The increasing PM concentration in the environment could lead to an increase in the amount of PM accumulation on the leaves of plants [35]. In this study, the PM concentration level was highest in winter, which could cause the highest PM accumulation on the needleleaf [36]. For the broadleaf, the PM accumulation on the leaf surface was higher in spring than in the other seasons due to the PM concentration in this season being higher than in summer and autumn. Under different environments, the PM concentration level, rainfall, and wind impact the amount of PM accumulation on leaves [37,38]. The high rainfall in summer and autumn could wash PM from a leaf, causing decreasing PM accumulation on the leaves of plants in the two seasons. The amount of PM accumulation in the 24 plant species tended to be higher in needleleaf than in broadleaf. Numerous studies have also shown that needleleaf are more effective in PM accumulation than broadleaf because of their high stomatal concentration and large amount of leaf wax [39,40]. The needleleaf secrete mucus oils on their leaf surfaces, reducing the amount of PM washed from leaves [41]. The trees, with their large leaf area, were considered the most effective PM accumulation plants. Although the crowns of the shrubs were much lower than the trees, nearing the ground helped shrubs accumulate PM closer to the ground [4]. In this study, the shrubs *E. japonicus* and *R. yedoense* showed more effective PM accumulation on the leaf surface than others.

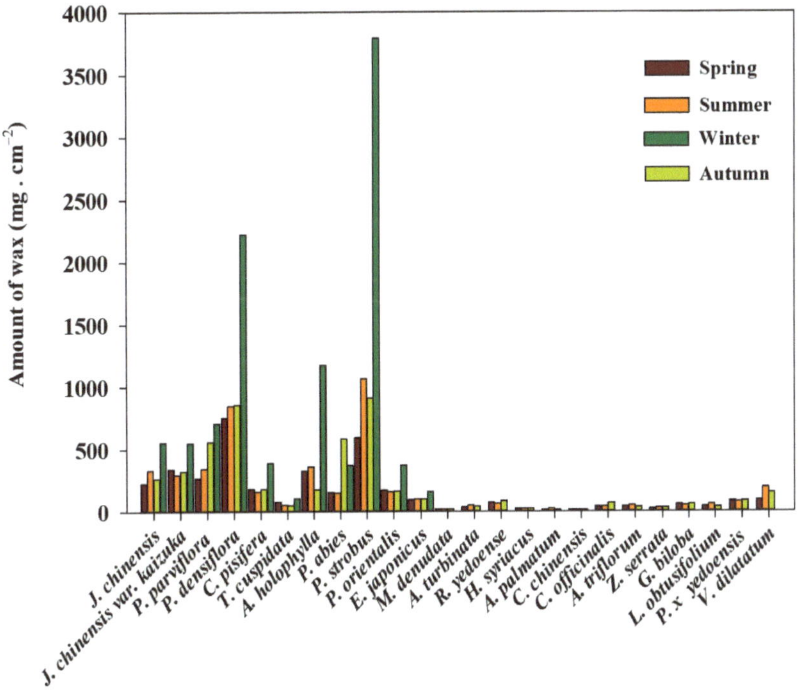

Figure 3. The amount of wax on 24 plant species in four seasons (spring, summer, autumn, and winter).

Among the ten needleleaf, *P. strobus* showed the highest PM accumulation on the leaf surface. The high stomatal concentration on the leaf could increase the PM accumulation on the leaf of *P. strobus*. Additionally, the amount of wax on *P. strobus* was the highest, causing an increasingly large amount of PM accumulation in this plant species. Other studies also showed that *P. parviflora* and *P. densiflora* were able to accumulate more PM than other plant species, which was the same as our result. Among the 14 broadleaf, the plant species that showed the most effective PM accumulation were *A. palmatum* and *E. japonicus* because of their leaf structures. Leaf structures, such as heave, groove, vein, chambers, bumps, glands, and secretions, impact PM deposition on leaves [42]. Many studies have shown that plants with rough leaves can accumulate more PM than plants with smooth leaves [39]. However, *E. japonicus* with smooth leaves showed high effective accumulation on leaves because of the plant's curled leaf edges, which could keep PM on the leaf apices despite the washing effect of rainfall. Moreover, the glands and secretions of the leaves were related to increasing PM accumulation in *E. japonicus*. For *A. palmatum*, the leaf fold structure led to increased PM accumulation. Conversely, *G. biloba*, with a rough leaf and high leaf area, showed the lowest PM accumulation. The long leaf petiole increased the wind's effective removal of PM from its leaves [43]. The ridged microstructure and dense wax tubules, due to the high self-cleaning ability of *G. biloba*, reduced the effective accumulation of PM in *G. bibola* [44]. The water repellency of the leaf prevented PM deposition on the leaf, which was also one of the reasons that led to decreasing PM accumulation on the leaf of *G. bibola*. The amount of wax on 24 plants differed due to their structural differences. The amount of wax in winter increased in response to the increasing PM in the atmosphere. The PM concentration level is positively correlated with the amount of wax in plants [45]. Leaf structures impacted the ability of PM to accumulate in the plants. However, other factors, such as leaf shape, canopy shape, and leaf height, need to be studied to better understand the complex relationships of plant species and PM accumulation.

4.2. The Leaf Traits of 24 Plant Species

In this study, the leaf traits of 24 plant species were analyzed. Among the leaf traits, the Chl a, Chl b, TChl, carotenoid, and RWC were different among plant species and seasons. We did not find any seasonal tendencies when comparing the Chl a and carotenoid contents of the 24 plant species. For the Chl b of 24 plant species, we found that Chl b tended to be lower in autumn than in summer; the species not following this trend were *P. strobus*, *Z. serrata*, and *L. obtusifolium*. The TChl of 24 plant species showed the same tendency as the Chl b, but *T. cuspidata*, *A. holophylla*, *P. strobus*, *Z. serrata*, and *L. obtusifolium* did not follow this trend. The RWC of the 24 plant species differed among the plants. In this study, the RWC of broadleaf did not show any seasonal tendencies. However, the RWC of needleleaf increased in autumn compared with summer. When comparing the RWC of needleleaf between autumn and winter, the RWC decreased in winter except in *P. strobus*. The pH differed among plant species. We found that the pH increased in autumn compared with summer for all plant species except *J. chinensis* and *P. parviflora*. The pH of 11 plant species was higher in winter than in other seasons. Additionally, the SLA of 24 plants differed among plant species. Among the four seasons, we found that the SLA of 11 plant species, except for *A. holophylla* and *P. strobus*, was lowest in winter. When comparing the SLA among the three seasons (spring, summer, and autumn), we did not find any seasonal tendencies (Tables 3 and 4).

Table 3. The leaf traits of 24 plant species in four different seasons (spring, summer, autumn, and winter).

	Seasons	Chl a ($mg{\cdot}g^{-1}$ FW)	Chl b ($mg{\cdot}g^{-1}$ FW)	TChl ($mg{\cdot}g^{-1}$ FW)	RWC (%)	pH	SLA ($cm^2{\cdot}g^{-1}$)
J. chinensis	Spring	0.056 ± 0.001	0.024 ± 0.001	0.080 ± 0.001	74.66 ± 1.15	5.08 ± 0.05	19.06 ± 0.60
	Summer	0.090 ± 0.003	0.030 ± 0.003	0.124 ± 0.004	73.45 ± 0.15	5.04 ± 0.02	24.87 ± 0.51
	Autumn	0.049 ± 0.001	0.020 ± 0.000	0.069 ± 0.002	79.28 ± 0.18	4.88 ± 0.03	20.97 ± 0.65
	Winter	0.066 ± 0.002	0.027 ± 0.002	0.094 ± 0.004	74.14 ± 0.58	5.70 ± 0.01	16.25 ± 0.32
J. chinensis var. *kaizuka*	Spring	0.052 ± 0.003	0.022 ± 0.002	0.078 ± 0.006	78.65 ± 0.04	4.75 ± 0.01	19.72 ± 0.96
	Summer	0.058 ± 0.001	0.026 ± 0.000	0.084 ± 0.001	77.93 ± 0.99	4.75 ± 0.01	18.77 ± 0.48
	Autumn	0.053 ± 0.002	0.023 ± 0.001	0.076 ± 0.004	91.87 ± 0.59	4.93 ± 0.05	19.40 ± 0.16
	Winter	0.055 ± 0.001	0.023 ± 0.002	0.078 ± 0.002	83.36 ± 1.12	5.77 ± 0.08	17.02 ± 0.60
P. parviflora	Spring	0.189 ± 0.007	0.060 ± 0.008	0.249 ± 0.015	71.99 ± 1.08	5.00 ± 0.03	9.65 ± 1.36
	Summer	0.117 ± 0.000	0.048 ± 0.000	0.166 ± 0.001	79.69 ± 1.47	5.09 ± 0.05	9.06 ± 0.97
	Autumn	0.119 ± 0.001	0.045 ± 0.000	0.165 ± 0.002	81.71 ± 0.49	4.93 ± 0.02	13.12 ± 0.70
	Winter	0.088 ± 0.000	0.039 ± 0.000	0.128 ± 0.000	76.85 ± 0.28	5.85 ± 0.04	9.02 ± 0.30

Table 3. *Cont.*

	Seasons	Chl a (mg·g^{-1} FW)	Chl b (mg·g^{-1} FW)	TChl (mg·g^{-1} FW)	RWC (%)	pH	SLA (cm^2·g^{-1})
P. densiflora	Spring	0.178 ± 0.004	0.075 ± 0.001	0.250 ± 0.007	77.58 ± 1.13	4.62 ± 0.03	14.34 ± 0.52
	Summer	0.122 ± 0.001	0.047 ± 0.001	0.184 ± 0.014	86.58 ± 1.26	4.59 ± 0.02	11.11 ± 0.67
	Autumn	0.106 ± 0.001	0.041 ± 0.001	0.146 ± 0.001	96.52 ± 1.10	4.85 ± 0.01	13.08 ± 1.24
	Winter	0.075 ± 0.004	0.032 ± 0.003	0.108 ± 0.007	93.19 ± 0.50	6.07 ± 0.05	9.23 ± 0.12
C. pisifera	Spring	0.157 ± 0.003	0.059 ± 0.001	0.216 ± 0.004	81.78 ± 0.80	4.90 ± 0.05	39.54 ± 2.07
	Summer	0.139 ± 0.007	0.056 ± 0.002	0.195 ± 0.009	80.96 ± 0.53	4.94 ± 0.03	58.19 ± 0.28
	Autumn	0.131 ± 0.001	0.051 ± 0.001	0.183 ± 0.001	84.92 ± 0.31	5.03 ± 0.01	42.99 ± 0.35
	Winter	0.089 ± 0.001	0.038 ± 0.001	0.127 ± 0.001	80.65 ± 1.66	6.04 ± 0.10	32.61 ± 0.43
T. cuspidata	Spring	0.124 ± 0.020	0.053 ± 0.008	0.140 ± 0.024	75.52 ± 0.54	5.21 ± 0.14	63.01 ± 1.89
	Summer	0.105 ± 0.005	0.045 ± 0.002	0.141 ± 0.006	77.25 ± 0.98	4.84 ± 0.07	55.18 ± 1.69
	Autumn	0.111 ± 0.003	0.042 ± 0.001	0.148 ± 0.005	79.31 ± 0.36	5.24 ± 0.01	58.30 ± 2.69
	Winter	0.105 ± 0.004	0.051 ± 0.002	0.156 ± 0.005	73.02 ± 1.14	5.69 ± 0.04	42.38 ± 0.77
A. holophylla	Spring	0.105 ± 0.009	0.044 ± 0.003	0.135 ± 0.011	74.88 ± 1.25	4.74 ± 0.03	28.94 ± 0.58
	Summer	0.113 ± 0.001	0.042 ± 0.000	0.156 ± 0.001	73.96 ± 0.24	4.74 ± 0.03	33.90 ± 1.49
	Autumn	0.114 ± 0.004	0.042 ± 0.001	0.157 ± 0.005	88.70 ± 0.57	4.93 ± 0.02	27.12 ± 0.59
	Winter	0.062 ± 0.002	0.030 ± 0.001	0.092 ± 0.002	77.56 ± 1.18	6.28 ± 0.07	28.10 ± 0.55
P. abies	Spring	0.103 ± 0.004	0.042 ± 0.003	0.128 ± 0.002	79.12 ± 1.07	4.19 ± 0.09	27.62 ± 1.04
	Summer	0.122 ± 0.003	0.046 ± 0.001	0.168 ± 0.004	73.40 ± 0.35	4.19 ± 0.09	31.79 ± 0.93
	Autumn	0.087 ± 0.001	0.033 ± 0.001	0.120 ± 0.002	83.60 ± 0.81	5.06 ± 0.04	33.57 ± 0.41
	Winter	0.101 ± 0.000	0.042 ± 0.002	0.142 ± 0.002	80.702 ± 0.35	5.84 ± 0.07	21.70 ± 0.41
P. strobus	Spring	0.094 ± 0.007	0.037 ± 0.003	0.147 ± 0.012	62.41 ± 0.53	4.84 ± 0.04	7.80 ± 0.86
	Summer	0.083 ± 0.001	0.035 ± 0.001	0.120 ± 0.003	72.78 ± 0.31	4.84 ± 0.04	3.78 ± 0.91
	Autumn	0.101 ± 0.004	0.042 ± 0.002	0.143 ± 0.006	73.44 ± 0.77	5.05 ± 0.02	1.27 ± 0.11
	Winter	0.084 ± 0.002	0.040 ± 0.000	0.124 ± 0.002	78.97 ± 0.46	5.54 ± 0.03	4.85 ± 0.52
P. orientalis	Spring	0.094 ± 0.002	0.040 ± 0.001	0.139 ± 0.003	78.17 ± 2.58	4.84 ± 0.02	49.33 ± 1.98
	Summer	0.105 ± 0.002	0.045 ± 0.001	0.150 ± 0.002	76.43 ± 0.20	4.84 ± 0.02	46.32 ± 1.88
	Autumn	0.096 ± 0.003	0.039 ± 0.001	0.135 ± 0.003	85.22 ± 1.06	5.38 ± 0.02	51.76 ± 0.49
	Winter	0.065 ± 0.005	0.029 ± 0.002	0.094 ± 0.008	75.84 ± 1.28	6.16 ± 0.15	39.36 ± 1.72
E. japonica	Spring	0.048 ± 0.003	0.024 ± 0.001	0.072 ± 0.002	67.96 ± 1.40	5.16 ± 0.01	81.74 ± 1.51
	Summer	0.071 ± 0.004	0.033 ± 0.001	0.103 ± 0.004	70.75 ± 0.85	5.16 ± 0.01	82.66 ± 4.59
	Autumn	0.039 ± 0.002	0.018 ± 0.001	0.057 ± 0.002	74.41 ± 0.59	5.31 ± 0.03	91.05 ± 2.82
	Winter	0.044 ± 0.006	0.021 ± 0.003	0.065 ± 0.009	60.96 ± 1.07	6.14 ± 0.03	77.15 ± 1.44
M. denudata	Spring	0.096 ± 0.005	0.038 ± 0.004	0.145 ± 0.014	67.29 ± 0.85	5.47 ± 0.01	179.51 ± 7.27
	Summer	0.111 ± 0.003	0.046 ± 0.002	0.156 ± 0.004	77.87 ± 0.31	5.47 ± 0.01	168.66 ± 6.71
	Autumn	0.086 ± 0.002	0.036 ± 0.001	0.122 ± 0.003	84.80 ± 0.13	6.14 ± 0.03	177.88 ± 10.6
	Winter	0	0	0	0	0	0
A. turbinata	Spring	0.214 ± 0.015	0.043 ± 0.005	0.271 ± 0.021	69.42 ± 0.35	5.17 ± 0.03	131.94 ± 2.31
	Summer	0.172 ± 0.000	0.073 ± 0.001	0.245 ± 0.001	72.26 ± 0.32	5.17 ± 0.03	145.68 ± 6.54
	Autumn	0.122 ± 0.003	0.047 ± 0.001	0.172 ± 0.005	80.75 ± 1.09	5.56 ± 0.05	123.81 ± 2.56
	Winter	0	0	0	0	0	0
R. yedoense	Spring	0.119 ± 0.026	0.029 ± 0.012	0.148 ± 0.037	77.55 ± 0.41	5.21 ± 0.04	165.65 ± 1.17
	Summer	0.165 ± 0.006	0.076 ± 0.005	0.261 ± 0.018	87.68 ± 1.78	5.18 ± 0.07	163.55 ± 2.48
	Autumn	0.165 ± 0.004	0.066 ± 0.001	0.230 ± 0.006	86.44 ± 0.40	5.47 ± 0.03	151.35 ± 1.82
	Winter	0	0	0	0	0	0
H. syriacus	Spring	0.128 ± 0.013	0.026 ± 0.004	0.154 ± 0.013	80.34 ± 0.65	5.63 ± 0.03	160.35 ± 4.51
	Summer	0.173 ± 0.012	0.071 ± 0.006	0.244 ± 0.017	78.41 ± 0.79	5.63 ± 0.03	147.46 ± 4.12
	Autumn	0.157 ± 0.002	0.071 ± 0.002	0.228 ± 0.003	81.45 ± 0.88	6.15 ± 0.02	151.61 ± 3.38
	Winter	0	0	0	0	0	0

Table 3. *Cont.*

	Seasons	Chl a ($mg{\cdot}g^{-1}$ FW)	Chl b ($mg{\cdot}g^{-1}$ FW)	TChl ($mg{\cdot}g^{-1}$ FW)	RWC (%)	pH	SLA ($cm^2{\cdot}g^{-1}$)
A. palmatum	Spring	0.075 ± 0.007	0.026 ± 0.012	0.070 ± 0.007	85.26 ± 0.19	4.71 ± 0.02	152.04 ± 5.05
	Summer	0.164 ± 0.015	0.071 ± 0.002	0.242 ± 0.012	91.30 ± 0.27	4.71 ± 0.02	193.06 ± 5.44
	Autumn	0.168 ± 0.003	0.068 ± 0.001	0.237 ± 0.003	92.00 ± 0.47	4.99 ± 0.09	201.17 ± 6.55
	Winter	0	0	0	0	0	0
C. chinensis	Spring	0.193 ± 0.008	0.054 ± 0.007	0.269 ± 0.015	61.45 ± 1.75	4.31 ± 0.05	159.38 ± 13.53
	Summer	0.191 ± 0.004	0.085 ± 0.002	0.274 ± 0.006	65.69 ± 0.34	4.42 ± 0.06	156.05 ± 17.76
	Autumn	0.166 ± 0.000	0.070 ± 0.001	0.234 ± 0.001	75.35 ± 0.50	5.44 ± 0.01	253.01 ± 21.21
	Winter	0	0	0	0	0	0
C. officinalis	Spring	0.127 ± 0.029	0.055 ± 0.014	0.182 ± 0.042	64.92 ± 0.16	5.84 ± 0.02	156.65 ± 2.40
	Summer	0.150 ± 0.005	0.065 ± 0.001	0.209 ± 0.003	73.81 ± 0.77	5.84 ± 0.02	210.40 ± 21.70
	Autumn	0.081 ± 0.006	0.036 ± 0.002	0.117 ± 0.008	73.75 ± 0.86	6.25 ± 0.04	121.68 ± 3.91
	Winter	0	0	0	0	0	0
A. triflorum	Spring	0.194 ± 0.019	0.070 ± 0.015	0.287 ± 0.035	68.34 ± 4.52	4.41 ± 0.02	214.34 ± 7.11
	Summer	0.217 ± 0.023	0.100 ± 0.009	0.317 ± 0.031	75.56 ± 0.85	4.39 ± 0.04	238.45 ± 9.92
	Autumn	0.152 ± 0.005	0.063 ± 0.002	0.212 ± 0.008	71.01 ± 0.31	5.81 ± 0.01	205.55 ± 8.00
	Winter	0	0	0	0	0	0
Z. serrata	Spring	0.177 ± 0.005	0.061 ± 0.002	0.237 ± 0.006	52.93 ± 3.92	5.15 ± 0.01	203.19 ± 0.53
	Summer	0.192 ± 0.018	0.078 ± 0.006	0.269 ± 0.023	61.44 ± 0.93	5.15 ± 0.01	174.44 ± 2.83
	Autumn	0.233 ± 0.030	0.102 ± 0.010	0.374 ± 0.035	56.92 ± 0.40	5.86 ± 0.02	224.08 ± 3.65
	Winter	0	0	0	0	0	0
G. biloba	Spring	0.093 ± 0.003	0.029 ± 0.002	0.122 ± 0.003	75.00 ± 0.06	5.00 ± 0.09	129.80 ± 5.15
	Summer	0.074 ± 0.004	0.034 ± 0.003	0.107 ± 0.006	72.13 ± 0.05	5.00 ± 0.09	131.70 ± 1.51
	Autumn	0.053 ± 0.002	0.024 ± 0.000	0.074 ± 0.001	80.03 ± 0.67	5.69 ± 0.03	156.01 ± 1.22
	Winter	0	0	0	0	0	0
L. obtusifolium	Spring	0.203 ± 0.005	0.070 ± 0.003	0.257 ± 0.017	69.74 ± 2.66	5.14 ± 0.02	122.36 ± 2.48
	Summer	0.174 ± 0.004	0.063 ± 0.002	0.237 ± 0.005	66.93 ± 0.71	5.14 ± 0.02	157.87 ± 13.67
	Autumn	0.156 ± 0.011	0.065 ± 0.003	0.240 ± 0.013	72.35 ± 3.70	5.34 ± 0.02	207.37 ± 3.50
	Winter	0	0	0	0	0	0
P. × *yedoensis*	Spring	0.137 ± 0.002	0.049 ± 0.002	0.186 ± 0.004	75.81 ± 0.22	5.03 ± 0.00	95.33 ± 1.98
	Summer	0.134 ± 0.005	0.055 ± 0.004	0.189 ± 0.008	76.55 ± 0.80	5.03 ± 0.00	110.75 ± 3.71
	Autumn	0.124 ± 0.019	0.042 ± 0.001	0.146 ± 0.004	81.83 ± 0.43	5.38 ± 0.01	120.18 ± 3.51
	Winter	0	0	0	0	0	0
V. dilatatum	Spring	0.164 ± 0.021	0.039 ± 0.006	0.203 ± 0.028	46.36 ± 1.82	5.60 ± 0.01	216.28 ± 11.07
	Summer	0.212 ± 0.010	0.091 ± 0.004	0.303 ± 0.014	64.15 ± 1.48	5.60 ± 0.01	251.32 ± 1.95
	Autumn	0.146 ± 0.007	0.063 ± 0.003	0.209 ± 0.010	62.67 ± 0.78	5.68 ± 0.01	302.54 ± 4.60
	Winter	0	0	0	0	0	0

Chl a: chlorophyll a concentration, Chl b: chlorophyll b concentration, TChl: total chlorophyll concentration, RWC: relative leaf water content, pH: leaf extract pH, and SLA: specific leaf area.

PM impacts plant growth and production by impacting its physiology and biological activities. Under stress from the environment, numerous changes in plants can be observed. However, the level of change depends on the response of the plants to the environment [26]. In this study, the leaf traits (chlorophyll, pH, RWC, and SLA) were analyzed to determine the influence of PM on the plant while the plant responded to environmental stress. Chlorophyll content signifies the photosynthesis process that determines plant growth and production [46]. PM accumulation on the leaf could block the stomata, leading to reduced stomatal conductance, which leads to reduced chlorophyll content in the plants [17]. Moreover, PM accumulation on the leaf prevented light absorption due to the decreasing effectiveness of photosynthesis [47]. PM could even lead to chlorosis (yellowing) in plants. In this study, the decreasing chlorophyll content of broadleaf caused them to change from green to yellow at the end of the growing season. We did not find any patterns among the chlorophyll content of the plants, but we suggest that the chlorophyll reduction in some plants caused PM to accumulate on the leaves. The increase in the chlorophyll content of other plants during the high PM concentration season showed plant tolerance to environmental stress [48]. In this study, we also found that the large sPM (10–100 μm), wPM (10–100 μm), and coarse wPM (2.5–10 μm) showed negative correlations with plant chlorophyll and carotenoid in summer, but, in winter, we found positive correlations between coarse wPM (2.5–10) and chlorophyll and carotenoid. In winter, the needleleaf and one broadleaf (*E. japonica*) were collected and analyzed.

Table 4. ANOVA analysis of PM accumulation on the leaf and leaf traits of 24 plant species in the four seasons (spring, summer, autumn, and winter).

	Species ($F_{23,192}$)	**Seasons** ($F_{3,192}$)	**Species × Season** ($F_{69,192}$)
sPM (10–100)	155.44 ***	20.46 ***	5.77 ***
sPM (2.5–10)	149.94 ***	47.48 ***	38.04 ***
wPM (10–100)	229.74 ***	34.54 ***	13.88 ***
wPM (2.5–10)	27.95 ***	9.71 ***	9.39 ***
Chl a	47.32 ***	750.56 ***	23.66 ***
Chl b	29.11 ***	498.36 ***	20.88 ***
TChl	48.89 ***	747.12 ***	27.2 ***
RWC	576.54 ***	8009.28 ***	297.33 ***
pH	882.41 ***	19134.2 ***	1415.78 ***
SLA	465.06 ***	2098.27 ***	84.38 **

Levels of significance: ** $p < 0.01$, *** $p < 0.001$. PM (10–100): particulate matter with diameter 10–100 μm, PM (2.5–10): particulate matter with diameter 2.5–10 μm, sPM: particulate matter on the leaf surface, wPM: particulate matter in the wax layer, Chl a: chlorophyll a concentration, Chl b: chlorophyll b concentration, TChl: total chlorophyll concentration, RWC: leaf relative water content, pH: leaf extract pH, SLA: specific leaf area.

The RWC reflected the status of water of plants, which plays a significant role in maintaining plant physiological balance. Plants with a high RWC could be more tolerant to stress from the environment, such as drought or air pollution [49]. PM locked on stomata led to a decreasing transpiration rate, which decreased the water pulled from plant roots and was a reason why minerals could not translate from plant roots for biosynthesis [6]. In this study, we suggested that the increase in PM concentration in winter caused a decrease in the RWC of needleleaf. In contrast, the increasing RWC of a plant in the high PM concentration season showed the high tolerance level of the plants to air pollution.

Further, pH is a sensitive indicator of stress in plant environments. Moreover, low pH reduces conversion of hexose sugar to ascorbic acid, which plays an important role in the tolerance level of plants [30]. Therefore, plants with a high pH have a greater ability to tolerate air pollution than plants with a low pH. Air pollutants lead to decreasing pH, which causes sensitive stomata due to reduced plant photosynthesis rates. Furthermore, acidic pollution, such as SO_2 and NO_2, is one reason for the decreasing pH of plants [50]. In this study, the increasing pH of needleleaf (in winter) and broadleaf (in autumn) may have caused the plants to respond to air pollution. We found a negative correlation between PM accumulation on leaves and pH in spring, summer, and autumn. Other factors, such as environmental soil pH, that could impact plant pH need to be studied to determine the correlation between pH and these factors.

SLA measures the thickness of the leaf. The changes in SLA mirror the changes in leaf structure and nutritional content. The thickness of leaves can help to increase effective light absorption [33]. However, PM accumulation on the leaf surface may increase the leaf's shape area and cause changes in the SLA of plants. Based on plant protective or adaptive mechanisms, SLA fluctuation levels vary [51]. In this study, we found that the SLA of the plant was different among species. Moreover, the SLA of needleleaf was lowest in winter, which caused a large amount of PM accumulation on the leaf surface of needleleaf during this season. We also found that PM accumulation was negatively correlated with plant SLA (Table 5).

Table 5. Pearson correlation analysis of the accumulation of different fractions of particulate matter on leaf and leaf traits of 24 plant species in the four seasons.

	PM Size	Chl a		Chl b		TChl		RWC		pH		SLA	
Spring	sPM (10–100)	−0.012		0.106		0.046		−0.037		−0.204		−0.554	***
	sPM (2.5–10)	0.035		0.110		0.069		0.141		−0.284	*	−0.548	***
	wPM (10–100)	−0.148		0.066		−0.070		0.047		−0.269	*	−0.633	***
	wPM (2.5–10)	0.087		0.095		0.127		−0.007		0.016		−0.154	
Summer	sPM (10–100)	−0.285	*	−0.295	*	−0.266	*	0.155		−0.148		−0.484	***
	sPM (2.5–10)	−0.196		−0.168		−0.183		0.314	*	−0.200		−0.381	***
	wPM (10–100)	−0.344	**	−0.359	**	−0.332	**	0.062		−0.244	*	−0.537	***
	wPM (2.5–10)	−0.306	**	−0.329	**	−0.300	*	0.138		−0.165		−0.471	***
Autumn	sPM (10–100)	−0.123		−0.136		−0.131		0.055		−0.313	**	−0.445	***
	sPM (2.5–10)	−0.175		−0.210		−0.181		0.182		−0.479	***	−0.579	***
	wPM (10–100)	−0.177		−0.214		−0.189		0.190		−0.477	***	−0.577	***
	wPM (2.5–10)	−0.134		−0.139		−0.137		−0.025		−0.258	*	−0.331	**
Winter	sPM (10–100)	0.231		0.287		0.252		0.200		−0.360		−0.559	***
	sPM (2.5–10)	0.177		0.250		0.202		0.107		−0.459		−0.472	**
	wPM (10–100)	0.142		0.209		0.165		0.269		−0.423		−0.545	***
	wPM (2.5–10)	0.149	***	0.217	***	0.172	***	0.214		−0.413		−0.492	**

Levels of significance: * $p < 0.05$, ** $p < 0.01$, *** $p < 0.001$. PM (10–100): particulate matter with diameter 10–100 μm, PM (2.5–10): particulate matter with diameter 2.5–10 μm, sPM: particulate matter on the leaf surface, wPM: particulate matter in the wax layer, Chl a: chlorophyll a concentration, Chl b: chlorophyll b concentration, TChl: total chlorophyll concentration, RWC: leaf relative water content, pH: leaf extract pH, SLA: specific leaf area.

5. Conclusions

In this study, we found that the amount of PM accumulation on the leaves of 24 plant species differed among the four seasons. In winter, the amount of PM accumulation increased with increasing PM concentration in the atmosphere. In summer and autumn, the low PM concentration and the high rainfall may have led to a reduction in the amount of PM accumulation on the plant leaves. In this study, needleleaf was more effective than broadleaf in accumulating PM. The shrubs demonstrated to be highly effective at reducing PM. Using both trees and shrubs can increase the effective PM accumulation in the urban area. The *P. strobus*, *P. parviflora*, *P. densiflora*, *E. japonicus*, and *A. palmatum* showed more effective PM accumulation than other plant species. The leaf traits differed regarding plant species and seasons. PM had a negative correlation with plant SLA. In summer, PM was negatively correlated with chlorophyll and carotenoids. Further, pH had a negative correlation with PM accumulation on the leaves in spring, summer, and autumn. PM accumulation impacted the leaf traits of the plant, but numerous other factors, such as temperature and soil, also impact leaf traits. More studies on the complex correlations among leaf PM accumulation, leaf traits, and environmental conditions are needed to effectively increase the use of plants to improve air quality.

Author Contributions: Conceptualization, S.-Y.K. and B.-J.P.; methodology, S.-Y.K. and B.-J.P.; investigation, H.-T.B. and U.O.; data analysis, H.-T.B. and U.O.; writing—review and editing, S.-Y.K. and B.-J.P.; funding acquisition B.-J.P. All authors have read and agreed to the published version of the manuscript.

Funding: This research was funded by the support of the R&D program for Forest Science Technology (Project No. 2019155B10-2021-001) provided by Korea Forest Service (Korea Forestry Promotion Institute), and this work was conducted during the research year of Chungbuk National University in 2021.

Institutional Review Board Statement: Not applicable.

Informed Consent Statement: Not applicable.

Data Availability Statement: All data supporting the conclusions of this article are included in this manuscript.

Conflicts of Interest: The authors declare no conflict of interest.

References

1. Doreswamy; Harishkumar, K.S.; Yogesh, K.M.; Ibrahim, G. Forecasting air pollution particulate matter ($PM_{2.5}$) using machine learning regression models. *Procedia Comput. Sci.* **2020**, *171*, 2057–2066. [CrossRef]
2. WHO. WHO's global air-quality guidelines. *Lancet* **2006**, *368*, 1302. [CrossRef]
3. Mukherjee, A.; Agrawal, M. World air particulate matter: Sources, distribution and health effects. *Environ. Chem. Lett.* **2017**, *15*, 283–309. [CrossRef]
4. Sæbø, A.; Popek, R.; Nawrot, B.; Hanslin, H.M.; Gawronska, H.; Gawronski, S.W. Plant species differences in particulate matter accumulation on leaf surfaces. *Sci. Total Environ.* **2012**, *427–428*, 347–354. [CrossRef]
5. Kwon, K.-J.; Urrintuya, O.; Kim, S.-Y.; Yang, J.-C.; Sung, J.-W.; Park, B.-J. Removal potential of particulate matter of 12 woody plant species for landscape planting. *J. People Plants Environ.* **2020**, *23*, 647–654. [CrossRef]
6. Bui, H.-T.; Ousuren, U.; Kwon, K.-J.; Kim, S.-Y.; Yang, J.-C.; Jeong, N.-R.; Park, B.-J. Assessment of air pollution tolerance and particulate matter accumulation of 11 woody plant species. *Atmosphere* **2021**, *12*, 1067. [CrossRef]
7. Bui, H.-T.; Odsuren, U.; Jeong, M.; Seo, J.-W.; Kim, S.-Y.; Park, B.-J. Evaluation of the air pollution tolerance index of 12 plant species growing in environments with different air pollution levels. *J. People Plants Environ.* **2022**, *25*, 23–31. [CrossRef]
8. Bui, H.-T.; Odsuren, U.; Kim, S.-Y.; Park, B.-J. Particulate matter accumulation and leaf traits of ten woody species growing with different air pollution conditions in Cheongju City, South Korea. *Atmosphere* **2022**, *13*, 1351. [CrossRef]
9. Wróblewska, K.; Jeong, B.R. Effectiveness of plants and green infrastructure utilization in ambient particulate matter removal. *Envrion. Sci. Eur.* **2021**, *33*, 110. [CrossRef]
10. Zhang, W.K.; Wang, B.; Niu, X. Study on the adsorption capacities for airborne particulates of landscape plant in different polluted regions in Beijing (China). *Int. J. Envrion. Res. Public Health* **2015**, *12*, 9623–9638. [CrossRef]
11. Zhang, W.; Wang, B.; Niu, X. Relationship between leaf surface characteristics and particle capturing capacities of different tree species in Beijing. *Forests* **2017**, *8*, 92. [CrossRef]
12. McDonald, A.G.; Bealey, W.J.; Fowler, D.; Dragosits, U.; Skiba, U.; Smith, R.I.; Donovan, R.G.; Brett, H.E.; Hewitt, C.N.; Nemitz, E. Quantifying the effect of urban tree planting on concentrations and depositions of PM10 in two UK conurbations. *Atmos. Environ.* **2007**, *41*, 8455–8467. [CrossRef]
13. Mo, L.; Ma, Z.; Xu, Y.; Sun, F.; Lun, X.; Liu, X.; Chen, J.; Yu, X. Assessing the capacity of plant species to accumulate particulate matter in Beijing, China. *PLoS ONE* **2015**, *10*, e0140664. [CrossRef] [PubMed]
14. Popek, R.; Gawrońska, H.; Wrochna, M.; Gawroński, S.W.; Sæbø, A. Particulate matter on foliage of 13 woody species: Deposition on surfaces and phytostabilisation in waxes—A 3-year study. *Int. J. Phytoremediation* **2013**, *15*, 245–256. [CrossRef] [PubMed]
15. Zhang, W.; Zhang, Z.; Meng, H.; Zhang, T. How does leaf surface micromorphology of different trees impact their ability to capture particulate matter? *Forests* **2018**, *9*, 681. [CrossRef]
16. Leonard, R.J.; McArthur, C.; Hochuli, D.F. Particulate matter deposition on roadside plants and the importance of leaf trait combination. *Urban For. Urban Green.* **2016**, *20*, 249–253. [CrossRef]
17. Przybysz, A.; Popek, R.; Gawrońska, H.; Grab, K.; Łoskot, K.; Wrochna, M.; Gawroński, S.W. Efficiency of photosynthetic apparatus of plants grown in sites differing in level of particulate matter. *Acta Sci. Pol. Hortorum Cultus* **2014**, *13*, 17–30.
18. Wang, H.; Shi, H.; Wang, Y. Effects of weather, time, and pollution level on the amount of particulate matter deposited on leaves of *Ligustrum lucidum*. *Sci. World J.* **2015**, *2015*, 935942. [CrossRef]
19. Chen, L.; Liu, C.; Zhang, L.; Zou, R.; Zhang, Z. Variation in tree species ability to capture and retain airborne fine particulate matter ($PM_{2.5}$). *Sci. Rep.* **2017**, *7*, 3206. [CrossRef]
20. Popek, R.; Haynes, A.; Przybysz, A.; Robinson, S.A. How much does weather matter? Effects of rain and wind on PM accumulation by four species of Australian native trees. *Atmosphere* **2019**, *10*, 633. [CrossRef]
21. Chaturvedi, R.K.; Prasad, S.; Pana, S.; Obaidullah, S.M.; Pandey, V.; Singh, H. Effects of dust load on the leaf attributes of the tree species growing along the roadside. *Environ. Monit. Assess.* **2013**, *185*, 383–391. [CrossRef] [PubMed]
22. Kong, K.; Yu, H.; Chen, M.; Piao, Z.; Dang, J.; Sui, Y. Effects of particle matters on plant: A review. *Phyton* **2019**, *88*, 367–378. [CrossRef]
23. Farmer, A.M. The effects of dust on vegetation—A review. *Environ. Pollut.* **1993**, *79*, 63–75. [CrossRef]
24. Mulenga, C.; Clarke, C.; Meincken, M. Physiological and growth responses to pollutant-induced biochemical changes in plants: A review. *Pollution, Pollution* **2020**, *6*, 827–848. [CrossRef]
25. He, C.; Qie, K.; Pott, R. Reduction of urban traffic-related particulate matter-leaf trait matters. *Environ. Sci. Pollut. Res.* **2020**, *27*, 5825–5844. [CrossRef] [PubMed]
26. Panda, L.R.L.; Aggarwal, R.K.; Bhardwaj, D.R. A review on air pollution tolerance index (APTI) and anticipated performance index (API). *Curr. World Environ.* **2018**, *13*, 55–65. [CrossRef]
27. Popek, R.; Przybysz, A.; Gawrońska, H.; Klamkowski, K.; Gawroński, S.W. Impact of particulate matter accumulation on the photosynthetic apparatus of roadside woody plants growing in the urban conditions. *Ecotoxicol. Environ. Saf.* **2018**, *163*, 56–62. [CrossRef]
28. Shannigrahi, A.S.; Fukushima, T.; Sharma, R.C. Anticipated air pollution tolerance of some plant species considered for green belt development in and around an industrial/urban area in India: An overview. *Int. J. Environ. Stud.* **2004**, *61*, 125–137. [CrossRef]

29. Ter, S.; Chettri, M.K.; Shakya, K. Air pollution tolerance index of some tree species of Pashupati and Budhanilkantha area, Kathmandu. *Amrit Res. J.* **2020**, *1*, 20–28. [CrossRef]
30. Dzierzanowski, K.; Popek, R.; Gawrońska, H.; Saebø, A.; Gawroński, S.W. Deposition of particulate matter of different size fractions on leaf surfaces and in waxes of urban forest species. *Int. J. Phytoremediation* **2011**, *13*, 1037–1046. [CrossRef]
31. Lichtenthaler, H.K. Chlorophylls and carotenoids: Pigments of photosynthetic biomembranes. In *Methods in Enzymology*; Elsevier: Amsterdam, The Netherlands, 1987; Volume 148, pp. 350–382.
32. Singh, S.; Rao, D.; Agrawal, M.; Pandey, J.; Naryan, D. Air pollution tolerance index of plants. *J. Environ. Manag.* **1991**, *32*, 45–55. [CrossRef]
33. Turner, N.C. Techniques and experimental approaches for the measurement of plant water status. *Plant Soil* **1981**, *58*, 339–366. [CrossRef]
34. Xu, X.; Yu, X.; Mo, L.; Xu, Y.; Bao, L.; Lun, X. Atmospheric particulate matter accumulation on trees: A comparison of boles, branches and leaves. *J. Clean. Prod.* **2019**, *226*, 349–356. [CrossRef]
35. Ahmad, I.; Abdullah, B.; Dole, J.M.; Shahid, M.; Ziaf, K. Evaluation of the air pollution tolerance index of ornamentals growing in an industrial area compared to a less polluted area. *Hortic. Environ. Biotechnol.* **2019**, *60*, 595–602. [CrossRef]
36. He, C.; Qiu, K.; Alahnad, A.; Pott, R. Particulate matter capturing capacity of roadside evergreen vegetation during the winter season. *Urban For. Urban Green.* **2020**, *48*, 126510. [CrossRef]
37. Beckett, K.P.; Freer-Smith, P.H.; Taylor, G. Particulate pollution capture by urban trees: Effect of species and windspeed. *Glob. Chang. Biol.* **2000**, *8*, 995–1003. [CrossRef]
38. Xu, X.; Zhang, Z.; Bao, L.; Mo, L.; Yu, X.; Fan, D.; Lun, X. Influence of rainfall duration and intensity on particulate matter removal from plant leaves. *Sci. Total Environ.* **2017**, *609*, 11–16. [CrossRef]
39. Jin, E.J.; Yoon, J.H.; Bae, E.J.; Jeong, B.R.; Yong, S.H.; Choi, M.S. Particulate matter removal ability of ten evergreen trees planted in Korea urban greening. *Forests* **2021**, *12*, 438. [CrossRef]
40. Li, Y.; Wang, S.; Chen, Q. Potential of thirteen urban greening plants to capture particulate matter on leaf surfaces across three levels of ambient atmospheric pollution. *Int. J. Environ. Res. Public Health* **2019**, *16*, 402. [CrossRef]
41. Liu, J.; Cao, Z.; Zou, S.; Liu, H.; Hai, X.; Wang, S.; Duan, J.; Xi, B.; Yan, G.; Zhang, S.; et al. An investigation of the leaf retention capacity, efficiency and mechanism for atmospheric particulate matter of five greening tree species in Beijing, China. *Sci. Total Environ.* **2018**, *616–617*, 417–426. [CrossRef]
42. Shi, J.; Zhang, G.; An, H.; Yin, W.; Xia, X. Quantifying the particulate matter accumulation on leaf surfaces of urban plants in Beijing, China. *Atmos. Pollut. Res.* **2017**, *8*, 836–842. [CrossRef]
43. Han, D.; Shen, H.; Duan, W.; Chen, L. A review on particulate matter removal capacity by urban forests at different scales. *Urban For. Urban Green.* **2020**, *48*, 126565. [CrossRef]
44. Chen, G.; Lin, L.; Zhang, Y.; Ma, K. Net particulate matter removal ability and efficiency of ten plant species in Beijing. *Urban For. Urban Green.* **2021**, *63*, 127230. [CrossRef]
45. Popek, R.; Łukowsk, A.; Karolewski, P. Particulate matter accumulation—Further differences between native *Prunus padus* and non-native *P. serotina*. *Dendrobiology* **2017**, *78*, 85–95. [CrossRef]
46. Tripathi, A.K.; Gautnam, M. Biochemical parameters of plants as indicators of air pollution. *J. Environ. Biol.* **2007**, *28*, 127–132. [PubMed]
47. Popek, R.; Łukowski, A.; Grabowski, M. Influence of particulate matter accumulation on photosynthetic apparatus of *Physocarpus opulifolius* and *Sorbaria sorbifolia*. *Pol. J. Environ. Stud.* **2018**, *27*, 2391–2396. [CrossRef]
48. Giri, S.; Shrivastava, D.; Deshmukh, K.; Dubey, P. Effect of air pollution on chlorophyll content of leaves. *Curr. Agric. Res. J.* **2013**, *1*, 93–98. [CrossRef]
49. Ghafari, S.; Kaviani, B.; Sedaghathoor, S.; Allahyari, M.S. Assessment of air pollution tolerance index (APTI) for some ornamental woody species in green space of humid temperature region (Rasht, Iran). *Environ. Dev. Sustain.* **2020**, *23*, 1579–1600. [CrossRef]
50. Rai, P.K.; Panda, L.L.S.; Chutia, B.M.; Singh, M.M. Comparative assessment of air pollution tolerance index (APTI) in the industrial (Rourkela) and non industrial area (Aizawl) of India: An ecomanagement approach. *Afr. J. Environ. Sci. Technol.* **2013**, *7*, 944–948. [CrossRef]
51. Wuytack, T.; Wuyts, K.; Van Dongen, S.; Baeten, L.; Kardel, F.; Verheyen, K.; Samson, R. The effect of air pollution and other environmental stressors on leaf fluctuating asymmetry and specific leaf area of *Salix ablba* L. *Environ. Pollut.* **2011**, *159*, 2405–2411. [CrossRef]

land

MDPI

Article

Micrometeorological and Hydraulic Properties of an Urban Green Space on a Warm Summer Day in a Mediterranean City (Attica–Greece)

Nikolaos D. Proutsos *, Alexandra D. Solomou , Michaela Petropoulou and Nikolaos E. Chatzipavlis

Institute of Mediterranean Forest Ecosystems-Hellenic Agricultural Organization "DEMETER", Terma Alkmanos, 11528 Athens, Greece
* Correspondence: np@fria.gr

Abstract: Urban Green Spaces (UGSs) are considered the most effective tool to mitigate Urban Heat Islands (UHIs). The optical properties of the materials and the vegetation types of the UGSs affect their surface temperatures, directly influencing their cooling ability. The hydraulic properties of urban soils are also affected by the vegetation coverage. The aim of this study is to investigate the temperature and reflected radiation (albedo) differences between UGS's elements, around noon on a warm summer day, in Greece. The results indicate that green elements have smaller surface temperatures and higher reflectance compared to the artificial or the dry bare soil, presenting differences with the direct air temperature (measured above the surfaces with unshielded thermometers) −5.5 °C (shrubs), −3.8 °C (grass), +7.8 °C or +8.7 °C (paved surfaces inside or outside the UGS), +10.8 °C (dry bare soil), +12.2 °C (concrete) and +12.5 °C (asphalt), and albedo values 0.14 (grass and shrubs), 0.15 (dry bare soil), 0.27 (concrete), 0.21 (asphalt) and 0.20 (paved surfaces). The tree shades also produce great surface differences. The unsaturated hydraulic conductivity of the urban soil is greater than the surfaces covered with grass compared to the shrub-covered or bare soil, presenting values of 27.6, 10.8 and 11.4 mm h^{-1}, respectively.

Keywords: urban heat island; urban climate; urban green spaces; optical properties; surface temperature; albedo; green infrastructure; LIFE GrIn project; Mediterranean city

1. Introduction

Urban Green Spaces (UGSs) provide multiple benefits for local communities [1]. Their importance has been enhanced and acknowledged by the public, especially during the last decades, mainly due to the expanding urbanization and, most recently, due to the restrictions imposed by the spread of the COVID-19 pandemic [2]. The most important benefit of urban vegetation is its ability to regulate urban climate by affecting several micrometeorological attributes of the urban environment that impact the exchange of the mass and energy fluxes, providing a cooling effect and relieving the impacts of the Urban Heat Island (UHI) phenomenon.

The characteristics of the UGS and the persisting weather conditions in each area can highly affect its ability to regulate urban climate. The seasonal radiation conditions and the size of the park are among the more influential parameters for the parks' cooling effect [3], whereas the shape of the green area, its coverage with trees and shrubs and grass-covered surfaces also affect UGSs' cooling effects [3]. Following up on the findings of other studies [4–8], it is concluded that the proportion of green areas within an urban setting is a key role factor in the distribution of the UHI intensity. According to Qui et al. [9], the intensity of the UHI phenomenon can be relieved by up to 1.57 °C in comparison with surrounding commercial areas, addressing evapotranspiration as an efficient attribute for UHI mitigation. Maimaitiyiming et al. [10] also suggest that the presence of vegetated areas lowers local temperatures, underling the importance of the quality characteristics of the green areas, i.e., the density of green patches might be positively linked with alleviation of the UHI but cannot compare in strength with the cooling effect that an uninterrupted large green space would offer, a fact also pointed out by Cao et al. [3], Li et al. [11] and Zhang et al. [12]. Qui et al. [13] reported that urban vegetation could amount to a 0.5–4.0 °C reduction in the air temperature of the neighboring area, whereas Mackey et al. [14] suggest that an NDVI (index indicative of vegetation cover) above 0.35 triggers a notable temperature drop. Jusuf et al. [15] observed a considerable correlation between land use and temperature data, as well as an obvious temperature reduction as the vegetation body increased. Similarly, Huang et al. [16] studied the cooling effects of several UGSs across Harbin, China, factoring in the features of the parks, and determined that the vegetal arrangement, including the size and shape of the parks, is positively linked to their cooling efficiency. They confirmed, using the NDVI, that higher vegetation cover translates into a less intense UHI phenomenon, a point also supported by Bao et al. [17].

Edmondson et al. [18] identified an alleviating effect of the UHI by vegetation cover, distinguishing the higher potential of woody plants compared to herbaceous coverage. They noted that the mean maximum surface temperature of the soil was mediated by about 5.7 °C more by the presence of trees and scrubs when compared to the effect of grass-covered surfaces. During summer, the estimated residual temperature under tree cover was found to be negative, whereas it was found to be positive in the case of non-woody coverage, and this relationship seems to be reversed in winter. However, they noted that, as also reported by Armson et al. [19] and Wu et al. [20], the existence of trees in urban environments, while it might offer benefits due to shading, can also hinder the free movement of air currents, thus impeding the release of long-wave radiation at nighttime, trapping heat.

The use of different materials in the city and also inside the UGSs may also affect their energy exchange characteristics. Mohajerani et al.'s [21] research on the contribution of different materials used in urban areas to the UHI phenomenon revealed asphalt's and concrete's great significance since their presence basically boosts the UHI. They note that asphalt's high heat capacity and low albedo in combination with its extensive use in cities largely influence the urban environment's thermal conditions. It has been reported that freshly paved asphalt concrete can capture up to 95% of the received solar radiation [22] thus easily surpassing surface temperatures of 60 °C on hot summer days. As commented by Grimmond [23], the artificial materials commonly used in urban surfaces differ substantially in their thermal, hydraulic and radiative properties from naturally occurring ones. They also state that modifying the albedo of a surface is an easier and more tangible solution than meddling with any of its other characteristics to mitigate UHI and that a suggested practice is to add a top, light-colored, low-albedo layer on paved surfaces to enhance its reflectance and promote the return of radiation back to the atmosphere.

Generally, solar radiation drives plants' physiological processes and productivity [24]. Solar light availability, especially at the photosynthetically active waveband [25–27], in conjunction with the plant's optical properties (reflectance, transmittance and absorbance), determines the energy and mass fluxes of the vegetation-covered surfaces and ecosystems [28]. The radiation reflectance of urban surfaces is considered a most influential factor for UHI [29]. The surfaces that comprise urban landscapes have lower reflectivity, larger heat capacity, and generally different thermal properties than others in less built-up areas, thus storing more solar energy, which, in turn, translates into higher surface temperatures [30,31]. Due to their high thermal inertia and heat flux, the surface materials that are commonly used in cities are greatly influenced by ambient temperatures [32]. In addition, the densely built space of the city does not allow the dispersal of long-wave radiation from the ground, effectively obstructing the otherwise naturally occurring reduction in temperature [33–35]. Baldinelli et al. [29] suggest that an increase in radiation reflectance in the city (urban albedo) would result in its cooling. Similarly, Erell et al. [36] state that the high albedo surfaces result in lowering air temperature but increase the thermal stress in sunny conditions due to radiant exchange.

Most studies conducted in cities analyze the impacts of UGSs on UHI by assessing differences in air temperature between highly built-up and open, green or not, spaces. Specifically for Athens—Greece, Giannopoulou et al. [37] identified five geographic zones in the city with different thermal balances. The authors explored the heat island phenomenon in 25 sites in Athens during the summer season and found higher air temperatures in the industrial western part and in the center of the city and lower in its northern and eastern parts. Their findings are in line with those by Livada et al. [38], who also mention that UHI was developed intensively in the central and western industrialized part of the city.

Skoulika et al. [39] studied the cooling island intensity of an urban park in Athens during summer and confirmed its important mitigation impact on its surroundings, whereas Zulia et al. [40] investigated the microclimatic conditions in the National Garden in the center of Athens during summer and found a clear influence of the park, being cooler compared to other monitored urban locations, with greater air temperature differences during the night in streets with high buildings and wide streets with low traffic, whereas the relative differences were higher during the daytime in streets with high anthropogenic heat during the day.

Melas et al. [41] examined the microclimatic conditions of small open spaces (small courtyards and backyards) and found that vegetated backyards produce stronger cool island patterns compared to non-vegetated spaces. Tsiros and Hoffman [42] also evaluated the cooling effect in a courtyard's garden during a hot weather summer period in Athens and found a well-defined and strong daytime cool island and a significant air temperature reduction compared to an urban square with low canopy coverage. In addition, Tsiros [43] studied the trees cooling effect in Athens by measuring air temperature under the canopy in five streets during the exceptionally hot weather period in 2007 and found that the trees' average cooling effect at noon varied from 0.5 to 1.6 °C.

UGSs also provide important services for flood prevention due to the high water absorption by the soil, which is highly affected by the type of urban vegetation cover. Hidayat et al. [44] mention that soil characteristics including hydraulic conductivity are affected by the forest canopy cover, with higher hydraulic conductivity and thus infiltration rate at high-density canopies. Luo et al. [45] reported the different hydraulic profiles of clay soils covered by different vegetation types or with no coverage. Measurements of the unsaturated hydraulic conductivity showed that planting Vetiver grass causes a significant rise in hydraulic conductivity values, probably due to the structure of its root system, whereas, conversely, the Bermuda grass cover reduced the conductivity compared to the reference bare soil studied. The same pattern was observed when it comes to the infiltration rate, with the Vetiver grass almost doubling its value, whereas the Bermuda grass cut it in half. Gadi et al. [46] inquired into the variation in the hydraulic properties of the soil in close proximity to a tree, taking into account the state of vegetation cover. When studying the hydraulic conductivity as a function of floral density, they discovered that densely vegetated soils differed by 33–99% from soils with poor vegetation cover. This can be attributed to the intensified suction related to the water uptake by roots, which in turn lessens the water flow through the soil [47,48]. Multiple studies have showcased the escalation potential of hydraulic conductivity as a response to the growth of the plant's root system [49–51], which is accredited to water's preferential flow around the roots [52]. Gadi et al. [46] also marked the significant influence of the presence of shredded leaves upon the soil surface, as the associated hydraulic conductivity was 49–100% higher than in portions where leaves were absent. Jarvis et al. [53] studied the effects of both land use and climate on the hydraulic conductivity values and found that arable sites have, on average, ca. 2–3 times smaller saturated water conductivity than natural vegetation, forests and perennial agriculture.

Galli et al. [54] measured the unsaturated hydraulic conductivity in degraded and rehabilitated urban green spaces and found increased values in rehabilitated soils, especially five years after the soil rehabilitation process. They also concluded that in spaces with an absence of soil and vegetation maintenance, the unsaturated hydraulic conductivity may decrease after 9–12 years. Vegetation improves the soil structure and increases the soil's unsaturated conductivity. In addition, Galli et al. [54] identified the time since soil rehabilitation, the soil compaction and the vegetation cover as the most critical factors affecting the unsaturated hydraulic conductivity. They suggested that the higher unsaturated hydraulic conductivity values in the examined rehabilitated green spaces are associated with the time after the vegetation introduction and the development of the plant's root system and the development of a coherent soil matrix with stable connections between pores and the establishment of pathways for the water movement.

Based on the above findings, it is concluded that the optical properties of the UGSs determine the microclimate of a city. Along with the UGSs' soil characteristics, the basic cycles of energy and water in the city are highly affected, impacting also the quality of life of the citizens. The aim of this study is to investigate the differences in the surface temperatures of the green and grey elements inside an urban park and extend our knowledge on the performance of UGSs on absorbing/reflecting radiant energy and infiltrating water, under the weather conditions persisting in a hot summer day at a city with a Mediterranean climate.

2. Materials and Methods

This work was implemented for the broader area of Athens (Attica) in the Municipality of Amaroussion. The long-term meteorological data for the region, derived by the nearest station of N. Philadelphia (38.05° N, 23.67° E, alt.: 12 m, operated by the Hellenic National Meteorological Service) indicate that the climate is semi-arid, according to UNEP's [55] aridity climate classification, which is based on Thornthwaite's [56,57] water balance approach, with an aridity index value of 0.44 for the period 1955–2019 [58,59], presenting a decadal trend to more arid conditions [60].

The study site is an urban green space (38.04° N, 23.80° E, alt.: 190 m), with a total area of 0.91 ha covered by different types of materials including both green (trees, shrubs and grass) and grey (paved corridors and concrete) elements, whereas the UGS is surrounded by asphalt roads (Figure 1).

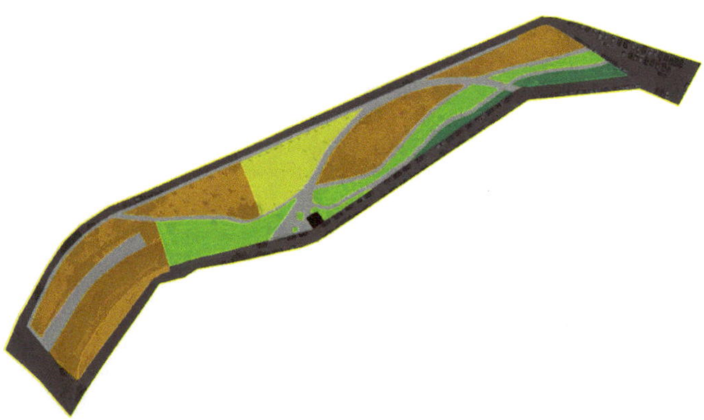

Figure 1. Different types of land cover (materials) in the UGS, i.e., asphalt (dark grey), paved surfaces (light grey), concrete (yellow), non-irrigated soil with trees (orange), shrubs (light green) and grass (dark green) (base map by Google Earth, Image © 2022, Google, Maxar Technologies, Westminster, CO, USA).

The vegetation in the UGS includes trees, shrubs and herbaceous plants. The trees are generally deciduous broad-leaved (e.g., *Tilia tomentosa*, *Acer negundo*, *Melia azedarach*, *Platanus orientalis*, *Morus alba*, *Cercis siliquastrum*, *Prunus cerasifera*, *Ailanthus altissima*) and randomly distributed in the UGS. Most of them are of small age (less than 5 years) and have not yet produced large canopies. The shrub-covered surfaces (Figure 1) host a variety of species (e.g., *Nerium oleander*, *Teucrium fruticans*, *Rosmarinus officinalis*, *Lavandula angustifolia*) in mixed patterns with herbaceous plants (e.g., *Calendula arvensis, Lactuca serriola, Matricaria recutita, Pallenis spinosa, Capsella bursa-pastoris, Solanum elaeagnifolium, Plantago lanceolata, Convolvulus arvensis*), whereas two surfaces were covered with grass (e.g., *Lolium perenne, Poa annua, Arundo donax, Cynodon dactylon, Eleusine indica*).

To study the micrometeorological environment inside the UGS, an automatic meteorological station was installed in 2019 by the Institute of Mediterranean Forest Ecosystems (IMFE), measuring several attributes including air temperature (at a height of 2 m above ground) and radiation attributes. For the study of the spatial variations in air and surface temperatures, portable sensors were used in order to take 258 point measurements at different points inside and outside the UGS. All measurements were taken during midday (from 13:00 to 15:00) of a hot summer day (23 June 2022) under clear sky conditions. A portable LP 471 RAD probe (Delta OHM), irradiance meter with cosine correction was used to measure shortwave solar radiation incoming and reflected flux densities within the spectral range of 400 to 1050 nm. In addition, an MI-210 infrared radiometer (Apogee Electronics) was used to measure surface temperatures above the different surfaces. The recordings of the surface temperatures were obtained by placing the sensor 40 cm above the surface facing downwards (zero angle). At the same height above the surfaces, the direct air temperature was measured with an HD 2301.0 handheld thermo-hygrometer (portable device with a Pt100 humidity/temperature combined sensor), considering that the substrate surface highly affects its above-measured air temperature [61] and thus the recordings of the portable device are more reliable compared to the temperatures recorded at the nearby meteorological station by a radiation-shielded thermometer. The sensor was directly exposed to radiation. To compare the differences in air temperature inside and outside the UGS, continuous 10-min temperature and relative humidity data, along with global solar radiation flux densities, were obtained by both IMFE's station and a nearby (270 m distance) automatic meteorological station (model: Davis Vantage Pro 2 Plus) installed by the Municipality of Amaroussion, on the roof of a one-floor (with height about 5 m) elementary school building.

The ability of the UGS to infiltrate water was assessed by measuring the unsaturated hydraulic conductivity at specific points inside the UGS, covered with bare soil, shrubs and grass (Figure 2). The field experiments were conducted with a minidisk infiltrometer (Meter Group) and the unsaturated hydraulic conductivity was estimated according to the method proposed by Zhang [62], applying the van Genuchten parameters obtained from Carsel and Parrish [63]. Before each infiltration experiment, the soil moisture was measured with a delta-t, SM150 sensor. In total, nine infiltration experiments were conducted above bare soil, shrub-covered and grass-covered surfaces.

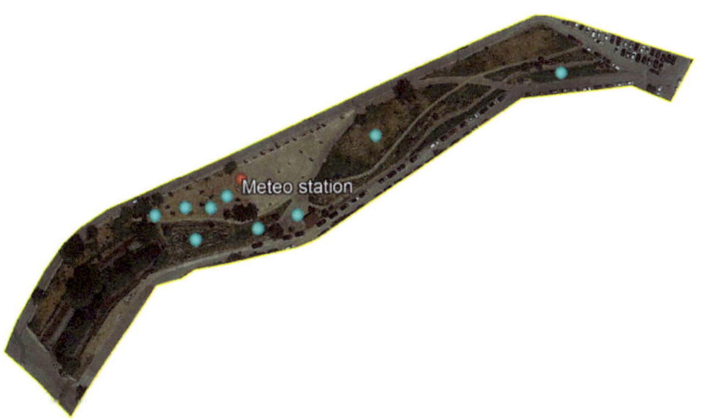

Figure 2. Points inside the UGS where infiltration measurements were taken for the estimation of the unsaturated hydraulic conductivity of the soil (4 points for dry bare soil, 3 points for shrubs and 1 point for grass) (base map by Google Earth, Image © 2022, Google, Maxar Technologies).

Zhang's [62] method is simple and reliable for measuring infiltration into dry soils. It uses cumulative infiltration (I) measurements plotted against time (t), and the results are fitted to the equation

$$I = C_1 \sqrt{t} + C_2\, t \tag{1}$$

where C_1 and C_2 (in mm s^{-1}) are factors related to soil sorptivity and hydraulic conductivity, respectively.

The soil hydraulic conductivity (K) is estimated by the equation:

$$K = \frac{C_1}{A} \tag{2}$$

where A is an estimator calculated by the following equation:

$$A = \begin{cases} \dfrac{11.65\,\left(n^{0.1}-1\right)\, e^{2.92(n-1.9)\ a\ h_0}}{(a\ r_0)^{0.91}}, & for\ n \geq 1.9 \\ \dfrac{11.65\,\left(n^{0.1}-1\right)\, e^{7.5(n-1.9)\ a\ h_0}}{(a\ r_0)^{0.91}}, & for\ n < 1.9 \end{cases} \tag{3}$$

where n and a are the van Genuchten parameters that differ for specific soils [62], r_0 is the infiltometer's disk radius, and h_0 is the suction applied at the disk's surface. In our experiments, the soil was loamy sand, r_0 = 2.25 cm and h_0 = −1 cm.

In the present study, spatial patterns of the attributes measured or calculated in the UGS were also produced in the form of contour maps, by using the Surfer® ver. 13 Software [64]. Point Kriging geostatistical gridding method was applied to each irregularly spaced point dataset to produce accurate grid values. The method is widely used and reliable, estimating the unknown grid values at all points across a well-defined spatial domain by using weighted averages of all known values around each grid point [65].

3. Results and Discussion

3.1. Temperature Attributes

During the warm summer day (23 June 2022) when measurements were taken, the air temperature inside the UGS was elevated, presenting a 24 h average value of 30.1 °C, which reached 36.0 °C at midday (13:00–15:00 h), whereas the soil temperature at 10 cm depth was even higher with a 24 h average 36.4 °C reaching 44.6 °C at midday (13:00–15:00 h). The respective air temperatures recorded on the concrete roof of the nearby building were similar and slightly higher (24 h average 30.3 °C and midday average at 13:00–15:00 h 36.3 °C). The 24 h temperatures were higher compared to the month's (June 2022) average by about 3.7 °C for air temperature and 2.8 °C for soil temperature in the UGS and by 4.2 °C for air temperature on the building's roof, indicating the hot conditions prevailing during the day when the measurement campaign was conducted.

The diurnal changes of the meteorological parameters during 23 June 2022 as recorded by both stations (inside the UGS and at the building's roof) along with the respective monthly average values for an average day of June 2022 are depicted in Figure 3. The air temperature and relative humidity in both stations appear to have similar values both during the day of the campaign (UGS 0.1 °C cooler than the roof) and during June 2022 (zero difference), especially during daytime (6:00–20:00 h). It is notable, however, that during the nighttime the temperature differences between the vegetation-covered UGS and the concrete-covered building's roof are maximized indicating cooler conditions in the UGS by about −0.9 °C in June 2022 becoming higher (−2.0 °C) during the campaign's day. The pattern of the relative humidity is the opposite compared to the respective one of temperature, presenting similar values for both stations at daytime and higher values (+2.4% in June 2022 and +6.4% on 23 June 2022) at nighttime.

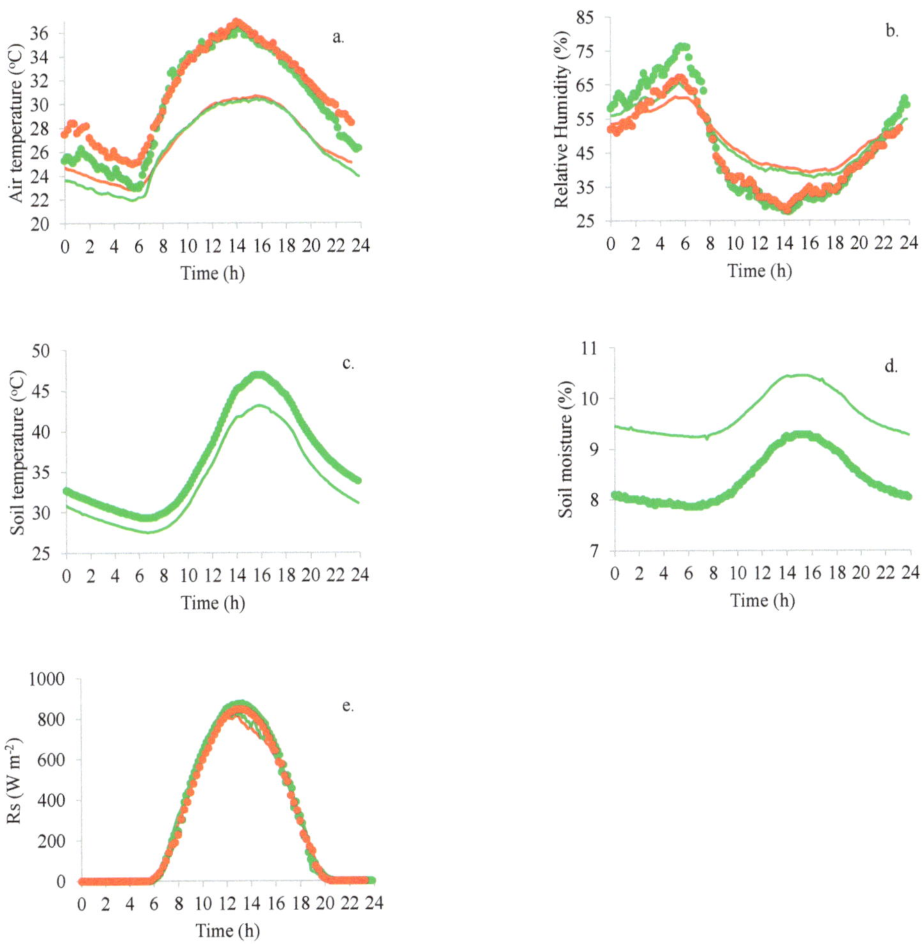

Figure 3. Average 10 min values of micrometeorological parameters: (**a**). air temperature, (**b**). relative humidity, (**c**). soil temperature, (**d**). soil moisture, (**e**). global solar radiation) measured inside the UGS (green) and on a nearby (270 m distance) roof of a building (red), during a warm summer day (23 June 2022, dots) and in all June 2022 (lines) in Amaroussion UGS.

Based on the above, the cooling effect of the UGS can be considered minor. However, this should be assessed in conjunction with the prevailing micrometeorological conditions during the very warm campaign day. The increased air temperature during the specific day, which reached 30.1 °C (compared to 26.4 °C for the average day of June 2022), presenting a maximum of 36.4 °C in association with the rather low relative humidity (daily average of 47.6% with a minimum 10 min value 27.6%), the cloudless sky conditions along with the high solar radiation flux density (24 h average 300.0 W m^{-2}, with maximum 10 min value 874.3 W m^{-2}) and the very high soil temperatures (daily average 36.2 °C with maximum 10 min value 46.9 °C), impose a very high water demand for the vegetation to sustain high growth rates. Such conditions, considering also the extremely reduced water availability in the rootzone (soil moisture 8.4% p.v.), define the reduced evapotranspiration rates and, thus, the less effective cooling ability of the plants to allocate water and energy from soil to the atmosphere. For the specific UGS, the municipality green managers apply deficit irrigation in an attempt to reduce water consumption, and this decreases the UGS's ability to produce cooling benefits to a maximum degree. The issue of reducing water consumption in UGSs, through plant selection or deficit irrigation, is critical and rather complicated, especially in the Mediterranean region, where water scarcity, on one hand, and the increased impact of UHI, on the other, impose contradicting criteria for UGS's design and management. The cooling effect is also associated with the size of the UGS, its shape and the vegetation architecture, which, in our case, consists of young trees with no fully developed canopies that enhance the openness of the park, resulting in a negative effect on the air temperature [66]. According to Jaganmohan et al. [67], the larger the area of an UGS, the more prominent the associated cooling, though manifested differently depending on the landscape type (forests vs. parks). In their research on several UGSs in Leipzig Germany, it was also indicated that the shape factor also affects their cooling ability. Moreover, the features of the green space are more significant than those of their surroundings when it comes to the UGS's cooling efficiency. Similarly, Monteiro et al. [68] assessed various sized UGSs in London, observing that very small UGSs (less than 0.8 ha) had no discernable cooling effect beyond their boundaries, but sizing up, one can note the drop in the air temperature of the surrounding environment. Furthermore, the tree canopy cover proved to be positively related to the cooling distance, but the degree of cooling was more strongly linked to the grass cover of the green space.

The general effect of the UGS was further assessed by analyzing surface temperature data taken above surfaces covered with different green and grey materials inside the park. The spatial patterns produced by the instant air temperature and surface temperature values in the UGS, after applying Kriging's interpolation, are depicted in Figure 4. The surface temperatures measured at several points of the UGS during midday (13:00–15:00) were, in general, higher for the artificial surfaces compared to the ones covered with natural elements. More specifically, surface temperatures above asphalt during midday were the highest, compared to all other materials, reaching an average of 54.4 °C, and were about +12.5 °C warmer compared to the direct air temperature measured just above the surface. The respective differences with the air and soil temperatures obtained by the meteorological station were even higher (18.2 and 8.4 °C, respectively). Similar was the behavior of the concrete-covered surface inside the park, which presented an average surface temperature of 52.9 °C, +12.8 °C higher than the above-point-measured direct air temperature, being +18.0 °C and +11.94 °C warmer than the air and soil temperatures measured by the meteo station, respectively. The internal paved corridors inside the UGS also presented high surface temperatures (average 51.4 °C) and they were also higher compared to the direct air temperatures above their surfaces (+8.7 °C), and much higher compared to the air and soil temperatures measured by the meteo station (15.2 and +5.4 °C, respectively). The respective temperature differences of the paved surfaces surrounding the UGS were also of the same magnitude since their average surface temperature was 49.6 °C, about +7.8, +13.8 and +3.3 °C warmer than the direct air temperature above the surfaces and the air or soil temperatures of the station inside the UGS, respectively. The above pattern is rather expected since the artificial elements absorb the radiant energy from the sun and store it in the form of heat, increasing their temperature.

The general pattern of the above-mentioned artificial surfaces, appears to be followed in the case of dry bare soil, which also presents high surface temperatures of the same magnitude (average 50.0 °C) during the hot summer midday, being +10.8 °C warmer compared to the direct air temperatures above the surface and +15.9 and +6.6 °C warmer than the air and soil temperatures of the station inside the UGS. This is also expected since the dry and non-vegetation-covered (bare) soil appears to behave as an artificial material, storing heat and increasing its temperature.

The effect of the live vegetation tissues is quite different since the absorbed solar energy does not transform into heat storage but is rather used to enhance the photosynthesis process. In our study, this seems to be true even under low soil water availability conditions. The natural surfaces covered with vegetation have lower surface temperatures compared to the artificial or dry bare soil-covered ones. More specifically, the surface temperature of the grass was measured on average to be 38.0 °C, i.e., −3.7 °C cooler compared to the direct temperature of the above air, −8.2 °C cooler than the soil temperature, but +2.2 °C warmer than the air temperature of the station inside the UGS.

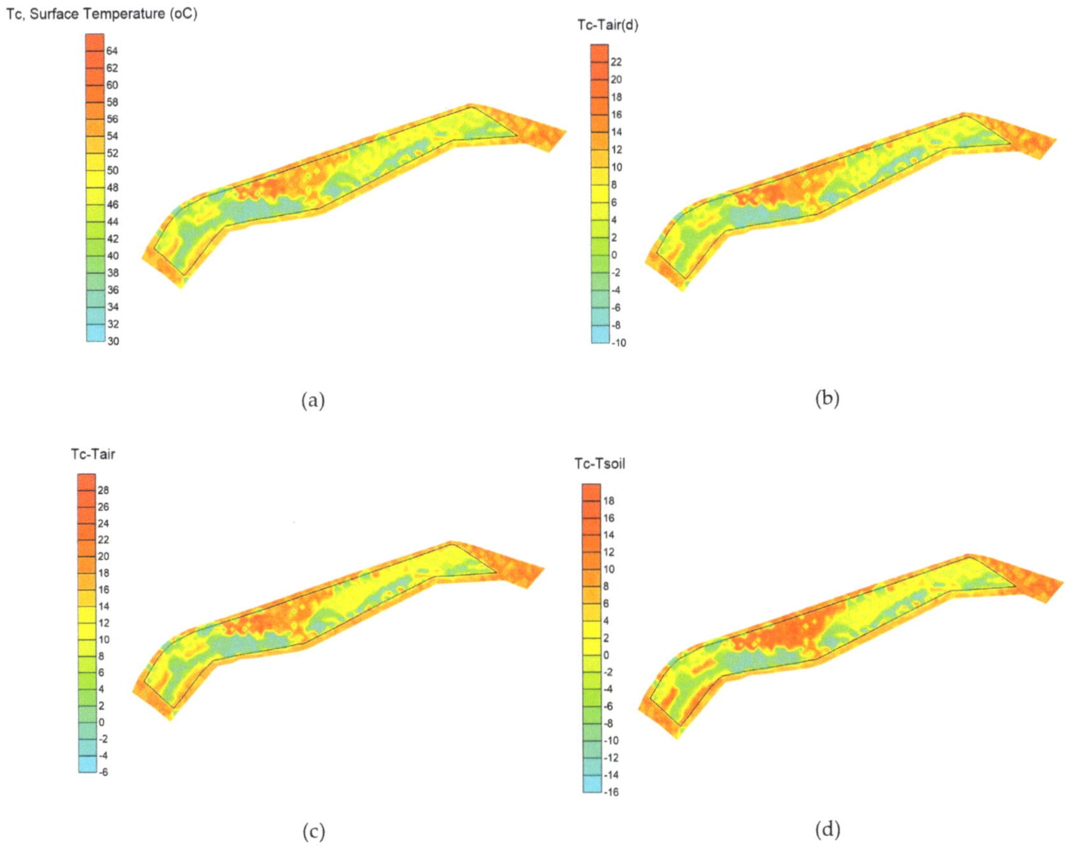

Figure 4. Spatial changes in (**a**) surface temperature in the UGS and differences (**b**) with direct air temperature measured above each point with sensor without radiation shield (**c**) with air temperature derived from the stationary meteorological station and (**d**) with soil temperature.

The cooling effect of shrubs was even more evident. Their surface temperatures were the lowest recorded (average 36.8 °C), and they were −5.5 °C cooler compared to the direct air temperatures above the surfaces, −0.2 °C cooler compared to the air temperature of the station and −9.9 °C lower than the soil temperature.

It should be noted that the above patterns describe the behavior of the natural and artificial elements of the UGS during the summer Mediterranean noon on a very warm day and under conditions of limited soil water availability. This is expected to have reduced the cooling effect of the vegetation elements of the UGS, since only small water quantities would have been available to the plants for evapotranspiration in order to regulate their tissues temperatures and perform photosynthesis, considering that evapotranspiration is demonstrated to be the most effective way to improve UHI [9,69–71].

According to the results of the study of Đekić et al. [72], a distinct variation between the temperatures of various surfaces in UGSs was identified, particularly in the high solar radiation hours—around noon. More specifically, they found that the temperature difference between the grass surface (coolest) and the dark asphalt varied between 9 and 19.7 °C, whereas in our study it was on average 16.3 °C. In addition, Đekić et al. [72]

also observed a difference of 10–24.5 °C between the hottest surface temperature and that of the ambient air, a range that is in line with the findings of our study where the respective difference between the asphalt and the ambient air temperature, measured by the meteorological station, was 18.2 °C. Similarly, Đekić et al. [72] mention that for the hotter period of measurements (mid-July to mid-August) the average maximum temperature for all artificial surfaces was well above 50 °C (average 52.1 °C in our study for all artificial materials ranging from 49.6 °C for the internal paved corridors to 54.4 °C for asphalt), while the grass-covered area more or less matched the temperature of the ambient environment, which is also in line with our study, since the temperature differences of grass and shrubs against ambient air temperature measured by our station were very close (38.0–35.8 °C and 36.7–36.1 °C, respectively.

During the measurement campaign, surface temperatures were also measured at shaded surfaces (below the tree canopies), to quantify how tree shades can reduce the increase in surface temperatures, especially above warm artificial surfaces and bare dry soil (Figure 4). The results indicate a rapid decrease in the surface temperature under shade in all cases. More specifically, the shaded asphalt was found to be −17.8 °C cooler than the unshaded, whereas similar differences were detected for the other surfaces (−15.2 °C for the internal paved corridors of the UGS, −11.6 °C for the concrete and −11.5 °C for the dry bare soil). The positive effect of shading is also portrayed by Đekić et al. [71] whereby even partly shaded asphalt surfaces were found to maintain lower temperatures than their counterparts that were fully exposed to sunlight.

3.2. Radiation Reflectance (Albedo)

The radiation reflectance of urban surfaces is considered the most influential factor for the Urban Heat Island effect [29]. The reflectance of the surfaces can highly affect their radiation budget and the radiative energy partitioning and exchange. The part of the incoming solar energy absorbed by the surfaces is determined by their reflectance characteristics including surface roughness and color. The shortwave radiation reflectance of a surface is the ratio of the reflected to the incoming global solar radiation flux density, and is commonly known as albedo.

The albedo changes above the different materials of the UGS of Amaroussion are depicted in Figure 5. Inside the UGS, the natural surfaces present, in general, lower reflectivity compared to the artificial surfaces, suggesting that they absorb more solar radiative energy. Specifically, the vegetation-covered surfaces, either with grass or shrubs, present albedos of about 0.14 and the dry bare soil appears to be of the same magnitude (0.15). On the other hand, the artificial surfaces' albedo is greater, being 0.27 for concrete, 0.21 for asphalt and 0.20 for the internal UGS's paved corridors.

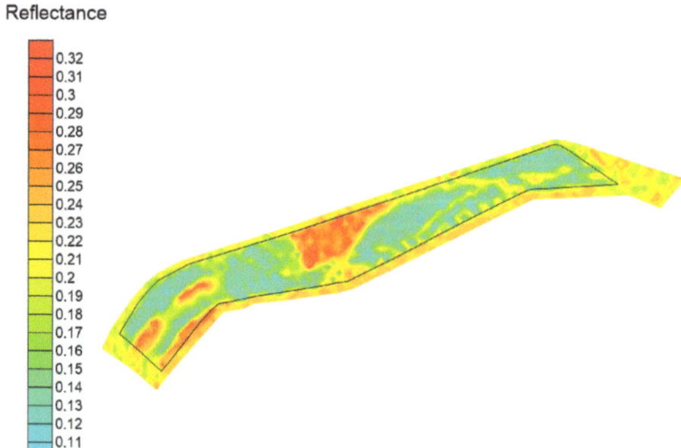

Figure 5. Spatial changes in the reflectance coefficient of shortwave solar radiation (albedo) inside the UGS.

Đekić et al. [72] cited the albedo values of some common urban artificial surfaces as well as those of green areas in order to accentuate their different influences on the city's thermal balance. Thereby, the albedo values' span for asphalt is 0.04–0.15 and the respective range for concrete is 0.10–0.35. As for the materials used for pedestrian routes, dark concrete tiles hold an albedo of 0.05–0.35 while white concrete tiles' albedo is considered to be around 0.70. Lawns and arbors have values of 0.25–0.30 and 0.15–0.18, respectively. Compared to our study, the values for asphalt and grass are quite different probably due to the different types of materials in the studies; however, the albedos of the other attributes are in line with our findings.

In general, high reflectance materials are suggested for use in UGS and cities in order to reduce radiation absorption and, thus, to prevent increasing surface temperatures, and they are considered as a cooling strategy to mitigate the UHI effect in cities [29]. It should be noted, however, that the reflected radiation in densely built cities with tall buildings is only partly reflected back to the atmosphere, mainly due to multiple radiation reflection and scattering between the surfaces, which finally trap the shortwave radiation in the building canopies, resulting in decreased urban albedo in the large scale of a city [73] with regard to the geometric structure of the buildings [74] and the scarcity of urban vegetation [75].

The reflectance of the materials inside the UGS differs at larger scales. Sugawara and Takamura [75] used radiometers to measure the surface albedo above two cities in Japan using a helicopter and found an albedo value of 0.12, whereas, in the nearby forest, it was higher (0.16). According to the authors, the lower city albedo is attributed to the roughness of the urban surface, which is enhanced by the city buildings and an increase in the absorbance of the urban surface. This appears to be contradicted when studying the optical properties of specific artificial materials since their surfaces are far smoother compared to the large scale of a city.

The part of the absorbed radiation by the artificial surfaces results in their temperature increase. In the case of the green elements, the absorbed radiation is mainly used for photosynthetic processes and, to a lesser degree, results in the increase in the tissues' temperatures. In our case, the lower reflectance of vegetation-covered surfaces results in greater sums of radiation absorbed by plants but, as already presented above, this did not lead to an increase in their temperature. It is worth noting that even the behavior of the dry bare soil is similar to the plant-covered surfaces, as regards its optical characteristics, but this increased radiation absorbance increased the heat storage in the soil, resulting in its higher temperatures.

3.3. Soil Water Infiltration

Water infiltration ability is a significant parameter in soil research and an important attribute for the UGSs, indicating the ability of the soil to absorb water. The unsaturated hydraulic conductivity is a measure of this soil's ability, highly associated with the specific soil mechanical characteristics and especially the coverage of its upper layer. In the UGS of the present study, nine infiltration experiments were conducted at different points in the UGS covered with shrubs, grass and dry bare soil. The results of the analysis are depicted in Figure 6, indicating that grass-covered surfaces present a higher ability to infiltrate water compared to dry bare soil and shrub-covered surfaces. The respective average unsaturated hydraulic conductivity values are 27.6 mm h^{-1} (or lit m^{-2} h^{-1}) for the grass, 10.8 mm h^{-1} for the shrubs and 11.4 mm h^{-1} for the dry bare soil.

Hidayat et al. [44] found that soil characteristics, including hydraulic conductivity, are affected by the forest canopy cover, with higher hydraulic conductivity at high-density canopies. The similar values for the bare soil and the shrubs in our UGS are probably attributed to the fact that the shrub-covered surface was only recently (1 year ago) planted and, thus, the short time slot passed was not enough to allow both the soil and the plants root to formulate water paths in the soil to enhance infiltration.

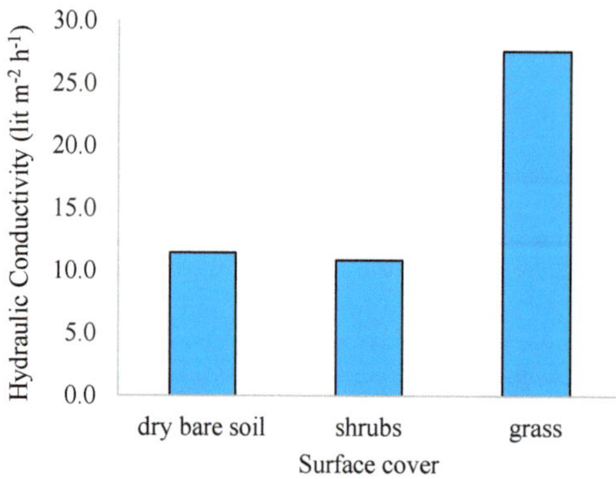

Figure 6. Unsaturated hydraulic conductivity values for soils with a different cover in the UGS.

Vegetation improves the soil structure and increases the soil's unsaturated conductivity. However, the time passing after the vegetation installation is critical. In a previous work by Galli et al. [54] on UGSs, the authors found increased unsaturated hydraulic conductivity values in rehabilitated soils, especially five years after the soil rehabilitation. They also identified that unsaturated hydraulic conductivity may decrease after 9–12 years in UGSs with no soil and vegetation maintenance. In our study, the soil at the spaces where shrubs were planted, was only maintained at the specific spots where the plants were established, whereas the experiments for the determination of the unsaturated hydraulic conductivity were conducted between the spots, where there was no soil maintenance.

On the other hand, the grass sections of the UGS were the only ones frequently irrigated and maintained, and within the first year of the grass installation, a dense root system was formulated and enhanced the soil's ability to infiltrate water. Galli et al. [54] identified the time after soil rehabilitation, the soil compaction and the vegetation cover as the most critical factors affecting the unsaturated hydraulic conductivity in the UGSs. They suggested that the higher unsaturated hydraulic conductivity values in the examined rehabilitated green spaces were associated with the time after the vegetation introduction. This is due to the fact that as the plants grow their root system, the development of a coherent soil matrix with stable connections between pores and the establishment of pathways for the water movement inside the soil, are enhanced. In addition, many studies [46,49–51] support that soil hydraulic conductivity is highly affected by the growth of the plant's roots, and this is accredited to water's preferential flow around the roots [52].

4. Conclusions

The positive impact of Urban Green Spaces (UGS) on the local climate is generally accepted, imposing a need to redesign our cities by enhancing green infrastructure in order to cope with climate change and establish resilient cities and neighborhoods. The positive effect of UGSs is attributed to their micrometeorological–optical characteristics and hydraulics properties that allow for the greater absorbance of solar radiation without increasing stored heat in the urban environment and, also, the increased water infiltration, which, in large scales, can reduce flooding phenomena.

In the present study, we present the findings of a campaign implemented in an UGS in Amaroussion city in Athens—Greece, where solar reflectance (albedo) and surface temperatures were measured above the different-type surfaces of the UGS during the midday of a warm summer day (23 June 2022), using portable radiometers and infrared thermometers. In addition, the soil hydraulic conductivity was estimated for bare soil and vegetation-covered surfaces in the UGS.

The results show that natural surfaces have lower reflection coefficients (0.14 for grass and shrub-covered surfaces and 0.15 for dry bare soil) compared to the artificial ones (albedo values 0.27 for concrete, 0.21 for asphalt and 0.20 for paved-covered surfaces). The surface temperatures of the natural elements are also cooler compared to the surrounding air (direct air temperatures measured above the surfaces with unshielded thermometers) presenting higher negative differences above vegetation −5.5 °C (for shrubs) and −3.8 °C (grass). The bare dry soil and the artificial surfaces are much warmer presenting positive surface to direct air temperature differences, with values of +7.8 °C (+8.7 °C) for the paved surfaces inside (outside of) the UGS, +10.8 °C for the dry bare soil, +12.2 °C for concrete and +12.5 °C for asphalt-covered surfaces. The above findings indicate that the vegetation-covered surfaces absorb higher solar radiation quantities that, however, do not lead to increases in the surfaces' temperatures. On the other hand, the artificial element's optical behavior allows higher radiation reflectance, and the absorbed solar radiation fluxes lead to higher surface temperatures enhancing the urban heat island effect.

The hydraulic conductivity of the unsaturated soil is higher at the grass-covered surfaces and less at dry bare soils and shrub-covered surfaces inside the UGS, indicating faster infiltration of the precipitation water, which is very important when assessing the performance of UGSs to prevent urban floods.

The above results refer to the micrometeorological attributes of a small urban green area, during the noon of a very warm summer day with clear sky conditions and high incoming radiation fluxes. The temporal (diurnal or seasonal) changes of the atmospheric environment are also important to be studied, in order to understand the behavior of the urban green infrastructures and their impact on formatting the urban climate. Additional sites inside the urban environment, especially in the Mediterranean, should also be studied to reach sound conclusions.

The study of additional urban green areas in the broader area of Athens, with different materials and green species composition and in different seasons are within the future research goals of the researchers of this work, in order to verify and quantify the micrometeorological characteristics of urban green areas and their effect on the Mediterranean urban climate.

Author Contributions: Conceptualization, N.D.P.; methodology, N.D.P. and A.D.S.; software, N.D.P.; validation, N.D.P. and M.P.; formal analysis, N.D.P., A.D.S. and N.E.C.; investigation, N.D.P., A.D.S., M.P. and N.E.C.; resources, N.D.P. and M.P.; data curation, N.D.P.; writing—original draft preparation, N.D.P., A.D.S. and M.P.; writing—review and editing, N.D.P., A.D.S., M.P. and N.E.C.; visualization, N.D.P.; supervision, N.D.P. All authors have read and agreed to the published version of the manuscript.

Funding: This research was funded by the LIFE GrIn project "Promoting urban Integration of GReen INfrastructure to improve climate governance in cities" LIFE17 GIC/GR/000029, which is co-funded by the European Commission under the Climate Change Action-Climate Change Governance and Information component of the LIFE Programme and the Greek Green Fund.

Data Availability Statement: The data presented in this study are available on request from the corresponding author.

Acknowledgments: The authors highly acknowledge the contribution of Nikolaos Papaioannou, for providing the June 2022 meteorological data from the automatic meteorological station of the Municipality of Amaroussion—Greece, installed on the roof of the 5th Elementary School of Amaroussion.

Conflicts of Interest: The authors declare no conflict of interest.

References

1. Kolimenakis, A.; Solomou, A.D.; Proutsos, N.; Avramidou, E.V.; Korakaki, E.; Karetsos, G.; Maroulis, G.; Papagiannis, E.; Tsagkari, K. The socioeconomic welfare of urban green areas and parks; a literature review of available evidence. *Sustainability* **2021**, *13*, 7863. [CrossRef]
2. Kolimenakis, A.S.; Proutsos, N.D.; Avramidou, E.V.; Korakaki, E.; Karetsos, G.; Kontogianni, A.B.; Kontos, K.; Georgiadis, C.; Maroulis, G.; Papagiannis, E.; et al. The socioeconomic importance of Urban Green Areas in the era of COVID-19: A case study of nationwide survey in Greece. *Land* 2022; *submitted paper-under revision*.

3. Cao, X.; Onishi, A.; Chen, J.; Imura, H. Quantifying the cool island intensity of urban parks using ASTER and IKONOS data. *Landsc. Urban Plan.* **2010**, *96*, 224–231. [CrossRef]
4. Chen, X.-L.; Zhao, H.-M.; Li, P.-X.; Yin, Z.-Y. Remote sensing image-based analysis of the relationship between urban heat island and land use/cover changes. *Remote Sens. Environ.* **2006**, *104*, 133–146. [CrossRef]
5. Tran, H.; Uchihama, D.; Ochi, S.; Yasuoka, Y. Assessment with satellite data of the urban heat island effects in Asian mega cities. *Int. J. Appl. Earth Obs. Geoinf.* **2006**, *8*, 34–48. [CrossRef]
6. Voogt, J.A.; Oke, T. Effects of urban surface geometry on remotely-sensed surface temperature. *Int. J. Remote Sens.* **1998**, *19*, 895–920. [CrossRef]
7. Weng, Q. Thermal infrared remote sensing for urban climate and environmental studies: Methods, applications, and trends. *ISPRS J. Photogramm. Remote Sens.* **2009**, *64*, 335–344. [CrossRef]
8. Weng, Q.; Lu, D.; Schubring, J. Estimation of land surface temperature–vegetation abundance relationship for urban heat island studies. *Remote Sens. Environ.* **2004**, *89*, 467–483. [CrossRef]
9. Qiu, G.Y.; Zou, Z.; Li, X.; Li, H.; Guo, Q.; Yan, C.; Tan, S. Experimental studies on the effects of green space and evapotranspiration on urban heat island in a subtropical megacity in China. *Habitat Int.* **2017**, *68*, 30–42. [CrossRef]
10. Maimaitiyiming, M.; Ghulam, A.; Tiyip, T.; Pla, F.; Latorre-Carmona, P.; Halik, Ü.; Sawut, M.; Caetano, M. Effects of green space spatial pattern on land surface temperature: Implications for sustainable urban planning and climate change adaptation. *ISPRS J. Photogramm. Remote Sens.* **2014**, *89*, 59–66. [CrossRef]
11. Li, X.; Zhou, W.; Ouyang, Z.; Xu, W.; Zheng, H. Spatial pattern of greenspace affects land surface temperature: Evidence from the heavily urbanized Beijing metropolitan area, China. *Landsc. Ecol.* **2012**, *27*, 887–898. [CrossRef]
12. Zhang, X.; Zhong, T.; Feng, X.; Wang, K. Estimation of the relationship between vegetation patches and urban land surface temperature with remote sensing. *Int. J. Remote Sens.* **2009**, *30*, 2105–2118. [CrossRef]
13. Qiu, G.-Y.; LI, H.-Y.; Zhang, Q.-T.; Wan, C.; Liang, X.-J.; Li, X.-Z. Effects of evapotranspiration on mitigation of urban temperature by vegetation and urban agriculture. *J. Integr. Agric.* **2013**, *12*, 1307–1315. [CrossRef]
14. Mackey, C.W.; Lee, X.; Smith, R.B. Remotely sensing the cooling effects of city scale efforts to reduce urban heat island. *Build. Environ.* **2012**, *49*, 348–358. [CrossRef]
15. Jusuf, S.K.; Wong, N.H.; Hagen, E.; Anggoro, R.; Hong, Y. The influence of land use on the urban heat island in Singapore. *Habitat Int.* **2007**, *31*, 232–242. [CrossRef]
16. Huang, M.; Cui, P.; He, X. Study of the cooling effects of urban green space in Harbin in terms of reducing the heat island effect. *Sustainability* **2018**, *10*, 1101. [CrossRef]
17. Bao, T.; Li, X.; Zhang, J.; Zhang, Y.; Tian, S. Assessing the distribution of urban green spaces and its anisotropic cooling distance on urban heat island pattern in Baotou, China. *ISPRS Int. J. Geo-Inf.* **2016**, *5*, 12. [CrossRef]
18. Edmondson, J.L.; Stott, I.; Davies, Z.G.; Gaston, K.J.; Leake, J.R. Soil surface temperatures reveal moderation of the urban heat island effect by trees and shrubs. *Sci. Rep.* **2016**, *6*, 33708. [CrossRef] [PubMed]
19. Armson, D.; Stringer, P.; Ennos, A. The effect of tree shade and grass on surface and globe temperatures in an urban area. *Urban For. Urban Green.* **2012**, *11*, 245–255. [CrossRef]
20. Wu, J.-H.; Tang, C.-S.; Shi, B.; Gao, L.; Jiang, H.-T.; Daniels, J.L. Effect of ground covers on soil temperature in urban and rural areas. *Environ. Eng. Geosci.* **2014**, *20*, 225–237. [CrossRef]
21. Mohajerani, A.; Bakaric, J.; Jeffrey-Bailey, T. The urban heat island effect, its causes, and mitigation, with reference to the thermal properties of asphalt concrete. *J. Environ. Manag.* **2017**, *197*, 522–538. [CrossRef]
22. Pomerantz, M.; Akbari, H.; Chang, S.-C.; Levinson, R.; Pon, B. Examples of Cooler Reflective Streets for Urban Heat-Island Mitigation: Portland Cement Concrete and Chip Seals. 2003. Available online: https://escholarship.org/content/qt53w2s92d/qt53w2s92d.pdf (accessed on 10 November 2022).
23. Grimmond, S. Urbanization and global environmental change: Local effects of urban warming. *Geogr. J.* **2007**, *173*, 83–88. Available online: https://www.jstor.org/stable/30113496 (accessed on 10 November 2022). [CrossRef]
24. Proutsos, N.; Liakatas, A.; Alexandris, S.; Tsiros, I. Carbon fluxes above a deciduous forest in Greece. *Atmósfera* **2017**, *30*, 311–322. [CrossRef]
25. Proutsos, N.; Alexandris, S.; Liakatas, A.; Nastos, P.; Tsiros, I.X. PAR and UVA composition of global solar radiation at a high altitude Mediterranean forest site. *Atmos. Res.* **2022**, *269*, 106039. [CrossRef]
26. Proutsos, N.; Liakatas, A.; Alexandris, S. Ratio of photosynthetically active to total incoming radiation above a Mediterranean deciduous oak forest. *Theor. Appl. Climatol.* **2019**, *137*, 2927–2939. [CrossRef]
27. Proutsos, N.D.; Liakatas, A.; Alexandris, S.G.; Tsiros, I.X.; Tigkas, D.; Halivopoulos, G. Atmospheric Factors Affecting Global Solar and Photosynthetically Active Radiation Relationship in a Mediterranean Forest Site. *Atmosphere* **2022**, *13*, 1207. [CrossRef]
28. Liakatas, A.; Proutsos, N.; Alexandris, S. Optical properties affecting the radiant energy of an oak forest. *Meteorol. Appl.* **2002**, *9*, 433–436. [CrossRef]
29. Baldinelli, G.; Bonafoni, S.; Rotili, A. Albedo retrieval from multispectral landsat 8 observation in urban environment: Algorithm validation by in situ measurements. *IEEE J. Sel. Top. Appl. Earth Obs. Remote Sens.* **2017**, *10*, 4504–4511. [CrossRef]
30. Jin, M.S.; Kessomkiat, W.; Pereira, G. Satellite-observed urbanization characters in Shanghai, China: Aerosols, urban heat island effect, and land–atmosphere interactions. *Remote Sens.* **2011**, *3*, 83–99. [CrossRef]
31. Kjelgren, R.; Montague, T. Urban tree transpiration over turf and asphalt surfaces. *Atmos. Environ.* **1998**, *32*, 35–41. [CrossRef]

32. Debbage, N.; Shepherd, J.M. The urban heat island effect and city contiguity. *Comput. Environ. Urban Syst.* **2015**, *54*, 181–194. [CrossRef]
33. Chudnovsky, A.; Ben-Dor, E.; Saaroni, H. Diurnal thermal behavior of selected urban objects using remote sensing measurements. *Energy Build.* **2004**, *36*, 1063–1074. [CrossRef]
34. Kalnay, E.; Cai, M. Impact of urbanization and land-use change on climate. *Nature* **2003**, *423*, 528–531. [CrossRef] [PubMed]
35. Kalnay, E.; Cai, M. Correction: Corrigendum: Impact of urbanization and land-use change on climate. *Nature* **2003**, *425*, 102. [CrossRef]
36. Erell, E.; Pearlmutter, D.; Boneh, D.; Kutiel, P.B. Effect of high-albedo materials on pedestrian heat stress in urban street canyons. *Urban Clim.* **2014**, *10*, 367–386. [CrossRef]
37. Giannopoulou, K.; Livada, I.; Santamouris, M.; Saliari, M.; Assimakopoulos, M.; Caouris, Y. On the characteristics of the summer urban heat island in Athens, Greece. *Sustain. Cities Soc.* **2011**, *1*, 16–28. [CrossRef]
38. Livada, I.; Santamouris, M.; Niachou, K.; Papanikolaou, N.; Mihalakakou, G. Determination of places in the great Athens area where the heat island effect is observed. *Theor. Appl. Climatol.* **2002**, *71*, 219–230. [CrossRef]
39. Skoulika, F.; Santamouris, M.; Kolokotsa, D.; Boemi, N. On the thermal characteristics and the mitigation potential of a medium size urban park in Athens, Greece. *Landsc. Urban Plan.* **2014**, *123*, 73–86. [CrossRef]
40. Zoulia, I.; Santamouris, M.; Dimoudi, A. Monitoring the effect of urban green areas on the heat island in Athens. *Environ. Monit. Assess.* **2009**, *156*, 275–292. [CrossRef]
41. Melas, E.; Tsiros, I.; Thoma, E.; Proutsos, N.; Pantavou, K.; Papadopoulos, G. An assessment of microclimatic conditions inside vegetated and non-vegetated small-scale open spaces in the Athens urban environment. In Proceedings of the 15th International Conference on Meteorology, Climatology and Atmospheric Physics—COMECAP 2021, Ioannina, Greece, 26–29 September 2021; pp. 269–273.
42. Tsiros, I.X.; Hoffman, M.E. Thermal and comfort conditions in a semi-closed rear wooded garden and its adjacent semi-open spaces in a Mediterranean climate (Athens) during summer. *Archit. Sci. Rev.* **2014**, *57*, 63–82. [CrossRef]
43. Tsiros, I.X. Assessment and energy implications of street air temperature cooling by shade tress in Athens (Greece) under extremely hot weather conditions. *Renew. Energy* **2010**, *35*, 1866–1869. [CrossRef]
44. Hidayat, Y.; Purwakusuma, W.; Wahjunie, E.D.; Baskoro, D.P.T.; Rachman, L.M.; Yusuf, S.M.; Adawiyah, R.M.; Syaepudin, I.; Siregar, M.M.R.; Isnaini, D.A. Characteristics of soil hydraulic conductivity in natural forest, agricultural land, and green open space area. *J. Pengelolaan Sumberd. Alam Dan Lingkung. J. Nat. Resour. Environ. Manag.* **2022**, *12*, 352–362. [CrossRef]
45. Luo, W.; Li, J.; Song, L.; Cheng, P.; Garg, A.; Zhang, L. Effects of vegetation on the hydraulic properties of soil covers: Four-years field experiments in Southern China. *Rhizosphere* **2020**, *16*, 100272. [CrossRef]
46. Gadi, V.K.; Tang, Y.-R.; Das, A.; Monga, C.; Garg, A.; Berretta, C.; Sahoo, L. Spatial and temporal variation of hydraulic conductivity and vegetation growth in green infrastructures using infiltrometer and visual technique. *Catena* **2017**, *155*, 20–29. [CrossRef]
47. Garg, A.; Leung, A.K.; Ng, C.W.W. Comparisons of soil suction induced by evapotranspiration and transpiration of *S. heptaphylla*. *Can. Geotech. J.* **2015**, *52*, 2149–2155. [CrossRef]
48. Leung, A.K.; Garg, A.; Coo, J.L.; Ng, C.W.W.; Hau, B. Effects of the roots of Cynodon dactylon and Schefflera heptaphylla on water infiltration rate and soil hydraulic conductivity. *Hydrol. Process.* **2015**, *29*, 3342–3354. [CrossRef]
49. Ghestem, M.; Sidle, R.C.; Stokes, A. The influence of plant root systems on subsurface flow: Implications for slope stability. *Bioscience* **2011**, *61*, 869–879. [CrossRef]
50. Mitchell, A.; Ellsworth, T.; Meek, B. Effect of root systems on preferential flow in swelling soil. *Commun. Soil Sci. Plant Anal.* **1995**, *26*, 2655–2666. [CrossRef]
51. Noordwijk, M.V.; Heinen, M.; Hairiah, K. Old tree root channels in acid soils in the humid tropics: Important for crop root penetration, water infiltration and nitrogen management. In *Plant-Soil Interactions at Low pH*; Springer: Berlin/Heidelberg, Germany, 1991; pp. 423–430.
52. Nieber, J.L.; Sidle, R.C. How do disconnected macropores in sloping soils facilitate preferential flow? *Hydrol. Process.* **2010**, *24*, 1582–1594. [CrossRef]
53. Jarvis, N.; Koestel, J.; Messing, I.; Moeys, J.; Lindahl, A. Influence of soil, land use and climatic factors on the hydraulic conductivity of soil. *Hydrol. Earth Syst. Sci.* **2013**, *17*, 5185–5195. [CrossRef]
54. Galli, A.; Peruzzi, C.; Beltrame, L.; Cislaghi, A.; Masseroni, D. Evaluating the infiltration capacity of degraded vs. rehabilitated urban greenspaces: Lessons learnt from a real-world Italian case study. *Sci. Total Environ.* **2021**, *787*, 147612. [CrossRef]
55. Arnold, E. *World Atlas of Desertification*; UNEP: London, UK, 1992.
56. Thornthwaite, C. Una aproximación para una clasificación racional del clima. *Geogr. Rev.* **1948**, *38*, 85–94. [CrossRef]
57. Thornthwaite, C.W.J.; Mather, J.R. *The Water Balance Climatology*; Drexel Institute of Technology, Laboratory of Climatology: Centerton, NJ, USA, 1955.
58. Proutsos, N.D.; Tsiros, I.X.; Nastos, P.; Tsaousidis, A. A note on some uncertainties associated with Thornthwaite's aridity index introduced by using different potential evapotranspiration methods. *Atmos. Res.* **2021**, *260*, 105727. [CrossRef]
59. Tsiros, I.X.; Nastos, P.; Proutsos, N.D.; Tsaousidis, A. Variability of the aridity index and related drought parameters in Greece using climatological data over the last century (1900–1997). *Atmos. Res.* **2020**, *240*, 104914. [CrossRef]

60. Proutsos, N.D.S.; Tigkas, D.A. Decadal variation of aridity and water balance attributes at the urban and peri-urban environment of Attica-Greece. In Proceedings of the HAICTA 2022: 10th International Conference on ICT in Agriculture, Food & Environment, Athens, Greece, 22–25 September 2022.
61. Alexandris, S.; Proutsos, N. How significant is the effect of the surface characteristics on the Reference Evapotranspiration estimates? *Agric. Water Manag.* **2020**, *237*, 106181. [CrossRef]
62. Zhang, R. Determination of soil sorptivity and hydraulic conductivity from the disk infiltrometer. *Soil Sci. Soc. Am. J.* **1997**, *61*, 1024–1030. [CrossRef]
63. Carsel, R.F.; Parrish, R.S. Developing joint probability distributions of soil water retention characteristics. *Water Resour. Res.* **1988**, *24*, 755–769. [CrossRef]
64. Golden Software LLC. Surfer® ver. 13. Available online: https://www.goldensoftware.com/products/surfer (accessed on 7 November 2022).
65. Salvati, L.; Zitti, M.; Di Bartolomei, R.; Perini, L. Climate aridity under changing conditions and implications for the agricultural sector: Italy as a case study. *Geogr. J.* **2012**, *2013*, 923173. [CrossRef]
66. Ha, J.; Lee, S.; Park, C. Temporal effects of environmental characteristics on urban air temperature: The influence of the sky view factor. *Sustainability* **2016**, *8*, 895. [CrossRef]
67. Jaganmohan, M.; Knapp, S.; Buchmann, C.M.; Schwarz, N. The bigger, the better? The influence of urban green space design on cooling effects for residential areas. *J. Environ. Qual.* **2016**, *45*, 134–145. [CrossRef] [PubMed]
68. Monteiro, M.V.; Doick, K.J.; Handley, P.; Peace, A. The impact of greenspace size on the extent of local nocturnal air temperature cooling in London. *Urban For. Urban Green.* **2016**, *16*, 160–169. [CrossRef]
69. Ca, V.T.; Asaeda, T.; Abu, E.M. Reductions in air conditioning energy caused by a nearby park. *Energy Build.* **1998**, *29*, 83–92. [CrossRef]
70. Tong, H.; Walton, A.; Sang, J.; Chan, J.C. Numerical simulation of the urban boundary layer over the complex terrain of Hong Kong. *Atmos. Environ.* **2005**, *39*, 3549–3563. [CrossRef]
71. Yu, C.; Hien, W.N. Thermal benefits of city parks. *Energy Build.* **2006**, *38*, 105–120. [CrossRef]
72. Đekić, J.P.; Mitković, P.B.; Dinić-Branković, M.M.; Igić, M.Z.; Đekić, P.S.; Mitković, M.P. The study of effects of greenery on temperature reduction in urban areas. *Therm. Sci.* **2018**, *22*, 988–1000. [CrossRef]
73. Oke, T.R. *Boundary Layer Climates*, 2nd ed.; Routledge: London, UK, 1987; p. 464. [CrossRef]
74. Aida, M. Urban albedo as a function of the urban structure—A model experiment. *Bound.-Layer Meteorol.* **1982**, *23*, 405–413. [CrossRef]
75. Sugawara, H.; Takamura, T. Surface albedo in cities: Case study in Sapporo and Tokyo, Japan. *Bound.-Layer Meteorol.* **2014**, *153*, 539–553. [CrossRef]

land

MDPI

Article

Forestry Bioeconomy Contribution on Socioeconomic Development: Evidence from Greece

Stavros Kalogiannidis [1], Dimitrios Kalfas [2,*], Efstratios Loizou [3] and Fotios Chatzitheodoridis [3,*]

[1] Department of Business Administration, University of Western Macedonia, 51100 Grevena, Greece
[2] Department of Agriculture, Faculty of Agricultural Sciences, University of Western Macedonia, 53100 Florina, Greece
[3] Department of Regional and Cross-Border Development, University of Western Macedonia, 50100 Kozani, Greece
* Correspondence: kalfdimi@otenet.gr (D.K.); fxtheodoridis@uowm.gr (F.C.)

Abstract: Forests are of utmost importance for sustainability because of their ongoing contributions to biodiversity protection, fertility management in agricultural areas, and the well-being of people. However, few studies have focused on the extent to which the bioeconomy of forests impacts a country's social and economic development. This study aimed to examine the bioeconomy contribution of forestry to social and economic development using Greece as a case study. Data was collected from 312 professionals in the forestry and finance sectors of Greece using a survey questionnaire. Forests are associated with direct and indirect contributions that impact human livelihood and contribute toward a country's economic development. However, the role of forestry in development is affected by policy-related and human-made challenges. The difficulties are primarily caused by shifts in how economic activity is distributed from the agricultural to the industrial to the service sectors, different government policies intended to increase forest cover, and in other instances, as a result of the role of global capital and trade. The forestry contributions to global commerce, national economies, employment, and family incomes remain consistent throughout all these patterns of loss, stabilization, and recovery. It was established that the bioeconomy can increase the benefits of forests by further exploiting forest wealth (biomass, resins) with the direct and indirect benefits for forest-related societies and local economies. In addition, the management and exploitation of forests by adopting bioeconomy practices, allows the attainment of important skills, knowledge, and new fields of entrepreneurship.

Keywords: forestry; bioeconomy; direct and indirect incomes from forests; social and economic development

1. Introduction

1.1. The Forestry Contribution and Value

Forests and multiple-use forest management are often essential to regional development [1]. In addition to producing wood, forests are valuable natural resources that also provide a variety of non-wood items, such as medicinal and aromatic plants, fruits, edible leaves, and game animals, that help boost people's incomes [2]. Forests offer different services relating to recreation, protection against soil erosion, biodiversity protection, preservation of water resources, and protection against climate change through carbon sequestration and the reduction of global warming [3]. The importance of forests in the modern era, especially their influence on climate change, has continued to attract significant attention. The economics of the forest industry is significantly impacted by climate change; hence, it is crucial to adjust forestry methods to combat it [4].

According to Fantechi and Fratesi (2022), forests help to provide job opportunities, add value to the GDP, and raise living standards; forests are essential drivers of regional development [5]. Additionally, the socioeconomic process of community development based on the forest sector has several facets. Reforestation and newly wooded regions are contributing to a larger regional development initiative, raising the value of forests to local economies [6]. Because the local populations choose to remain in the place and take advantage of the new jobs generated by forestry activities, forest products and services significantly impact less-favored areas [7]. Furthermore, the income from the forest's natural resources is crucial to the poor. Reduced income disparities and the potential to lessen socioeconomic disparities among households dependent on forests but with varying economic status are two of the most crucial functions of forests in regional development [3]. Recently, Cheng et al. (2019) created a systematic map protocol for forests' role in reducing poverty [2]. Afforestation programs are also used to enhance forest acreage, which helps to reduce inequality and spur economic development in rural regions [8].

Ballas et al. (2017) claimed the global forest sector is in a phase of creative destruction, which can be attributed to the decline in the protection and proper management of traditional forest products and the emergence of new production opportunities, such as wood products, thus impacting the economy [9]. Resources from the national forests are regarded as a source of commodities and services. One of the primary goals of national forest policy across the globe has always been the sustainable utilization of these resources. The forest industry also can boost national economies. Harvested wood products from forests and other forested areas are a significant part of the productive function. The amount of wood taken shows how valuable forest resources are to local economies and societies [10].

Eurostat (2020) indicates that although forests are a valuable natural resource, the European Union's forestry sector lacks a Common Policy [10]. To offer a compelling framework for the national forest policies of the member states, the European Union Commission produced the EU Forest Strategy in 2013. Kupec et al. (2022) indicated that some of the barriers to implementing a standard EU-based forest policy is that it is cross-sectorial and consequently interferes with other policies at the European level, including those related to agriculture, rural development, the environment, energy, and the climate change, among others, and lacks efficient coordination mechanisms [11]. In addition, establishing a comprehensive framework for the EU's forest policy must consider the forest value chain's extensive coverage of intersecting sectoral interests and policy tools [12].

As a Mediterranean country, Greece has favorable agro-climatic conditions for producing and collecting medicinal and aromatic plants. The growing demand for these raw materials, which traditional recipes can explain and the shift observed towards a healthy diet, increased the cultivation of these plants, which was non-existent and the needs were covered by over-exploitation and the irreversible damage to wild populations within forests [13,14]. Furthermore, Greece has a large number of plants (>7000) with 22% of them being endemic and contributing to forest biodiversity [15–17].

Forests help in sustainability because of their influence on biodiversity, agricultural areas, and the standards of living of people who rely on them. However, very few studies have examined how much the forestry development based on a bioeconomy affects a country's social and economic growth. It is therefore important to investigate this contribution of forestry to social and economic development.

1.2. Purpose of the Study

The study's main purpose was to investigate the bioeconomy contribution of forests' social and economic development, using evidence from Greece. This objective was analyzed based on two specific objectives:

1. To establish the relationship between dimensions of forest bioeconomy and economic benefits for forest-related societies and local economies.
2. To investigate the relationship between management and exploitation of forests and to explore the development of new skills, knowledge, and new fields of entrepreneurship by the local population related to forestry exploitation.

1.3. Research Questions

- What is the relationship between dimensions of forest bioeconomy and economic benefits for forest-related societies and local economies?
- What is the relationship between management and exploitation of forests and developing new skills, knowledge, and new fields of entrepreneurship?

1.4. Research Hypotheses

Hypothesis 1 (H1.) *The bioeconomy could increase the benefits provided by forests as it could further exploit forest wealth (biomass, resins) with economic benefits for forest-related societies and local economies.*

Hypothesis 2 (H2.) *Those involved in the management and exploitation of forests, by adopting bioeconomy practices, will develop new skills, knowledge, and new fields of entrepreneurship.*

1.5. Significance of the Study

The study findings will provide key insights into the contribution of forest bioeconomy towards social and economic development. In this case, new knowledge will be generated about the relationship between dimensions of forest bioeconomy and economic benefits for forest-related societies and local economies and the effect of management and exploitation of forests using bioeconomy practices on the development of new skills, knowledge, and new fields of entrepreneurship. The study has a significant academic contribution as future researchers can utilize this study to make more informed conclusions in the same or related area of study.

2. Literature Review

2.1. Economic Contribution of Forestry

The forestry sector employs a significant portion of the world's population and provides a primary, secondary, or alternative source of income [18], especially the SMEs dealing in forest products contribute significantly to the economy in terms of employment and income [4]. According to Masiero et al. (2016), the forestry sector creates employment opportunities for many people, and the production and trading of wood fuel employ tens of thousands of workers, many of whom work informally [7,19].

SMEs' contribution to employment is stable or expanding, notably in the US domestic wood furniture sub-sector, in contrast to worldwide declining employment in wood processing. In the US, SMEs dealing in forest products account for 37.4% of all solid wood products processing industry employment [4].

Li et al. (2019) state that in 2011, the global forestry sector directly employed more than 18.21 million people and created more than 45.15 million jobs through direct, indirect and induced effects. The direct contribution of the global forestry sector amounted to more than USD 539 billion, and the total contribution of more than USD 1298 billion to the global GDP, always through direct, indirect and induced effects [20].

In Table 1, the forestry sector output was higher in Sweden, Germany, and France, three European economies that have historically relied on the forestry industry. On the other hand, Greece comes almost last (excluding Cyprus, Luxembourg, and Malta). Greece is one of the European nations with the lowest productivity in primary round wood production [10,21].

Greece, together with the Netherlands, is ranked 25th among the 29 countries of the European Continent that provide sufficient data on the contribution of the forestry sector to their Gross Domestic Product (0.04%) (Table 1) with data for the year 2019. Two Baltic countries, Latvia and Estonia, have the largest contribution of the forestry sector to their Gross Domestic Product (4.56% and 3.85%, respectively). In total, eleven (11) countries have more than 1% contribution of forests to GDP [21]. The research of Tsiaras et al. (2021) reaches similar conclusions with data for the year 2016 [22].

Table 1. Output of forestry and GDP in the year 2019 for selected European countries.

Country	Forestry Output 2019 (Million €)	GDP-Based Market Prices (Million €)	%
Belgium	407.1	478,645.0	0.09
Bulgaria	697.51	61,558.5	1.13
Czechia	2720.8	225,613.5	1.21
Denmark	562.27	309,526.4	0.18
Germany (until 1990, former territory of the FRG)	6947.38	3,473,260.0	0.20
Estonia	1069.1	27,764.7	3.85
Ireland	173.9	356,704.6	0.05
Greece	76.7	183,351.2	0.04
Spain	1941.05	1,245,513.0	0.16
France	6485.81	2,437,635.0	0.27
Croatia	328.31	55,644.4	0.59
Italy	2457.1	1,796,648.5	0.14
Cyprus	4.46	23,176.2	0.02
Latvia	1398.6	30,678.6	4.56
Lithuania	561.8	48,908.2	1.15
Luxembourg	19.99	62,373.6	0.03
Hungary	584.6	146,526.1	0.40
Malta	0	14,047.9	0.00
Netherlands	350	813,055.0	0.04
Austria	1966.91	397,169.5	0.50
Poland	5332.44	532,504.7	1.00
Portugal	1306.31	214,374.6	0.61
Romania	2507.91	224,178.6	1.12
Slovenia	547.63	48,533.1	1.13
Slovakia	1127.4	94,437.5	1.19
Finland	5,745	239,858.0	2.40
Sweden	9571.73	476,869.5	2.01
Norway	1405.77	361,734.6	0.39
Switzerland	864.17	644,443.2	0.13

Source: Authors' own work, based on Eurostat [(https://ec.europa.eu/eurostat/statistics-explained/index.php?title=Forests,_forestry_and_logging#Forests_in_the_EU (accessed on 10 November 2022))].

In Greece, forest production is concentrated in the Regional Unit of Drama located in the northeast of the country and in the Region of Eastern Macedonia and Thrace, where we find the large forest and transitional forest areas, as well as pastures. The second-most valuable regions of Greece in terms of forest production are the Regional Units of Grevena and Florina in the northwest of the country, in the Region of Western Macedonia in Northern Greece. Furthermore, the geomorphological relief and the weather conditions favor the production of energy from the exploitation of wind and solar radiation, with wind generation mainly located above the upper forest line [23–25].

The global economic crisis significantly impacted the forestry sector in Greece, whereby forest consumption levels per person were greatly reduced, and employment and the total output of the forestry sector were drastically reduced between 2008 and 2017 [26]. Greece's predominant trait is that most of its forestland is found in regions with steep mountains and slopes, which usually makes harvesting very difficult. For various management and ecological reasons, wood production and quality are often constrained. Karametou and Apostolopoulos (2010) also listed Greece as one of the EU countries with the lowest productivity [27]. The economic growth of rural regions and the well-being of the Mediterranean region's urban inhabitants depends on the forest ecosystems' variety of forest products and services [19].

FAOSTAT (2022) indicates that several factors, such as organizational problems, limited funding for forest management, issues with following forest law, and ambiguous ownership of forest land, as well as governance and bureaucracy issues, all affect the removal of wood. Consequently, Greece's national forest industry has had a trade imbalance in forest products. To meet local demand, Greece imports more forest products than it exports [28]. Figure 1 depicts the trade balance for forest products in Greece over the last 18 years and shows that the imbalance is still one of the industry's most pressing problems.

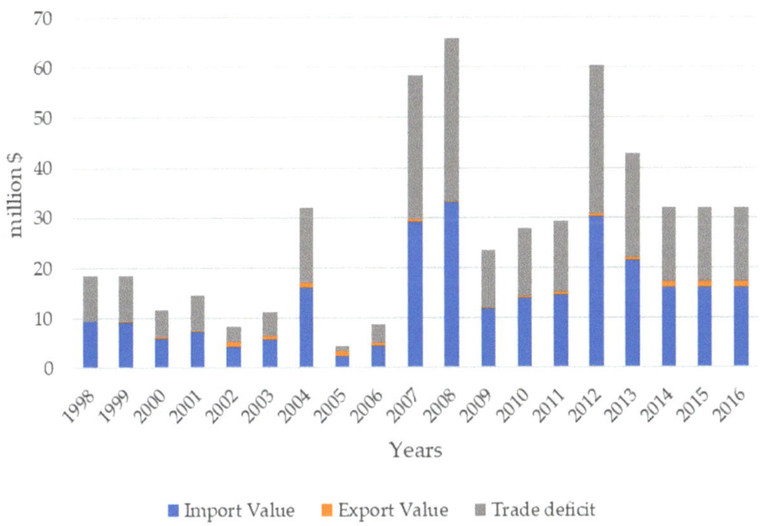

Figure 1. The trade balance and deficit for the Greek forest sector. Source: Authors' own work, based on FAO data [https://www.fao.org/faostat/en/#home (accessed on 24 September 2022)].

According to FAOSTAT (2022) data, Figure 2 was created to demonstrate the employment in forestry and logging between 1992 to 2020 in Greece. The course of the employment curve in these two decades seems to have reduced significantly in the Greek forestry and logging sector, and the whole activity as a result of the employee reduction [28].

Greece's National Forest Strategy made an effort to incorporate the forestry strategy of the EU and its key priorities while taking into account the various qualities of Greek forests, such as their protective role, multiple functions, significant contribution to the ecosystem, and efforts to produce innovative forestry and products with added value. Additionally, the National Forest Strategy of Greece seeks to address two significant environmental problems in Greece: the restoration of forest ecosystems and the absence of effective forest management, which are two of the biggest inefficiencies for Greek forests [26,29].

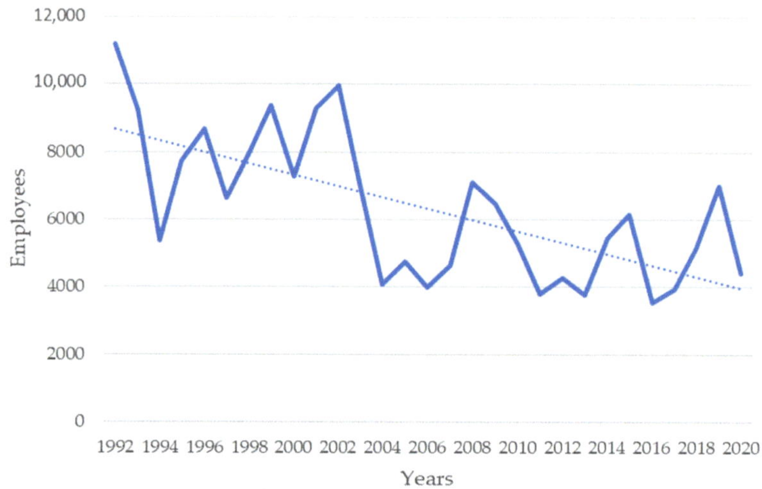

Figure 2. Employment in forestry and logging for Greece. Source: Authors' own work, based on FAO data [https://www.fao.org/faostat/en/#home (accessed on 24 September 2022)].

2.2. Forestry's Direct and Indirect Income Contribution

The high levels of economic advantages that forests continue to provide to individuals, businesses, and governments served as the first motivation for protective laws and regulations. According to Masiero et al. (2016), the forest sector contributed more than USD 450 billion to national earnings in 2008, accounting for close to 1% of the global GDP and formal employment 0.4% of the total working force worldwide [7]. Additionally, forests provide chances for informal jobs, alternative sources of income, and economic value reservoirs that lessen family income shocks, especially in rural regions of developing nations. However, there are no valid statistical data on a global or national scale to allow for aggregate estimations of the non-industrial economic benefits of forests [9]. Globally and locally, the forest sector has contributed significantly to formal direct employment and continues to do so. In addition, since 1990, economic diversification and liberalization of the economy have increased, proving positive for the trade of forest products but sometimes negative for long-term employment. How the forest sector influences economic development, as well as the changing ways in which it continues to impact global political, economic, and social development, are all highlighted by these changing dynamics [30].

Aggestam and Pülzl (2018) indicate that even when there is an absence of sales income of any forest products, different indirect economic benefits can be enjoyed from forests by stakeholders. Second, indirect economic benefits nearly always contribute more to total family income than direct [12]. The connection between forests' direct and indirect income contributions to the total family income varies. The most significant direct and indirect income sources were firewood, construction supplies, and forest foods. However, other commodities like fiber and herbal remedies also rank highly in terms of relative value to the family [1]. Spanos et al. (2021) revealed that firewood accounts for over 36% of the direct income category's forest revenue, housebuilding materials account for about 36%, and forest foods and herbal remedies account for 25%. The significance of timber is insignificant. Even when the revenue proportions differ somewhat, the indirect income category's order of importance remains the same [26].

2.3. Role of Forests in Eradication of Poverty and Social Contribution

Forests have a far more nuanced function in lowering long-term poverty and assisting people in escaping it than was once believed. Initially, it was thought that poverty would be decreased by identifying forest goods and increasing their production. However, only selling wood would do that, and even if governments were willing to let the world's poorest people become loggers, timber production requires too much cash [2].

In many circumstances, trees are valued for their welfare benefits rather than for the income they may provide. Studies demonstrate how much woods support local lives. They are good days for both men and women, for wealthy and poorer people, and not simply under challenging times. The livelihood benefits associated with forests are very important in uplifting the household incomes of different people, which further helps to boost the national GDP [1].

Aggestam and Pülzl (2018) showed how rare it is for people living in remote locations to escape poverty quickly. Before poverty can be reduced, it is often necessary to alter the relationship between agriculture and reliance on the forest [31]. The forest also plays a part in helping some families get by during hard times at home as the primary breadwinners establish a foothold as labor migrants to cities for employment opportunities or to get more money to put into the farm [12]. Moreover, the authors revealed that community forestry organizations are still respected since they provide beneficial advantages [12]. Women and their female offspring increasingly manage the local forests and generate the local economy [1,26,32].

Cheng et al. (2019) revealed that people often discover that a dual strategy is the greatest approach to harnessing the synergy between agriculture and the forest. Investing in cattle and utilizing the forest as feed in tropical dry forests is often the simplest way to escape poverty [2]. Multistory forest gardens attest to the pattern that has prevailed across Southeast Asia, and this tactic is now being used in Papua. Some people are concerned by the changes in the tree species that make up forests, but forest function is preserved in each instance. In addition, chances for poverty reduction exist while maintaining or even improving the forest cover. In the case of post-conflict rebuilding, the forest has been able to temporarily pick up the slack while families return to their previous lifestyles and start to search for methods to save money for the future. In all these ways, woodlands assist locals in finding detours away from poverty. These ideas about "direct revenue from trees," prevalent ten to fifteen years ago, are pretty different [8,32].

The stabilization and control of soil erosion are significant benefits of forests. Studies have shown that forest growth stabilizes soils and prevents sedimentation and erosion. The estimated values for soil stabilization mainly account for the expenses of sedimentation. The prices vary from $1.94 per ton in Tennessee to $5.5 million yearly in Oregon's Willamette Valley. In Tucson, Arizona, 500,000 mesquite plants are anticipated to lessen runoff, which would otherwise need the $90,000 building of detention ponds. Forests also help to enhance air quality. Because trees capture airborne dust, the environment and people's health are improved. Only one study on the importance of trees' contributions to air quality is discussed in this essay. According to the findings of that research, Tucson, Arizona, plans to plant 500,000 mesquite trees, which, when fully grown, would remove 6500 tonnes of particulate matter yearly. An alternate dust management method in Tucson costs $1.5 million. Therefore, each tree is worth $4.16 in terms of air quality [33].

Forests are crucial for carbon sequestration and climate regulation. By retaining moisture and cooling the earth's surface, trees contribute to climate control. According to Fantechi and Fratesi (2022), benefits from climate control provided by U.S. woods amount to $18.5 billion annually [5]. According to studies conducted in metropolitan areas, 100,000 correctly positioned, mature trees in American cities might save $2 billion in heating and cooling expenses [7]. Additionally, trees absorb atmospheric carbon dioxide, which slows global warming. According to the U.S. Forest Service, these carbon sequestration services result in benefits of $65 per ton, or $3 4 billion per year, for all U.S. forests [19].

Pilli & Grassi (2021) indicate that forests are crucial in preserving biodiversity [34]. Numerous factors contribute to the importance of biological variety, including its capacity to produce valuable pharmaceuticals, its function as a genetic resource bank that can be used to selectively breed plants and animals, and its involvement in natural pest and disease management. Although there have been few studies on the worth of biological variety in forest ecosystems, it is predicted that utilizing chemical pesticides to replace the natural pest control services provided by all-natural ecosystems would cost Greece's agriculture USD 54 billion yearly [7].

Tourism and recreation forests are well-liked sites for outdoor leisure because of their scenic beauty and recreational features. According to Krieger (2001), recreational activities in national forests alone boost our country's GDP by USD 110 billion annually. Regionally, the proximity of population centers and the distinctive qualities of a region's forest resources influence the economic impact of forest-based recreation. The estimated yearly economic effect of entertainment impacted by forests ranges from USD 6 billion in the Southern Appalachians to USD 736 million in Montana [35]. Numerous studies have calculated the benefit of outdoor recreation using wilderness-related areas that are untamed and unloaded.

In addition, trees have a significant impact on regional microclimates and perhaps worldwide climate. The environment around trees is impacted by temperature, humidity, moisture availability, and lighting changes. The ability of the trees to raise relative humidity and regulate soil and air temperatures, two variables crucial for better crop development, plays a role in the success of many agroforestry systems [19].

Recent studies also imply that trees may affect rainfall patterns, surface reflectance, and other meteorological factors, which may impact climate [9,11,36]. One aspect is how clearing trees alters how sunlight is reflected from the earth's surface. The leaves, branches, and tree trunks in a living forest absorb sunlight. When a forest is cleared, reflectivity rises, and heat absorption decreases on the land. Additionally, less solar energy is utilized in deforested regions to evaporate moisture from plant and tree leaves. This causes more climatic variations, raising daytime temperatures and reducing nighttime ones. Furthermore, forests play a significant role in the carbon cycle [26,37]. When forests are cleared and burnt, their carbon is released into the atmosphere, increasing the level of atmospheric carbon dioxide, one of the leading causes of the greenhouse effect-induced global warming [33]. Living trees provide the opposite function by absorbing carbon dioxide from the atmosphere. Reforestation on a large scale has been advocated as a critical strategy for reducing anticipated global warming. However, if afforestation were to reduce the levels of carbon dioxide in the atmosphere significantly, it would need to be done on a continental scale [12].

2.4. The Role of Ecosystems Services and the Landsenses Ecology

If sustainable land management is sought over time, regional policy must include economic and environmental aspects, which reflect the conditions prevailing in the specific geographical area [38–41]. The study of ecosystem services over the last two to three decades has changed how the concept of nature conservation is generally viewed, changed the rationale for ecosystem management and the wider policy for natural ecosystems (e.g., forests, lakes, etc.). A few years ago, protecting the environment was the priority of governance at all levels (regional, national, European and global), but today the preservation of natural ecosystems and the restoration of disturbed ones are at the tip of the spear [38,42–45].

The ecosystem services of natural forests, without excluding urban and peri-urban forests, play an essential role in adjacent populations' economic and social cohesion [46,47]. The assessment of ecosystem services, introduced more recently, is constantly developing since many ecosystem services (e.g., protection from landslides, floods, strong winds, etc.) cannot be easily measured [48–51]. Therefore, assessing ecosystem services requires the contribution of many scientific disciplines (economics, ecology, statistics, geography, mathematics, computers, etc.) [52–55].

For the quantification of ecosystem services, international standards have been created that de facto use geometric methods and Geographic Information Systems (GIS) [56–58]. The quantification results are considered in decisions related to spatial planning on land and sea [59,60], but also with land use in agriculture and forestry [61–64].

Scientists who deal with ecosystem services study existing management practices, but also synergies that develop between man and nature and to evaluate the policies applied at all levels, having sustainability as the background of the study [65–67].

A relatively more recent approach is "Landsenses ecology" and "Landsenseology", which is defined as the scientific discipline that investigates the planning, construction, and management of land use for sustainable development [68–72]. "Landenses" is based on ecological principles but also the analysis of physical factors, senses, perceptions, and socio-economic conditions [68,69,73]. A new approach could not lack technology, the Internet of Things (IoT), GIS, intelligent systems, and artificial intelligence as part of earth sensing [69,70,72,74,75].

2.5. Bioeconomy in Europe

The EU Bioeconomy Strategy encouraged many member states to adopt such projects [76]. Bioeconomy focuses on producing renewable biological resources and converting these resources and their waste into value-added products such as food, feed, bio-based products, or bio-energy [77,78]. The circular economy, which appeared in the European Union's revised bioeconomy strategy, is a model of production and consumption that focuses on preserving the value of products, materials and resources for as long as possible, minimizing waste production [79]. Therefore, the integration of both, i.e., the circular bioeconomy, is intended to represent a sustainable economic and social model [80,81], bringing together many existing economic sectors, including the primary sector (agriculture, forestry, fisheries and aquaculture), the bio-based industrial sector (food, textiles, textiles, paper, chemicals, pharmaceuticals) and the service sector (consulting, logistics, trade, transport) [79]. As the core concepts of the bioeconomy and circular economy overlap in their attempt to reconcile economic, environmental, and social goals through the development of a sustainable economy, this search included documents published after 2018, as well as green economy and green growth strategies [82,83].

Further objectives of the bioeconomy strategies are to promote energy security, to green the energy industry, and to contribute to rural development. To strengthen the agricultural and forestry sectors, since they are the prominent bio-mass resources, through the development and application of biotechnology, biotechnology strategies aim to promote economic growth, healthcare, and environmental security [84,85]. They relate technology advancements to social progress while promoting socioeconomic well-being, the green economy or green growth plans and adopt a comprehensive approach to supporting low-carbon, resource-efficient, and resilient development approaches [86,87].

The Greece bioeconomy strategy focuses on technology and economics and places a lot less emphasis on the social aspects of a bioeconomy transition, placing a lot more emphasis on job development [88]. The fact that this approach primarily omits the utilization of forest resources and only sometimes discusses the significance of rural areas when highlighting employment creation in the biofuel and agricultural industries is of special relevance [82].

Hodge et al. (2017) revealed that the most comprehensive socioeconomic perspective is provided by green economy policies, which recognize the value of the forestry and agroforestry industries in achieving their objectives [89]. This is not surprising given that green economy strategies are more extensive than bioeconomy strategies and include a wider range of social and disciplinary viewpoints. It is interesting to note that different stakeholder categories were engaged in the strategies' design [90].

Recent initiatives, such as the European Green Deal [91], confirm the expected role of the circular bioeconomy in the European Union of the future and in each region. In particular, it can be seen how regions and Member States are starting to implement circular bioeconomy planning (strategies, action plans) to promote the development of this sector, largely as a result of the political impetus given at a higher level. Many corresponding policy documents address primarily the agricultural and forest sectors while highlighting the significance of research and innovation programs as the pillars of a knowledge-based transition towards a sustainable bioeconomy [92].

3. Methodology

3.1. Research Design, Study Area, Target Population and Data Collection

3.1.1. Research Design

The study utilized a quantitative research methodology based on the cross-sectional survey design. The cross-sectional research design depends on an in-depth investigation of a group or event to explore the causes of different underlying principles associated with the research problem or topic of study. The cross-sectional research design made it easy to focus on specific aspects of forestry in Greece, the dimensions of forest bioeconomy such as direct and indirect income contributions of forests, management and exploitation of forests and their effect on social and economic development.

3.1.2. Research Population and Sample Size

The research targeted professionals in the forestry and finance sector of Greece. The research population included workers in primary forest production (loggers, transporters, 4000 resin collectors, but also 4600 public employees in the country's public forestry service) [28,93]. In addition, workers in the private sector who related to forest production (e.g., sawmills and sale of firewood) as well as design-construction companies mainly in the tertiary sector, but also board members from forest management companies were included in the research population, which were estimated at 5200 employees. The total research population was estimated to comprise 13,800 employees and professionals related to direct and indirect forestry in Greece. From this research population, we estimated a total sample of 312 professionals. The purposive sampling technique helped in the selection of the survey sample.

3.1.3. Data Collection

A well-structured online questionnaire was used in the collection of data. Data were only collected after obtaining informed consent from the participants, conforming to their willingness to participate in the study. The data gathered helped establish relationships between this study's variables to answer the research questions. The questionnaire contained questions about forestry and social and economic results from the specific activity. A sample of 312 study participants, mainly from the forestry areas of Greece, was employed in the investigation. The study was carried out between 5–25 September 2022.

The sample size was determined after assessing survey reliability (P = 99.7%) and precision (km 26.76). S^2 = 16,254.46 and s = 127.53 were estimated for each respondent using a preliminary (or pilot) sample of 50 people. The value of z is determined by the desired degree of confidence (P). A value of z = 3 is often used when calculating the samples. This corresponds to a confidence interval of P = 99.7%. We use the values N = 13,800, s = 127.53, z = 3, and d = 24.00 (the desired precision d was chosen arbitrarily to represent half the confidence interval, giving the confidence interval 11.5% "air") [47,94–96]. Equation (1) calculates that the minimum sample size should be 311.86 or 312 people.

$$n = \frac{N(zs)^2}{Nd^2 + (zs)^2} \tag{1}$$

Calculation of the minimum sample of respondents.

$$n = \frac{13,800\ (3 * 127.53)^2}{13,800 * 24.00^2 + (3 * 127.53)^2} \Leftrightarrow n = 311.86$$

The following are some of the important questions from the survey questionnaire. At the start, there were questions to identify the respondents' profiles, such as gender, degree of education, and time in the forestry sector.

This was followed by questions examining the impact of forest cash revenue on social and economic development. "Timber sales are a great source of income for many people in the forestry sector; farm trees grown as cash crops can provide people with brushwood for both cooking and selling in the market; hunting and trading game are extremely profitable forest-based enterprises in forest-endowed countries; and forest products contribute greatly to the economic transformation of households," the questions stated. For the issue, the response scale was 4: SD—severely disagree, D—disagree, U—undecided, A—agree, and SA—strongly agree.

This was followed by questions designed to elicit information on the results of indirect revenue from forests. Forests provide soil nutrients and forage for crops and livestock, which greatly contributes to agriculture; People can earn a living through employment, processing, and trade of forest products and energy; Forests provide several opportunities for recreation and spiritual renewal in most communities; The majority of forest income is non-cash and includes food, fuel, fodder, and construction materials, as well as herbal medicine. The response scale for the question was 4: SD stands ofr for severely disagree, D for disagree, U for uncertain, A for agree, and SA for strongly agree.

This was followed by questions aimed to elicit locals' perspectives on critical aspects of forest bioeconomy:

- Income, both direct and indirect;
- Products derived from bioenergy;
- Goods for the consumer;
- Industrial goods.

This was followed by questions meant to elicit communities' perspectives on critical areas of forest management and exploitation:

- Preventing forest overexploitation;
- Chemical management in the forest;
- Forest zone management based on policy;
- Forest fire prevention and control;
- Correct timber harvesting;
- Reforestation of forest land.

This was followed by questions designed to elicit information on the effects of forests on social and economic development. Forests could regulate the climate through carbon storage, which contributes to a high quality of life; trees are typically produced as an insurance policy against bad times and as an investment for the future; trees may be cut down to provide cash for emergencies or to pay for equipment or real estate; many nations have historically benefited from increased food security because the money made from tree cultivation is used to purchase food; trees may be cut down to provide cash for emergencies or to pay for equipment or real estate; trees may be cut down to provide cash for emergencies. The response scale for the question was 4: SD is for severely disagree, D stands for disagree, U stands for uncertain, A stands for agree, and SA stands for strongly agree.

3.2. Data Analysis

The quantitative data was coded and then analyzed using the Statistical Package for Social Sciences (SPSS) software. Tables were utilized to display the study findings, and frequencies and percentages were relied on in interpreting the results. The total predictive power of the various independent factors on the study's dependent variable was determined using regression analysis. In this instance, calculating various predictive values requires the use of a multiple regression model.

$$Y = \beta_0 + \beta_1 X_1 + \beta_2 X_2 + \varepsilon \ldots\ldots\ldots\ldots$$

where;

Y = Social and economic development;
β_0 = Constant (coefficient of intercept);
X_1 = Aspects of forestry bioeconomy;

X_2 = Management and exploitation of forests;

ε = Represents the error term in the multiple regression model.

The study's hypotheses are assessed based on the 5% (0.05) level of significance. Table 2 describes and measures variables and their a priori expectations.

Table 2. Description and measurement of variables and their a priori expectations.

Variable	Description	Measurement	A Priori Expectation
	Dependent variable		
Social and economic development	Economic benefits for forest-related societies and local economies New skills, knowledge, and new fields of entrepreneurship	1 = yes, 0 = otherwise	+/−
	Independent variables		
Dimensions of forest bioeconomy	Cash income from forests, non-cash income from forests	1 = yes, 0 = otherwise	+
Management and exploitation of forests	Bioeconomy practices of management	1 = yes, 0 = otherwise	+

Source: Authors' own work (2022).

3.3. Ethical Considerations

Informed consent was obtained to confirm the willingness of the sample to participate in the study. This was in addition to protecting the respondents' data with a high level of secrecy and privacy. Respondents were also allowed to interpret the various opinion questions to respond to inquiries. This made it easier to get general responses to certain inquiries.

4. Results

Results obtained after analysis using SPSS are presented in this section.

4.1. Univariate Analysis

This section focuses on the presentation and general interpretation of the results.

Most survey participants (59.9%) were male, and 40.1% were female. Most participants (42.3%) had a bachelor's degree, followed by 31.1% with postgraduate studies degrees. Most participants (49%) had spent over 10 years in forestry, and only 8.7% had spent below five years in this sector (Table 3).

Table 3. Demographic data of study participants.

Characteristics	Frequency	Percentage (%)
	Gender	
Male	186	59.6
Female	126	40.4
	Education level	
Diploma	62	19.9
Bachelor's	142	45.5
Master's	97	31.1
Ph.D.	11	3.5
	Duration in the forestry sector	
Below 5 years	27	8.7
5–10 years	132	42.3
Above 10 years	153	49.0
Total	312	100

Source: Authors' work (2022).

4.2. Descriptive Statistics

The study also sought to explore the effect of cash income from forests on social and economic development, and the findings are presented in Table 4.

Table 4. Results on direct income from forests.

	SD	D	U	A	SA
	%	%	%	%	%
Timber sales are a great source of income to many people in the forestry sector	7.6	11.3	2.6	53.7	25.4
Farm trees that are grown as cash crops can provide people with brushwood for both cooking and for selling in the market	3.0	2.7	5.8	62.8	25.6
Hunting and trading game are extremely lucrative forest-based enterprises in forest-endowed countries	11.8	20.2	4.4	50.9	4.6
Forestry products contribute greatly to the economic transformation of households	10.3	4.7	11.5	28.2	45.3

Key: SD—strongly disagree, D—disagree, U—undecided, A—agree, SA—strongly agree. Source: Primary Data (2022).

The results in Table 4 indicate that 53.1% of respondents agreed that Timber sales are a great source of income for people in the forestry sector. A percentage of 62.8% of respondents also agreed that Farm trees that are grown as direct income crops can provide indirectly through brushwood for both cooking and for selling in the market. A total of 50.9% agreed that Hunting and trading games are extremely lucrative forest-based enterprises in forest-endowed countries. In addition, 45.3% of the respondents strongly agreed that Forestry products contribute greatly to the economic transformation of households.

The study also sought to explore the effect of indirect income from forests on social and economic development; the findings are presented in Table 5.

Table 5. Results of indirect income from forests.

	SD	D	U	A	SA
	%	%	%	%	%
Forests provide soil nutrients and forage for crops and livestock, which contributes greatly to agriculture	4.6	18.4	3.7	49.2	24.1
People are able to earn a living through employment, processing, and trade of forest products and energy	8.6	12.6	9.4	11.9	57.5
Forests provide several opportunities for recreation and spiritual renewal in most communities	5.9	7.7	10.2	43.2	33.1
Most of the income from forests is non-cash and cuts across food, fuel, fodder, and construction materials, as well as herbal medicine	3.8	4.3	20.2	47.9	23.9
Forests are a great source of shelter, livelihoods, water, food, and fuel security for both humans and animals	5.7	8.9	13.2	60.2	12.1

Key: SD—strongly disagree, D—disagree, U—undecided, A—agree, SA—strongly agree. Source: Primary Data (2022).

The results in Table 5 show that 49.2% of participants agreed that forests provide soil nutrients and forage for crops and livestock, significantly contributing to agriculture. A percentage of 57.5% of respondents strongly agreed that people can earn a living through employment, processing, and trade of forest products and energy. A percentage of 43.2% of the study participants agreed that forests provide several opportunities for recreation and spiritual renewal in most communities. Furthermore, 47.9% agreed that most of the income from forests is indirect and cuts across food, fuel, fodder, construction materials, and herbal medicine. Furthermore, 60.2% of the study participants agreed with the fact that forests are a great source of shelter, livelihoods, water, food, and fuel security for both humans and animals.

The study established the key dimensions of forest bioeconomy, and the results are presented in Figure 3.

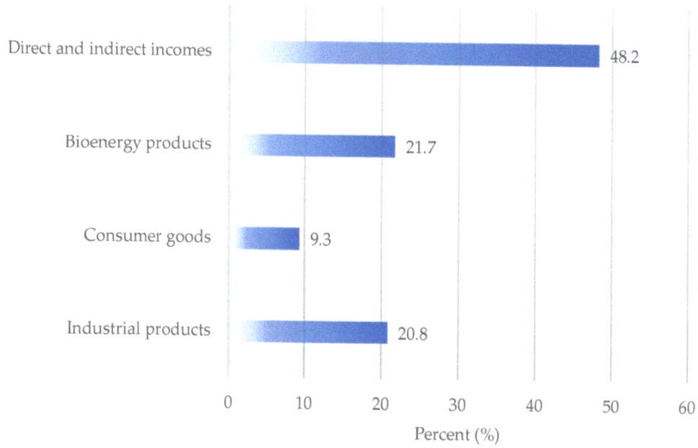

Figure 3. The dimensions of forest bioeconomy. Source: Authors' own work (2022).

Most of the participants (48.2%) mentioned direct and indirect income contributions as the major dimension of the forest bioeconomy, followed by bioenergy products (21.7%), then industrial products (20.8%), and the least number of respondents (9.3%) mentioned consumer goods from forests, such as herbal medicine.

This study established the key aspects in managing and exploiting forests, and the results are presented in Figure 4.

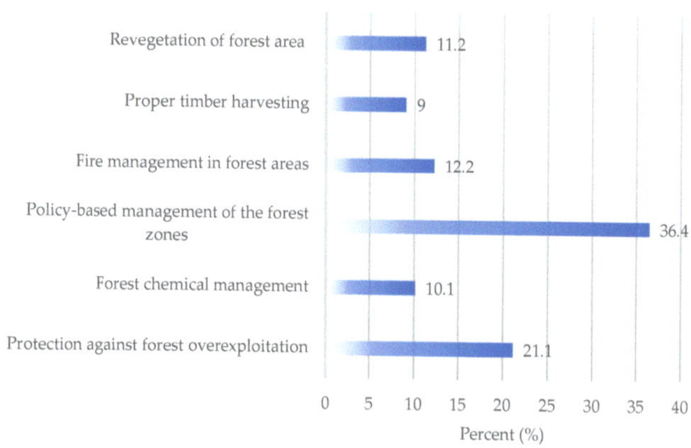

Figure 4. Key aspects in the management and exploitation of forests. Source: Authors' own work (2022).

From Figure 4, policy-based management of forest zones was selected by the largest percentage of participants (36.4%) as a key aspect of the management and exploitation of forests, followed by protection against forest overexploitation (21.1%), fire management in forest areas (12.2%), revegetation of forest areas (11.2%), and the least number of participants (9.0%) cited proper timber harvesting as a key aspect of management and exploitation of forests.

Table 6 presents the findings concerning the perspective of social and economic development in relation to forestry.

Table 6. Results on social and economic development related to forests.

	SD	D	U	A	SA
	%	%	%	%	%
Forests could regulate the climate through carbon storage, which contributes to a high quality of life	5.2	3.5	6.6	38.4	46.3
Trees are typically produced as an insurance policy against bad times and as an investment for the future	2.6	6.8	17.1	47.0	26.5
Trees may be cut down to provide cash for emergencies or to pay for equipment or real estate.	4.0	6.0	7.7	51.5	30.8
Many nations have historically benefited from increased food security because the money made from tree cultivation is used to purchase food, additional agricultural land, machinery, and inputs.	1.9	4.7	6.3	55.6	31.6

Key: SD—strongly disagree, D—disagree, U—undecided, A—agree, SA—strongly agree. Source: Primary Data (2022).

According to Table 6, the largest number of participants (46.3%) strongly agreed that forests could regulate the climate through carbon storage, contributing to the high quality of life. A percentage of 47.0% of respondents agreed that trees are typically produced as an insurance policy against bad times and as an investment for the future. The respondents (51.5%) agreed that trees may be cut down to provide income for emergencies or to pay for equipment or real estate. A percentage of 55.6% of the respondents agreed that many nations have historically benefited from increased food security because the money made from tree cultivation is used to purchase food, additional agricultural land, machinery, and inputs.

4.3. Regression Analysis

The relationship between forest bioeconomy and social and economic development was established using regression analysis as presented in the subsequent tables (Tables 7–9).

Table 7. Model Summary.

Model	R	R-Square	Adjusted R-Square	Std. Error of the Estimate
	0.798 a	0.786	0.684	0.10214

a—Predictors: (Constant), aspects of forestry bioeconomy, management, and exploitation of forests.

Table 8. ANOVA.

	Sum of Squares	Df.	Mean Square	F	Sig.
Regression	76.204	2	28.031	73.261	0.014
Residual	71.051	310	0.413		
Total	147.255	312			

Dependent variable: social and economic development. Predictors: (Constant), dimensions of forest bioeconomy, management, and exploitation of forests.

Table 9. Coefficients.

Model	Unstandardized Coefficients		Standardized Coefficients	T	Sig.
	B	Std. Error	Beta		
(Constant)	0.588	0.126		1.941	0.210
Dimensions of forest bioeconomy	0.168	0.054	0.371	1.124	0.024
Management and exploitation of forests	0.042	0.072	0.062	0.817	0.011

Dependent Variable: Social and economic development.

The dependent variable is social and economic development. The independent variable is regressed against the dependent variable obtaining a R^2 value of 0.673. This indicates that the independent variables jointly explain 78.6% of the variation in the dependent variable (social and economic development). The regression results also confirm that the study's independent variables do not influence 21.4% of the changes.

The F-statistic of 73.261 at prob. (Sig) = 0.014 at 5% significance level means that there is a statistically significant linear relationship between the independent variables (dimensions of forest bioeconomy, management, and exploitation of forests) and the dependent variable (social and economic development) as a whole.

The results in Table 9 confirm a relationship between forestry measured in terms of forest bioeconomy, management and exploitation of forests, and social and economic development since $p < 0.05$.

Hypotheses Testing

Since the significance level of 0.024 is less than 0.05%, we confirm that dimensions of forest bioeconomy, such as direct and indirect income from forests, have a positive effect on social and economic development. Therefore, we accept hypothesis H1 and conclude that the bioeconomy could increase the benefits of forests as it could further exploit forest wealth (biomass, resins) with economic benefits for forest-related societies and local economies.

In addition, there is a relationship between the management and exploitation of forests and social and economic development since the significance level of 0.011 is less than 0.05%. This indicates that the management and exploitation of forests help develop new skills, knowledge, and new fields of entrepreneurship. Therefore, we accept H2 and conclude that those involved in the management and exploitation of forests, by adopting bioeconomy practices, will develop new skills, knowledge, and fields of entrepreneurship.

5. Discussion

This study investigated the bioeconomy contribution of forestry on social and economic development. The study confirmed a positive relationship between dimensions of forest bioeconomy and social and economic development. It is clear that the bioeconomy can increase the benefits of forests as it could further exploit forest wealth (biomass, resins) with direct and indirect benefits for forest-related societies and local economies. In addition, the management and exploitation of forests by adopting bioeconomy practices allows the attainment of important skills, knowledge, and new fields of entrepreneurship. Globally and throughout many locations, the forest sector has contributed significantly to formal and direct employment and continues to do so. It is important to note that as economic diversification and liberalization have increased, these consequences have decreased proportionally. Trade in forest products has also grown in importance. These dynamic natures show how the forest sector has aided the economy in the previous era and how it plays a key role in global economic, political, and social development. They also show the importance of keeping and developing "real-time" data sets to map these changes [3,33].

The study showed that forests could regulate the climate through carbon storage, contributing to a high quality of life. Managi et al. (2019) also noted that by helping to preserve the natural conditions required for agricultural production, forests and trees play a crucial role in ensuring global food security. They balance soil temperatures, stop erosion, improve the ability of the land to hold water and stabilize the soil. With the removal of tree cover and the resultant loss of millions of hectares of fertile land, the significance of these consequences has often been overlooked in the past [1]. Additionally, the resource foundation for agriculture continues to be weakened by soil erosion and land degradation as forests are cut down, exposing the land to direct wind and rain assault. Industrialized and developing nations employ trees as windbreaks to cover crops, stop erosion, and save the soil. Trees help to protect crops, water supplies, soils, and towns and increase agricultural output by reducing wind speeds [2,19,26].

The study showed that forests provide soil nutrients and forage for crops and livestock, contributing greatly to agriculture. This agrees with Karametou and Apostolopoulos (2010), who argued that trees stabilize dunes and prevent the spread of deserts in arid and semi-arid regions of the globe so that crops may be cultivated there. In many arid and semi-arid environments, shelter belts provide fuelwood, food, and fodder, shielding crops from the wind's wrath and protecting them from grazing animals [27]. Additionally, the belts lessen the pace at which crops lose water via evapotranspiration. Thus, the crops use less water. As salt barriers along coastlines, trees may enable cropping closer to the water [4,22,31]. In addition to protecting against wave damage during storms, these salt barriers also lessen the likelihood of floods and bodily harm from tidal surges to inland regions. Greece scores poorly in this area in terms of raw numbers. Furthermore, Greece is ranked third from the bottom among EU nations in terms of how much its forest sector contributes to the GDP, with a meager 0.05 percent. This is because the EU countries are ranked according to how much their forest sectors contribute to their respective country's GDPs. Only Cyprus and the Netherlands do worse than Greece in terms of economic performance [8,22,27]. Likewise, Greece performs poorly regarding the economic activities related to forestry's gross value added.

The national forest sector in Greece has a significant trade imbalance in forest products over time, which has a negative impact on the industry's ability to contribute to the national economy. On the other hand, the sector has viable and attractive new growth potential, including non-wood forest products and forest services.

Another barrier for the forest industry is the general inclination of Greek administrations to cut spending throughout the years of the economic crisis. Growth in the forest sector's GDP contribution has been further hampered by the Green Fund, one of the largest investors in Greece's forest industry, cutting its financing by around half between 2011 and 2015 [26]. The nation's poor performance in the EU's most recent Regional Competitiveness Index worsens the issue. Despite the obstacles above, there are still many possibilities for development in Greece's forest industry. The National Forest Strategy's most current law in Greece gives Greece's forest industry a fantastic chance to increase its share of the country's GDP [8,27]. The recently enacted National Forest Strategy of Greece adopts the Mediterranean forestry model, ideally adapted to the local circumstances, and enhances the numerous functions of forests [22]. Its promotion of collaboration with rural communities, which results in regional development and employment possibilities and, in turn, may improve the general contribution of the forest industry to the nation's GDP, is one of its essential features. However, addressing the Greek forest industry's structural issues will take time. On the contrary, they need extensive policy adjustments, reprogramming of forest money, encouragement of fresh investments, and limitations on realizing the NFP's new, expansive goal [26].

6. Conclusions

This study confirmed that forestry has a significant influence on the social and economic transformation of a country. Both direct and indirect benefits generated by people and the government from forests greatly influence the social well-being of people and the economic transformation of a country. Governments must modify current forest policies and regulations to accomplish these new goals. Production and environmental and developmental objectives are covered in national forest policies, which provide a basic overview of a government's strategy in relation to forest management. These policies often seek to increase profits and foreign currency from wood and timber while ensuring the availability of raw materials for significant forest-based businesses. Many nations have implemented laws granting exclusive use of forest land and wood reserves to governments and commercial companies to accomplish these goals.

To unlock the potential of the forest bioeconomy and move towards sustainable development, governments must adopt a strong sustainability approach on the base of bioeconomy practices and also integrate innovations in forest activities that result in the valorization of biomass and the production of value-added goods and services. In other words, a transition to sustainability and a transition to "new" forest bioeconomy techniques activities are required. Failure to transition to an innovative bioeconomy and to challenge traditional forest activities will result in missed opportunities for socioeconomic development and inefficient resource use.

6.1. Recommendations

In national forests, local agroforestry programs should be created to generate a variety of goods such as bushmeat, fuelwood, and traditional medicines, among other forest foods. These programs should also be implemented. This may be accomplished by setting aside forest areas to serve as animals' homes or to cultivate regionally valuable crops. Alternatively, this goal can be accomplished by planting rows of these crops in government plantations.

Numerous people cut down trees and produce forest products to generate income for themselves. These activities can become more profitable and sustainable if favorable forest policies and government regulations exist. This would improve the means of subsistence and the food security of the impacted communities. Those individuals who depend on these activities the most, often those without land or otherwise disadvantaged, benefit the most from this change.

6.2. Limitations and Future Research

Even today, the COVID-19 pandemic introduced many limitations to this research; it was very difficult for the survey to be conducted face-to-face with the selected sample. In this type of research, where the researchers would gain more than the direct transfer of the participants' experience, the interview's limitations due to the pandemic were significant. Another limitation was the impossibility of collecting quantitative and economic data (due to reliability from a distance) from the forestry stakeholders for methodological support of the research and technical, economic analysis. In future research, the team intends to collect primary quantitative data related to the forestry activities to conduct techno-economic analyses.

The current study focused on the influences of forestry and especially the forestry bioeconomy on socioeconomic development and, for this purpose, used evidence from Greece. Future research should emphasize the role of government and EU policies in promoting the forestry bioeconomy and how contributes to the GDP of a region or a country. Moreover, the EU can set the rules for forest exploitation in a sustainable way and with the circular bioeconomy as a vehicle. For this reason, in our future research, we will examine those conditions that will govern planning at the European and national levels through international experts and the Delphi approach. The ultimate goal is to formulate forest environmental and circular economy programs that will ensure sustainability, promote circular bioeconomy forestry products and the economic result for local communities connected to the forest.

Author Contributions: Conceptualization, S.K. and F.C.; methodology, S.K. and D.K.; software, D.K.; validation, F.C. and E.L.; formal analysis, S.K.; investigation, S.K. and D.K.; resources, S.K., E.L. and D.K.; data curation, F.C. and E.L.; writing—original draft preparation, S.K.; writing—review and editing, D.K.; visualization, E.L. and D.K.; supervision, F.C.; project administration, F.C. and E.L.; funding acquisition, F.C. and E.L. All authors have read and agreed to the published version of the manuscript.

Funding: This research received no external funding.

Institutional Review Board Statement: Not applicable.

Informed Consent Statement: Informed consent was obtained from all subjects involved in the study.

Data Availability Statement: Data available on request.

Conflicts of Interest: The authors declare no conflict of interest.

References

1. Managi, S.; Wang, J.; Zhang, L. Research progress on monitoring and assessment of forestry area for improving forest management in China. *For. Econ. Rev.* **2019**, *1*, 57–70. [CrossRef]
2. Cheng, S.H.; MacLeod, K.; Ahlroth, S.; Onder, S.; Perge, E.; Shyamsundar, P.; Rana, P.; Garside, R.; Kristjanson, P.; McKinnon, M.C.; et al. A systematic map of evidence on the contribution of forests to poverty alleviation. *Environ. Evid.* **2019**, *8*, 3. [CrossRef]
3. Kupěák, V.; Šmída, Z. Forestry and wood sector and profitability development in the wood-processing industry of the Czech Republic. *J. For. Sci.* **2015**, *61*, 244–249. [CrossRef]
4. Antonarakis, A.S.; Pacca, L.; Antoniades, A. The effect of financial crises on deforestation: A global and regional panel data analysis. *Sustain. Sci.* **2022**, *17*, 1037–1057. [CrossRef] [PubMed]
5. Fantechi, F.; Fratesi, U. Measuring competitiveness differentials inside the same region: A propensity-score matching approach. *Soc. Indic. Res.* **2022**. [CrossRef]
6. United Nations. *Special Edition: Progress towards the Sustainable Development Goals*; United Nations: New York, NY, USA, 2019.
7. Masiero, M.; Pettenella, D.M.; Secco, L. From failure to value: Economic valuation for a selected set of products and services from Mediterranean forests. *For. Syst.* **2016**, *25*, 1–16. [CrossRef]
8. Maheras, G. *Forests Fires in Greece. The Analysis of the Phenomenon Affecting Both Natural And Human Environment. The Role of Sustainable Development in Controlling Fire Effects*; Lund University: Lund, Sweden, 2002.
9. Ballas, D.; Dorling, D.; Hennig, B. Analysing the regional geography of poverty, austerity and inequality in Europe: A human cartographic perspective. *Reg. Stud.* **2017**, *51*, 174–185. [CrossRef]
10. Eurostat Urban and Rural Living in the EU. Available online: https://ec.europa.eu/eurostat/web/products-eurostat-news/-/edn-20200207-1 (accessed on 22 September 2022).
11. Kupec, P.; Marková, J.; Pelikán, P.; Brychtová, M.; Autratová, S.; Fialová, J. Urban Parks Hydrological Regime in the Context of Climate Change— A Case Study of Štěpánka Forest Park (Mladá Boleslav, Czech Republic). *Land* **2022**, *11*, 412. [CrossRef]
12. Aggestam, F.; Pülzl, H. Coordinating the Uncoordinated: The EU Forest Strategy. *Forests* **2018**, *9*, 125. [CrossRef]
13. Taghouti, I.; Cristobal, R.; Brenko, A.; Stara, K.; Markos, N.; Chapelet, B.; Hamrouni, L.; Buršić, D.; Bonet, J.-A. The Market Evolution of Medicinal and Aromatic Plants: A Global Supply Chain Analysis and an Application of the Delphi Method in the Mediterranean Area. *Forests* **2022**, *13*, 808. [CrossRef]
14. Martinez de Arano, I.; Maltoni, S.; Picardo, A.; Mutke, S. *Non-Wood Forest Products for People, Nature and the Green Economy. Recommendations for Policy Priorities in Europe. A White Paper Based on Lessons Learned from around the Mediterranean*; Giessen, L., Adams, S., Martinez de Arano, I., Eds.; The European Forest Institute and the Food and Agriculture Organization of the United Nations: Joensuu, Finland, 2021; ISBN 9789251347270.
15. Solomou, A.; Martinos, K.; Skoufogianni, E.; Danalatos, N.G. Medicinal and Aromatic Plants Diversity in Greece and Their Future Prospects: A Review. *Agric. Sci.* **2016**, *4*, 9–20. [CrossRef]
16. Cheminal, A.; Kokkoris, I.P.; Strid, A.; Dimopoulos, P. Medicinal and Aromatic Lamiaceae Plants in Greece: Linking Diversity and Distribution Patterns with Ecosystem Services. *Forests* **2020**, *11*, 661. [CrossRef]
17. Kougioumoutzis, K.; Kokkoris, I.P.; Panitsa, M.; Kallimanis, A.; Strid, A.; Dimopoulos, P. Plant Endemism Centres and Biodiversity Hotspots in Greece. *Biology* **2021**, *10*, 72. [CrossRef] [PubMed]
18. Fitzgerald, J.; Lindner, M. (Eds.) *Adapting to Climate Change in European Forests—Results of the MOTIVE Project*; Pensoft Publishers: Sofia, Bulgaria, 2013.
19. Schulz, T.; Lieberherr, E.; Zabel, A. How national bioeconomy strategies address governance challenges arising from forest-related trade-offs. *J. Environ. Policy Plan.* **2022**, *24*, 123–136. [CrossRef]
20. Li, Y.; Mei, B.; Linhares-Juvenal, T. The economic contribution of the world's forest sector. *For. Policy Econ.* **2019**, *100*, 236–253. [CrossRef]
21. Eurostat Forests, Forestry and Logging. Available online: https://ec.europa.eu/eurostat/statistics-explained/index.php?title=Forests,_forestry_and_logging#Forests_in_the_EU (accessed on 10 November 2022).

22. Tsiaras, S.; Koulelis, P.; Tsiroukis, A.; Spanos, I. The contribution of forests in regional development: The role of National Forest Strategy in Greece. *Mibes Trans.* **2021**, *14*, 110–122.
23. Ioannou, K.; Tsantopoulos, G.; Arabatzis, G.; Andreopoulou, Z.; Zafeiriou, E. A Spatial Decision Support System Framework for the Evaluation of Biomass Energy Production Locations: Case Study in the Regional Unit of Drama, Greece. *Sustainability* **2018**, *10*, 531. [CrossRef]
24. Arabatzis, G.; Kyriakopoulos, G.; Tsialis, P. Typology of regional units based on RES plants: The case of Greece. *Renew. Sustain. Energy Rev.* **2017**, *78*, 1424–1434. [CrossRef]
25. Koundouri, P.; Kountouris, Y.; Remoundou, K. Valuing a wind farm construction: A contingent valuation study in Greece. *Energy Policy* **2009**, *37*, 1939–1944. [CrossRef]
26. Spanos, K.A.; Skouteri, A.; Gaitanis, D.; Petrakis, P.V.; Meliadis, I.; Michopoulos, P.; Solomou, A.; Koulelis, P.; Avramidou, E.V. Forests of Greece, Their Multiple Functions and Uses, Sustainable Management and Biodiversity Conservation in the Face of Climate Change. *Open J. Ecol.* **2021**, *11*, 374–406. [CrossRef]
27. Karametou, P.; Apostolopoulos, C. The causal nexus between social capital and local development in mountain rural Greece. *Int. J. Soc. Inq.* **2010**, *3*, 29–66.
28. FAOSTAT FAOSTAT. Available online: https://www.fao.org/faostat/en/#home (accessed on 24 September 2022).
29. Hoover, K.; Riddle, A.A. *National Forest System Management: Overview, Appropriations, and Issues for Congress*, 13th ed.; Congressional Research Service, U.S. Congress, House Committee: Washington, DC, USA, 2019.
30. Agrawal, A.; Cashore, B.; Hardin, R.; Shepherd, G.; Benson, C.; Miller, D. Economic contributions of forests. *Backgr. Pap.* **2013**, *1*, 1–127.
31. Annoni, P.; Dijkstra, L. *EU Regional Competitiveness Index 2019*; Dijkstra, L., Ed.; Publications Office of the European Union: Luxembourg, 2019.
32. OECD. *The Economic Significance of Natural Resources: Key Points for Reformers in Eastern Europe, Caucasus and Central Asia*; OECD: Paris, France, 2011; pp. 1–42. Available online: http://www.oecd.org/env/outreach/2011_AB_Economic%20significance%20of%20NR%20in%20EECCA_ENG.pdf (accessed on 29 September 2022).
33. Lele, U.; Karsenty, A.; Benson, C.; Fétiveau, J.; Agarwal, M.; Goswani, S. Changing roles of forests and their cross-sectorial linkages in the course of economic development. In Proceedings of the United Nations Forum for Forests, Istanbul, Turkey, 8–19 April 2013; United Nations Forum for Forests, Ed.; United Nations: Istanbul, Turkey, 2013; p. 172.
34. Pilli, R.; Grassi, G. *Provision of Technical and Scientific Support to DG ESTAT in Relation to EU Land Footprint Estimates and Gap-Filling Techniques for European Forest Accounts (LAFO)*; Publications Office of the European Union: Luxembourg, 2021; pp. 1–46. [CrossRef]
35. Krieger, D.J. *Economic Value of Forest Ecosystem Services: A Review*; Kloepfer, D., Ed.; The Wilderness Society: Washington, DC, USA, 2001.
36. Huber, N.; Bugmann, H.; Cailleret, M.; Bircher, N.; Lafond, V. Stand-scale climate change impacts on forests over large areas: Transient responses and projection uncertainties. *Ecol. Appl.* **2021**, *31*, e02313. [CrossRef] [PubMed]
37. Campos Arce, J.J. Forests, inclusive and sustainable economic growth and employment. In Proceedings of the Fourteenth Session of the United Nations Forum on Forests, New York, NY, USA, 6–10 May 2019; United Nations Forum on Forests: New York, NY, USA, 2019; pp. 1–58.
38. Liu, S.; Costanza, R.; Farber, S.; Troy, A. Valuing ecosystem services. Theory, practice, and the need for a transdisciplinary synthesis. *Ann. N. Y. Acad. Sci.* **2010**, *1185*, 54–78. [CrossRef] [PubMed]
39. Odgaard, M.V.; Turner, K.G.; Bøcher, P.K.; Svenning, J.-C.; Dalgaard, T. A multi-criteria, ecosystem-service value method used to assess catchment suitability for potential wetland reconstruction in Denmark. *Ecol. Indic.* **2017**, *77*, 151–165. [CrossRef]
40. Kalfas, D.G.; Zagkas, D.T.; Raptis, D.I.; Zagkas, T.D. The multifunctionality of the natural environment through the basic ecosystem services in the Florina region, Greece. *Int. J. Sustain. Dev. World Ecol.* **2019**, *26*, 57–68. [CrossRef]
41. Liu, Y.; Hou, X.; Li, X.; Song, B.; Wang, C. Assessing and predicting changes in ecosystem service values based on land use/cover change in the Bohai Rim coastal zone. *Ecol. Indic.* **2020**, *111*, 106004. [CrossRef]
42. Song, X.; Yang, G.; Yan, C.; Duan, H.; Liu, G.; Zhu, Y. Driving forces behind land use and cover change in the Qinghai-Tibetan Plateau: A case study of the source region of the Yellow River, Qinghai Province, China. *Environ. Earth Sci.* **2009**, *59*, 793. [CrossRef]
43. Reid, W.V.; Mooney, H.A.; Cropper, A.; Capistrano, D.; Carpenter, S.R.; Chopra, K.; Dasgupta, P.; Dietz, T.; Du-raiappah, A.K.; Hassan, R.; et al. Monika. Millennium Ecosystem Assessment. In *Ecosystems and Human Well-Being: Synthesis*; Island Press: Washington, DC, USA, 2005; Volume 5, ISBN 1597260401.
44. Carpenter, S.R.; Mooney, H.A.; Agard, J.; Capistrano, D.; Defries, R.S.; Diaz, S.; Dietz, T.; Duraiappah, A.K.; Oteng-Yeboah, A.; Pereira, H.M.; et al. Science for managing ecosystem services: Beyond the Millennium Ecosystem Assessment. *Proc. Natl. Acad. Sci. USA* **2009**, *106*, 1305–1312. [CrossRef]
45. Palmer, M.A.; Filoso, S.; Fanelli, R.M. From ecosystems to ecosystem services: Stream restoration as ecological engineering. *Ecol. Eng.* **2014**, *65*, 62–70. [CrossRef]
46. Gottero, E.; Cassatella, C.; Larcher, F. Planning Peri-Urban Open Spaces: Methods and Tools for Interpretation and Classification. *Land* **2021**, *10*, 802. [CrossRef]
47. Kalfas, D.; Zagkas, D.; Dragozi, E.; Zagkas, T. Estimating value of the ecosystem services in the urban and peri-urban green of a town Florina-Greece, using the CVM. *Int. J. Sustain. Dev. World Ecol.* **2020**, *27*, 310–321. [CrossRef]

48. Crossman, N.D.; Burkhard, B.; Nedkov, S.; Willemen, L.; Petz, K.; Palomo, I.; Drakou, E.G.; Martín-Lopez, B.; McPhearson, T.; Boyanova, K.; et al. A blueprint for mapping and modelling ecosystem services. *Ecosyst. Serv.* **2013**, *4*, 4–14. [CrossRef]
49. Turner, K.G.; Odgaard, M.V.; Bøcher, P.K.; Dalgaard, T.; Svenning, J.-C.C. Bundling ecosystem services in Denmark: Trade-offs and synergies in a cultural landscape. *Landsc. Urban Plan.* **2014**, *125*, 89–104. [CrossRef]
50. Andrew, M.E.; Wulder, M.A.; Nelson, T.A.; Coops, N.C. Spatial data, analysis approaches, and information needs for spatial ecosystem service assessments: A review. *GIScience Remote Sens.* **2015**, *52*, 344–373. [CrossRef]
51. Kay, S.; Crous-Duran, J.; Ferreiro-Domínguez, N.; García de Jalón, S.; Graves, A.; Moreno, G.; Mosquera-Losada, M.R.; Palma, J.H.N.; Roces-Díaz, J.V.; Santiago-Freijanes, J.J.; et al. Spatial similarities between European agroforestry systems and ecosystem services at the landscape scale. *Agrofor. Syst.* **2017**, *92*, 1075–1089. [CrossRef]
52. Costanza, R.; Kubiszewski, I. The authorship structure of "ecosystem services" as a transdisciplinary field of scholarship. *Ecosyst. Serv.* **2012**, *1*, 16–25. [CrossRef]
53. de Groot, R.S.; Alkemade, R.; Braat, L.; Hein, L.; Willemen, L. Challenges in integrating the concept of ecosystem services and values in landscape planning, management and decision making. *Ecol. Complex.* **2010**, *7*, 260–272. [CrossRef]
54. Fisher, B.; Turner, R.K.; Morling, P. Defining and classifying ecosystem services for decision making. *Ecol. Econ.* **2009**, *68*, 643–653. [CrossRef]
55. Farber, S.; Costanza, R.; Childers, D.L.; Erickson, J.; Gross, K.; Grove, M.; Hopkinson, C.S.; Kahn, J.; Pinceti, S.; Troy, A.; et al. Linking Ecology and Economics for Ecosystem Management. *Bioscience* **2006**, *56*, 121–134. [CrossRef]
56. Martinez-Harms, M.J.; Bryan, B.A.; Balvanera, P.; Law, E.A.; Rhodes, J.R.; Possingham, H.P.; Wilson, K.A. Making decisions for managing ecosystem services. *Biol. Conserv.* **2015**, *184*, 229–238. [CrossRef]
57. Maes, J.; Egoh, B.; Willemen, L.; Liquete, C.; Vihervaara, P.; Schägner, J.P.; Grizzetti, B.; Drakou, E.G.; La Notte, A.; Zulian, G.; et al. Mapping ecosystem services for policy support and decision making in the European Union. *Ecosyst. Serv.* **2012**, *1*, 31–39. [CrossRef]
58. Mancinelli, R.; Di Felice, V.; Karkalis, K.; Bari, S.; Radicetti, E.; Campiglia, E. Assessment of the state of agroecosystem sustainability using landscape indicators: A comparative study of three rural areas in Greece. *Int. J. Sustain. Dev. World Ecol.* **2018**, *25*, 35–46. [CrossRef]
59. Outeiro, L.; Häussermann, V.; Viddi, F.; Hucke-Gaete, R.; Försterra, G.; Oyarzo, H.; Kosiel, K.; Villasante, S. Using ecosystem services mapping for marine spatial planning in southern Chile under scenario assessment. *Ecosyst. Serv.* **2015**, *16*, 341–353. [CrossRef]
60. Von Haaren, C.; Albert, C. Integrating ecosystem services and environmental planning: Limitations and synergies. *Int. J. Biodivers. Sci. Ecosyst. Serv. Manag.* **2011**, *7*, 150–167. [CrossRef]
61. Doré, T.; Makowski, D.; Malézieux, E.; Munier-Jolain, N.; Tchamitchian, M.; Tittonell, P. Facing up to the paradigm of ecological intensification in agronomy: Revisiting methods, concepts and knowledge. *Eur. J. Agron.* **2011**, *34*, 197–210. [CrossRef]
62. Grêt-Regamey, A.; Weibel, B.; Kienast, F.; Rabe, S.-E.; Zulian, G. A tiered approach for mapping ecosystem services. *Ecosyst. Serv.* **2015**, *13*, 16–27. [CrossRef]
63. Soussana, J.-F.; Fereres, E.; Long, S.P.; Mohren, F.G.; Pandya-Lorch, R.; Peltonen-Sainio, P.; Porter, J.R.; Rosswall, T.; von Braun, J. A European science plan to sustainably increase food security under climate change. *Glob. Chang. Biol.* **2012**, *18*, 3269–3271. [CrossRef]
64. Lavorel, S.; Bayer, A.; Bondeau, A.; Lautenbach, S.; Ruiz-frau, A.; Schulp, N.; Seppelt, R.; Verburg, P.; Van Teeffelen, A.; Vannier, C.; et al. Pathways to bridge the biophysical realism gap in ecosystem services mapping approaches. *Ecol. Indic.* **2017**, *74*, 241–260. [CrossRef]
65. Costanza, R.; D'Arge, R.; de Groot, R.; Farber, S.; Grasso, M.; Hannon, B.; Limburg, K.; Naeem, S.; O'Neill, R.V.; Paruelo, J.; et al. The value of the world's ecosystem services and natural capital. *Nature* **1997**, *387*, 253–260. [CrossRef]
66. Costanza, R.; de Groot, R.; Braat, L.; Kubiszewski, I.; Fioramonti, L.; Sutton, P.; Farber, S.; Grasso, M. Twenty years of ecosystem services: How far have we come and how far do we still need to go? *Ecosyst. Serv.* **2017**, *28*, 1–16. [CrossRef]
67. Pártl, A.; Vačkář, D.; Loučková, B.; Lorencová, E.K. A spatial analysis of integrated risk: Vulnerability of ecosystem services provisioning to different hazards in the Czech Republic. *Nat. Hazards* **2017**, *89*, 1185–1204. [CrossRef]
68. Zhao, J.; Liu, X.; Dong, R.; Shao, G. Landsenses ecology and ecological planning toward sustainable development. *Int. J. Sustain. Dev. World Ecol.* **2016**, *23*, 293–297. [CrossRef]
69. Zhao, J.; Yan, Y.; Deng, H.; Liu, G.; Dai, L.; Tang, L.; Shi, L.; Shao, G. Remarks about landsenses ecology and ecosystem services. *Int. J. Sustain. Dev. World Ecol.* **2020**, *27*, 196–201. [CrossRef]
70. Zheng, S.; Cui, K.; Sun, S.; Wu, J.; Qiu, Q.; Tian, Y. Planning and design based on landsenses ecology: The case study of Chongming Island Landsenses Ecol-industrial Park. *Int. J. Sustain. Dev. World Ecol.* **2020**, *27*, 435–442. [CrossRef]
71. Tian, Y.; Guo, Z.; Zhong, W.; Qiao, Y.; Qin, J. A design of ecological restoration and eco-revetment construction for the riparian zone of Xianghe Segment of China's Grand Canal. *Int. J. Sustain. Dev. World Ecol.* **2016**, *23*, 333–342. [CrossRef]
72. Kalfas, D.G.; Zagkas, D.T.; Dragozi, E.I.; Melfou, K.K. An approach of landsenses ecology and landsenseology in Greece. *Int. J. Sustain. Dev. World Ecol.* **2021**, *28*, 677–692. [CrossRef]
73. Linwei, H.; Longyu, S.; Fengmei, Y.; Xue-qin, X.; Lijie, G. Method for the evaluation of residents' perceptions of their community based on landsenses ecology. *J. Clean. Prod.* **2021**, *281*, 124048. [CrossRef]

74. Dong, R.; Yu, T.; Ma, H.; Ren, Y. Soundscape planning for the Xianghe Segment of China's Grand Canal based on landsenses ecology. *Int. J. Sustain. Dev. World Ecol.* **2016**, *23*, 343–350. [CrossRef]
75. Zhang, L.; Huang, G.; Li, Y.; Bao, S. Advancement trajectory of emerging landsenses ecology for sustainability research and implementation. *Int. J. Sustain. Dev. World Ecol.* **2022**, *29*, 641–652. [CrossRef]
76. Stephenson, P.J.; Damerell, A. Bioeconomy and Circular Economy Approaches Need to Enhance the Focus on Biodiversity to Achieve Sustainability. *Sustainability* **2022**, *14*, 10643. [CrossRef]
77. European Commission, Directorate-General for Research and Innovation. *Innovating for Sustainable Growth. A Bioeconomy for Europe*; Publications Office of the European Union: Brussels, Belgium, 2012.
78. Papadopoulou, C.-I.; Loizou, E.; Melfou, K.; Chatzitheodoridis, F. The Knowledge Based Agricultural Bioeconomy: A Bibliometric Network Analysis. *Energies* **2021**, *14*, 6823. [CrossRef]
79. European Commission. *A Sustainable Bioeconomy for Europe: Strengthening the Connection between Economy, Society and the Environment*; European Commission: Brussels, Belgium, 2018.
80. Carus, M.; Dammer, L. The Circular Bioeconomy—Concepts, Opportunities, and Limitations. *Ind. Biotechnol.* **2018**, *14*, 83–91. [CrossRef]
81. Kardung, M.; Cingiz, K.; Costenoble, O.; Delahaye, R.; Heijman, W.; Lovrić, M.; van Leeuwen, M.; M'Barek, R.; van Meijl, H.; Piotrowski, S.; et al. Development of the Circular Bioeconomy: Drivers and Indicators. *Sustainability* **2021**, *13*, 413. [CrossRef]
82. Lovrić, M.; Lovrić, N.; Mavsar, R. Mapping forest-based bioeconomy research in Europe. *For. Policy Econ.* **2020**, *110*, 101874. [CrossRef]
83. Papadopoulou, C.-I.; Loizou, E.; Chatzitheodoridis, F. Priorities in Bioeconomy Strategies: A Systematic Literature Review. *Energies* **2022**, *15*, 7258. [CrossRef]
84. Spies, M.; Zuberi, M.; Mählis, M.; Zakirova, A.; Alff, H.; Raab, C. Towards a participatory systems approach to managing complex bioeconomy interventions in the agrarian sector. *Sustain. Prod. Consum.* **2022**, *31*, 557–568. [CrossRef]
85. Mihova, T.B.; Alexieva, V.N. Business communities—A factor of industry and bioeconomy development. *IOP Conf. Ser. Mater. Sci. Eng.* **2020**, *878*, 012070. [CrossRef]
86. Ranacher, L.; Wallin, I.; Valsta, L.; Kleinschmit, D. Social dimensions of a forest-based bioeconomy: A summary and synthesis. *Ambio* **2020**, *49*, 1851–1859. [CrossRef]
87. Lillemets, J.; Fertő, I.; Viira, A.-H. The socioeconomic impacts of the CAP: Systematic literature review. *Land Use Policy* **2022**, *114*, 105968. [CrossRef]
88. Papadopoulou, E.; Vaitsas, K.; Fallas, I.; Tsipas, G.; Chrissafis, K.; Bikiaris, D.; Kottaridi, C.; Vorgias, K.E. Bio-economy in Greece: Current trends and the road ahead. *EuroBiotech J.* **2018**, *2*, 137–145. [CrossRef]
89. Hodge, D.; Brukas, V.; Giurca, A. Forests in a bioeconomy: Bridge, boundary or divide? *Scand. J. For. Res.* **2017**, *32*, 582–587. [CrossRef]
90. Jurga, P.; Loizou, E.; Rozakis, S. Comparing Bioeconomy Potential at National vs. Regional Level Employing Input-Output Modeling. *Energies* **2021**, *14*, 1714. [CrossRef]
91. European Commission—Green Deal European Commission—Green Deal. Available online: https://ec.europa.eu/info/strategy/priorities-2019-2024/european-green-deal_en (accessed on 31 August 2022).
92. Backhouse, M.; Lehmann, R.; Lorenzen, K.; Puder, J.; Rodríguez, F.; Tittor, A. (Eds.) *Contextualizing the Bioeconomy in an Unequal World: Biomass Sourcing and Global Socio-Ecological Inequalities BT—Bioeconomy and Global Inequalities: Socio-Ecological Perspectives on Biomass Sourcing and Production*; Springer International Publishing: Cham, Switzerland, 2021; pp. 3–22, ISBN 978-3-030-68944-5.
93. MEE-YPEN Ministry of Environment and Energy. Available online: https://ypen.gov.gr/ (accessed on 23 October 2022).
94. Kalfas, D.; Chatzitheodoridis, F.; Loizou, E.; Melfou, K. Willingness to Pay for Urban and Suburban Green. *Sustainability* **2022**, *14*, 2332. [CrossRef]
95. Kalogiannidis, S.; Toska, E.; Chatzitheodoridis, F.; Kalfas, D. Using School Systems as a Hub for Risk and Disaster Management: A Case Study of Greece. *Risks* **2022**, *10*, 89. [CrossRef]
96. Kalogiannidis, S.; Kalfas, D.; Chatzitheodoridis, F.; Papaevangelou, O. Role of Crop-Protection Technologies in Sustainable Agricultural Productivity and Management. *Land* **2022**, *11*, 1680. [CrossRef]

land

Article

Public Perceptions of the Socioeconomic Importance of Urban Green Areas in the Era of COVID-19: A Case Study of a Nationwide Survey in Greece

Antonios Kolimenakis [1], Alexandra D. Solomou [1,*], Nikolaos Proutsos [1], Evangelia V. Avramidou [1], Evangelia Korakaki [1], Georgios Karetsos [1], Aimilia B. Kontogianni [2], Konstantinos Kontos [2], Christos Georgiadis [1], Georgios Maroulis [1], Eleftherios Papagiannis [1], Konstantinos Lagouvardos [3] and Konstantinia Tsagkari [1]

[1] Institute of Mediterranean Forest Ecosystems-Hellenic Agricultural Organization "DEMETER," Terma Alkmanos, 11528 Athens, Greece
[2] Homeotech Co. Environmental Management Company, 102 Egeou Str., Kalamaria, 55133 Thessaloniki, Greece
[3] Institute for Environmental Research and Sustainable Development, National Observatory of Athens, 11810 Athens, Greece
* Correspondence: solomou@fria.gr

Abstract: Considering the emerging challenges posed by the spread of COVID-19, this study was designed to evaluate citizens' perceptions of the role of urban green areas in the era of COVID-19 in Greece. The evaluation was based on the implementation of an electronic questionnaire survey through the Google Forms platform, which was conducted nationwide. The survey was conducted in 2020 and 735 responses were collected in total based on 14 structured questions. Among the key findings of the study, of notable importance is that citizens considered urban green areas as an important means of improving public health, while citizens were willing to accept an increase of EUR1 to EUR20 in their municipal taxes for improving the services offered by the urban green areas. Results indicate that in a period of both climatic and public health crises, healthy and green urban environments can play a seminal role for alleviating and mitigating different challenges and impacts, while at the same time ensuring sustainability of urban ecosystems. A certain necessity arises for investigating the socioeconomic importance of urban green areas both from an ecosystemic and public health perspective considering the novel challenges of COVID-19 to public policy and decision making.

Keywords: urban green areas; socioeconomic welfare; COVID-19 pandemic

1. Introduction

The interdependence and interconnection of the urban and natural environment of the city can be observed in urban green areas [1]. This connection is of the utmost importance for life in cities and for their inhabitants, and therefore the importance of urban green spaces for the well-being of city dwellers should be recognized [1].

The association of urban green areas (UGAs) with public health is highly acknowledged in global policies and reports, indicating that urban green spaces can provide important socioeconomic and health benefits [2,3]. The recent COVID-2019 pandemic further underlined the significance and the role of UGAs and urged the need for rethinking the design and organization of modern cities to more resilient and sustainable schemes. The WHO's manifesto for a healthy recovery from COVID-19 [4], outlines "Building healthy and liveable cities" as one of the six main prescriptions for a green and healthy recovery from COVID-19. Specifically, recommendations relate to "pedestrianizing streets and massively expanding cycle lanes—enabling "physically distant" transport during the crisis, and enhancing economic activity and quality of life afterwards," which outlines the tandem socioeconomic and health benefits that can arise from sustainable UGAs, also in line with the recent recognition of the "right to a healthy, clean and sustainable environment" of the COP27 decision [5].

However, the positive effects of urban green spaces on citizens' well-being cannot be fully acknowledged and expressed in monetary terms. This is due to the fact that the existence of urban green space does not provide direct economic benefits to the citizens. Urban green infrastructure has positive effects on citizens' well-being through the provision of essential ecosystem services, i.e., "the benefits people obtain from ecosystems" [6] thus increasing their quality of life. Ecosystems services such as carbon sequestration (regulating ecosystem service) or even the aesthetic pleasure (cultural ecosystem service) are factors that indirectly influence the citizens' well-being and should be taken into consideration. However, it will be also illustrated that the existence of urban green spaces is seen as an amenity and contributes to the increase of property values that are located in the vicinity of those areas [6].

It is characteristic that in the many case studies, where the effect of urban green spaces on citizens' well-being is estimated, the "hypothetical market" method is employed. The reason is obvious, since there is no real market for goods and services those spaces may provide. Therefore, the economic value of such urban spaces is extracted through the creation of hypothetical market conditions and is expressed through the willingness to pay or other indirect positive effects, such as carbon sequestration, are also estimated, as these contribute to an ameliorated quality of life in urban areas [7]. Additionally, the increased value of properties located near the area of green urban spaces can be regarded as an indirect monetary indicator that highlights the need for such spaces.

Such estimation is of primary importance. With the help of monetary valuation, the main challenges regarding the governance of the urban green space can be identified. More specifically, these challenges include the increase in development pressure due to population growth and economic constraints on the municipal budget, the loss of expertise and the low awareness of the green benefits to various factors through insufficient communication [8,9].

Several studies have been carried out throughout the last years that attempted to underline the critical importance of UGAs by highlighting the positive effects theses area can have on citizens' well-being [2,3,10,11]. Nevertheless, it is often an arduous task to codify those benefits to human welfare by devising one single metric or indicator that will be based on monetary values [12]. Undoubtedly, UGAs' positively influence human well-being, as they provide specific ecosystem services. Some of those ecosystem services have an indirect impact on citizens' well-being. However, their role should not be undermined at any circumstance and they should be carefully researched and adequately assessed [13]. Additionally, the multidimensional role of UGAs has been underlined in the research, stressing their importance in urban areas [14]. Obviously, important aspects of citizens' well-being such as health preservation and recreation can be influenced by the planning, realization and maintenance of UGAs, such as parks and forests [14].

Concerning the Greek peninsula, the recent rapid warming trends have been shown to have an impact on the viability and growth of vegetation on urban and natural settings [15]. In addition, recent studies have evaluated the changes in aridity in the last century in Greece and identified that the recent climatic period is characterized by more arid conditions compared to the past, suggesting also reduced water availability for the natural and urban vegetation, with significant impacts on plant growth, which is also supported by other studies on Greece [16,17]. The warming trends and the changes to more arid conditions are even rapidly occurring in Greek cities [18–24]. Under such climate conditions, the green infrastructure in Greek cities is already coping with climate change and are considered a regulatory tool to mitigating urban climate and the urban heat island (UHI) phenomenon imposing an urging need for sustainable urban planning. However, in Greek cities, the availability of green spaces is generally scarce, especially in densely populated and built-up areas such as Athens. Specifically for the city of Athens, Giannopoulou et al. [25] and Livada et al. [26] explored the UHI phenomenon in 25 sites inside the city during the summer period and suggested the division of the city in five geographical zones considering their thermal balance characteristics. Both studies identified that the industrial western and central parts of the city had higher air temperatures compared to the northern and eastern parts, underlining the positive cooling effect of green spaces, which is also addressed in other studies conducted in Athens, either in urban parks [27–29], small courtyards [30,31], or single trees [32].

According to recently conducted studies on the impact of the COVID-19 pandemic in cities [33], the main findings indicate that overall, the health and well-being in cities worsened due to the COVID-19 pandemic, while the built environment has contributed to COVID-19-related changes in health and well-being, with denser neighborhoods being particularly linked to lower well-being during COVID-19 [34,35]. In a study conducted in Mexico City, it was highlighted that UGA use has served as an option to decrease the effects of stress and isolation caused by COVID-19. More specifically UGAs have served as a "coping mechanism" that has increased citizens' physical and mental well-being [36]. Nevertheless, in low- and middle-income neighborhoods where there is a lack of UGAs, the problem of access has been highlighted as a crucial factor that has hampered citizens' well-being in those areas [36].

Furthermore, a general trend concerning visits to UGAs has been observed. In an extreme case, a threefold increase in recreational use of outdoor spaces in and around Oslo, Norway has been found [37,38]. A similar but considerably lower trend can be found in other European cities. In Bonn, Germany, visitors' visits to urban forests nearly doubled during March 2020 [39]. In the UK, researchers found evidence of a radical substitution of leisure time for recreation in available UGAs coupled with a drastic decrease of car use by 47% [40]. On the contrary, a 13.1% decrease in visits to UGAs has been observed in Poland, which was merely the result of a regulation banning the use of UGAs. However, over 50% of the respondents admitted that visits to UGAs have a considerable positive effect on their well-being [41]. Similarly, a study conducted in six countries (Croatia, Israel, Italy, Slovenia, Lithuania, and Spain) revealed similar conclusions. Characteristically, 64% of the respondents in Spain and in Italy limited their visits due to the government restrictions. This happened to a lesser extent in the other countries. Nevertheless, COVID-19 curfews and access restrictions to UGAs highlighted the physical, cognitive and emotional need that could by fulfilled by UGAs. Characteristically, respondents underlined that during the COVID-19 period they missed their park-related activity, e.g., exercising outdoors, meeting other people observing nature, and breathing fresh air [42].

2. Methods

Considering the emerging challenges posed by the spread of COVID-19, the current study was designed to evaluate citizens' perception of the role of urban green areas iduring the COVID-19 pandemic in Greece.

The purpose of this study was to elucidate:

- Citizens' considerations of urban green areas as an important means for improving public health;
- Citizens' perception in regard to urban green areas being hotspots for improving public health;
- A trade-off question between different economic preferences of a service in the form of a public good, access to urban green areas vs free access to home internet for each month of lockdown measures;
- Citizens' ranking of favorite urban ecosystem services in the era of lockdown.

The study's structure was based on the following steps:

- Design of the questionnaire by the research team;
- Conducting the survey through a web questionnaire;
- Evaluation and analysis of results by the research team.

The evaluation was based on the implementation of a web survey, uploaded on a Google Forms questionnaire, which was conducted at a national scale and distributed through a frequently visited meteorological website (www.meteo.gr, accessed on 11 October 2022) run by the National Observatory of Athens. This website is a popular meteorological data website with a high number of daily visitors, and for the purposes of the survey a special banner appeared randomly to visitors on the home page, from which visitors followed a link to the web survey. The survey was conducted from 25 November to 15 December 2020, with a total of 735 responses collected within this period from all over the country. It should be noted that during the first year of the COVID-19 outbreak, a set of lockdown measures was imposed by the Greek authorities starting in March 2020, while the measures were gradually eased from May 2020 onward.

The survey was based on 14 questions that were structured on the following main pillars:

- Citizens' considerations and perceptions of urban green areas in relation to the COVID-19 pandemic using different scaling options based on each question;
- Citizens' appraisal of ecosystem services provided by the UGAs in their municipality by using a grade from 1 to 5;
- Citizens' perceived socioecological benefits linked to the importance of UGAs, ranked on a four-point scale;
- Citizens' willingness to pay for selected ecosystem services through an increase in the municipal tax for improving UGAs;
- Demographic data.

The participants were all inhabitants of the Greek peninsula, and during the COVID-2019 quarantine had to follow the limitations in transportation imposed by the Greek government. It should be noted that the climate of Greece is generally mild, allowing visits to parks and green areas most of the days of the year. More specifically, Greece's climate is Mediterranean and according to Thornthwaite's aridity classification [43,44] is humid in most areas, but there are also regions with subhumid or subarid climate [15,45].

3. Respondents' Profile

Regarding the respondents' characteristics (Figure 1), the questionnaire was filled predominantly by males (71%). Furthermore, the respondents' level of education was high or very high, as 48.4% were university graduates and 34.6% had master's degrees or PhDs.

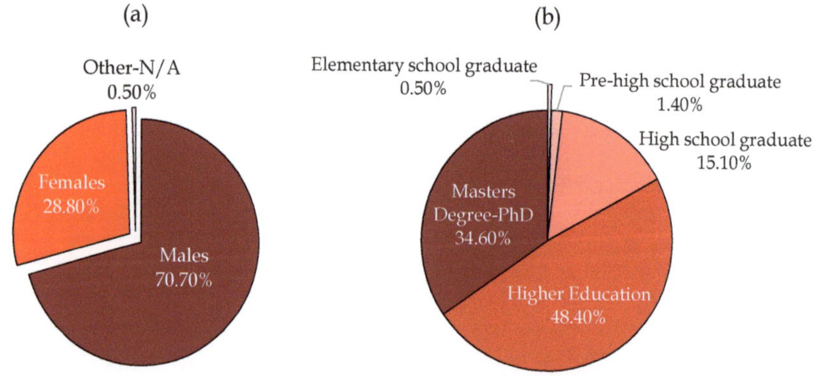

Figure 1. Distribution of the participants by (**a**) gender and (**b**) educational level.

As far as occupation is concerned (Figure 2), 33% of the respondents were employed, while 22% were self-employed, followed by civil servants (18%), pensioners (12%) and unemployed (9%).

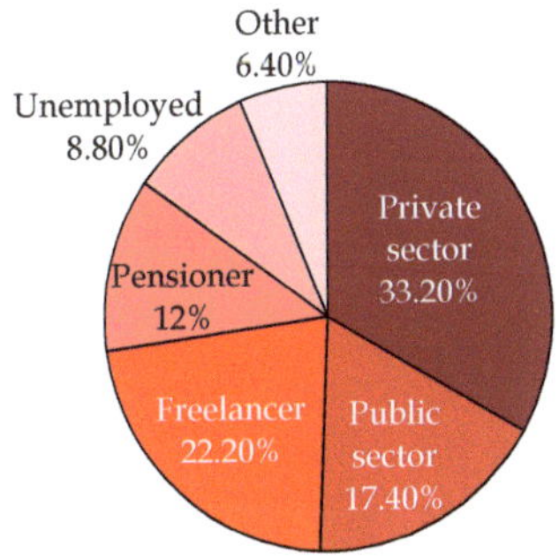

Figure 2. Occupation of the participants.

The composition of the respondents' households varied principally between 2–4 persons (79.2% of the responses). The annual household income (Figure 3) was divided between those having an income more than EUR20,000 (44%) and those on less than EUR20,000 (56%). It should be noted, however, that the participants were almost equally distributed between the different household income classes, presenting percentages within a small range (from 16% with household income of EUR5,000–10,000 to 26% with household income greater than EUR25,000). This distribution indicates that the interest of the citizens in UGAs is independent of their financial status.

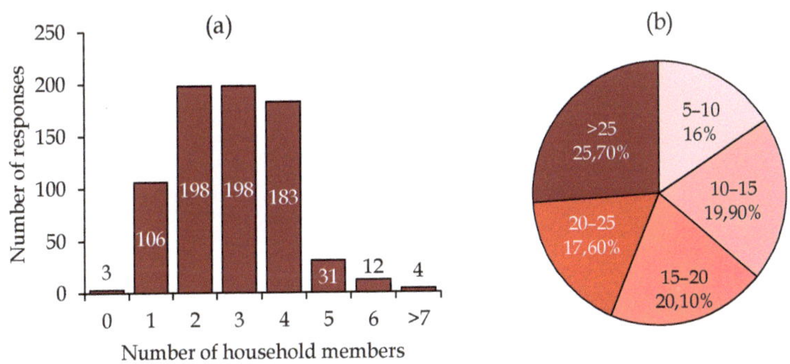

Figure 3. (**a**) Household members and (**b**) annual household income (in EUR1000s) of the participants.

Concerning the participants' origins, it should be noted that all were citizens with permanent residence in Greece. Their distribution in the 13 Greek prefectures indicated that all prefectures were represented. The great majority of the responders (65.3%) were living in the broader area of Athens (prefecture of Attica), whereas the prefecture of Central Macedonia (hosting the second-largest city of the country, i.e., Thessaloniki) also showed a high percentage (10.5 %). All other prefecture percentages were significantly smaller, varying from 0.5% for the Ionian Islands to 5.2% for Crete.

4. Results

4.1. Willingness to Pay for Ecosystem Services Related to Urban Green Spaces

The 735 participants of the survey were asked to appraise different ecosystem service functions provided by the UGAs in their municipality using a grade from 1 to 5. The average scores for each service function are depicted in Figure 4. The participants valued most (4.36, sd: 1.00) the upgrade of the aesthetic value of the urban landscape and to a similar degree (score 4.35, sd: 1.05) the climate regulation services and the noise and air pollution reduction (score 4.28, sd: 1.09) that the green infrastructures provide in the city. The score of the landscape aesthetic values presented no statistical difference in scores for the services of climate regulation ($p = 0.696$) and the noise and air pollution reduction ($p = 0.089$), but were significantly different ($p < 0.001$) for all other services. The recreation activities and the conservation/enhancement of the biodiversity appear to be less important for the participants with scores 3.76 (sd: 1.20) and 3.87 (sd: 1.25), respectively, which presented no significant difference ($p = 0.116$). Citizens evaluated the recreation activities and also the function of UGAs as a play space for children with lower scores (3.76 and 4.02, respectively) compared to landscape aesthetics or the UGAs' ability to regulate the local urban climate, suggesting that the average participant indicated an environmentally sensitive profile. The relatively low score for the function of UGAs as biodiversity hotspots in the city is probably explained by the fact that the citizens' idea for a UGA is that it should provide security and enhanced aesthetic surroundings to the visitors and that a great variety of floristic or fauna species would probably not be a priority.

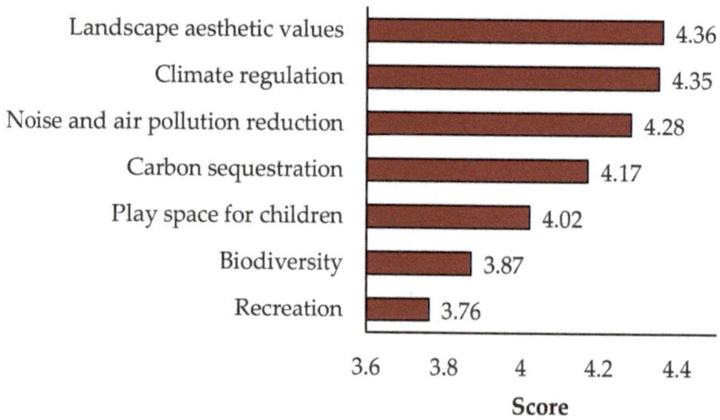

Figure 4. Average scores for the appraisal of the UGA services based on the evaluation of the participants (n = 735), using a 5-degree scale.

The participants also ranked the importance of UGAs on enhancing socioecological benefits based on four given options, and the average scores are presented in Figure 5 using a 4-degree ranking scale. In all cases, the score differences were statistically different ($p < 0.001$). Securing public health and natural environment gained the highest score (3.52/4.00) among the benefits of UGAs, whereas the production of food, drinking water or other basic materials had the lowest score, i.e., 2.48. This was expected, considering the impact of COVID-19 in public health and local communities. However, the relatively low score for enhancing social relations and social networks is notable, especially during the restrictions imposed for restraining the spread of COVID-19.

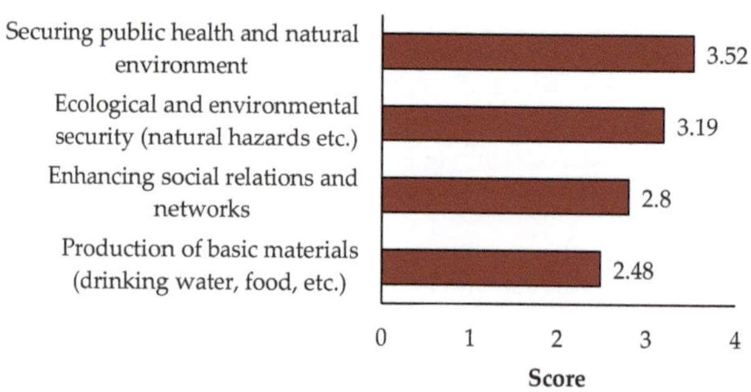

Figure 5. Average scores for assessing UGA importance for specific benefits using a 4-degree scale.

The question on the respondents' willingness to accept an increase in municipal taxes for specific services (Figure 6) revealed that 27% would be willing to pay for the amelioration and enhancement of green urban spaces for public health purposes. Surely, this question was greatly influenced by the COVID-19 restrictions, i.e., curfews, as well as the discussions regarding a lower spread of COVID-19 in open spaces. Interestingly, a fair share of the respondents would be willing to accept an increase in taxes for urban green spaces as a measure against climate change (20%) or for environmental conservation (18%). Then, taking into consideration the fact that urban green spaces are mainly dedicated to recreation activities, only 16% of the respondents were willing to additionally pay for that service.

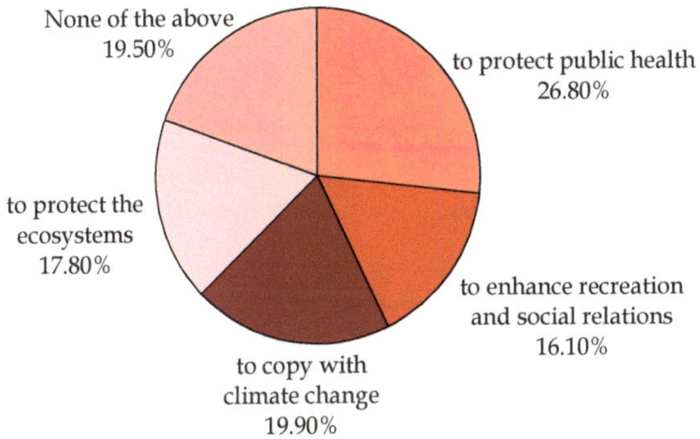

Figure 6. Responses on services accepted for an increase in the municipal tax for improving UGAs.

The level of the additional tax increase varied among the respondents (Figure 7). On the one hand, 26.2% were willing to pay the highest amount possible (EUR10–20 per year) or the second-highest amount (EUR7.5–10 per year). On the other hand, 23.5% were willing to pay the lowest possible amount (EUR1–2.5 per year).

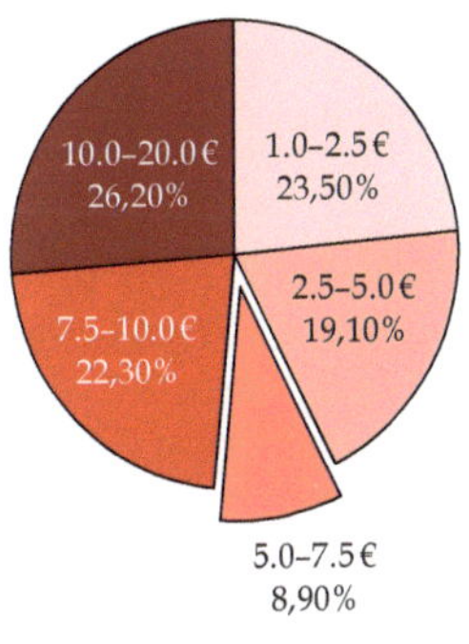

Figure 7. Responses on the monetary level of municipal tax increase (euros per year) citizens accept per year (household level).

4.2. COVID-19 and Urban Green Spaces

Between the "dilemma" among small and numerous or large and few urban green spaces (Figure 8), there is a slight preference towards the first option, i.e., many small urban green areas (56.7%). This is surely based on the fact that the second option of large urban green spaces is not realizable in urban centers in Greece, where open space is limited. Apart from that, due to COVID-19 concerns, the option of many small urban green spaces would facilitate the dispersion of the residents in urban centers and could prevent the spread of the virus.

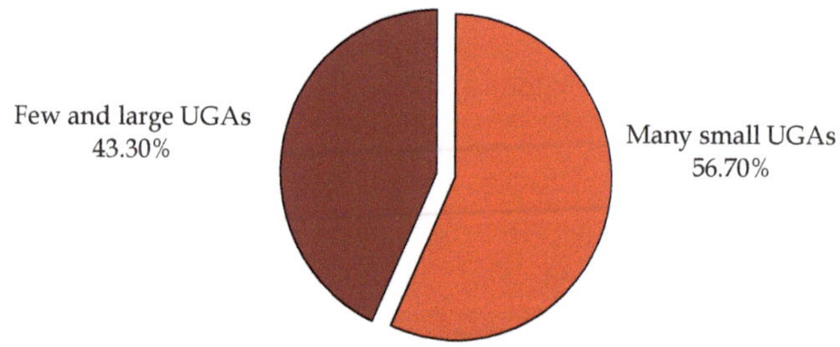

Figure 8. Responses on the perception of safety of small vs. large UGAs in regard to experience from the restriction measures (quarantine) to minimize the spread of COVID-19 (research question: "Based on your experience from the restriction measures (quarantine) to minimize the spread of COVID-19, do you think it is safer to have a lot small UGAs in your municipality, or few and large ones?").

The visit frequency to urban green areas prior to the COVID-19 outbreak and the subsequent restrictions was varied (Figure 9). In general, 34.5% would visit an urban green space more than once a week, while 23.9% of the respondents would visit it only once a week.

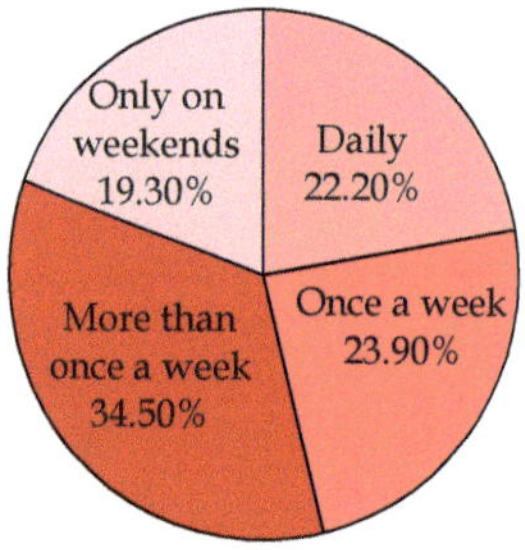

Figure 9. Responses on the frequency of visits of the UGAs before the beginning (March 2020) of COVID-19 restrictions (research question: "How often did you visit the UGAs of your municipality, before the beginning (March 2020) of the COVID-19 restrictions?").

In general, it is obvious that COVID-19 restrictions had an effect on the visit frequency (Figure 10), as 37.7% answered that these had a moderate effect. Nevertheless, 89% of the respondents underlined that urban green areas are critical for the improvement of public health. By analyzing the answers of all respondents, the average score of 4.85/5.00 also suggests that the citizens' perception of UGAs is highly associated with public health, further enforcing the strong bond between citizens' welfare and green infrastructure.

Regarding the perception change due to COVID-19 (Figure 11), 39% of the respondents admitted that this had positively altered their perception in regard to urban green areas being hotspots for improving public health. In general, 78.7% of the respondents acknowledged a moderate (3) to maximum (5) change in their perception of urban green areas. Also, 114 respondents (16% of the total number of participants) answered that their perception of UGAs was less (grade 1) affected regarding the improvement in human public health after the quarantine. It is interesting to note that the great majority (103) of this latest group of citizens also considers that UGAs are of maximum importance for the improvement in human health.

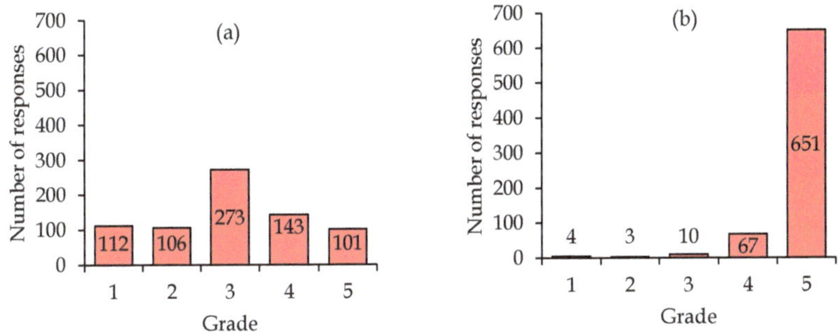

Figure 10. Responses on (**a**) the change of frequency of visits in the UGAs change after the end (May 2020) of COVID-19 restrictions (research question: "How has the frequency of your visits in the UGAs changed after the end (May 2020) of the COVID-19 restrictions?" and (**b**) importance of UGAs, regarding the improvement of human health (research question: "How important do you consider UGAs to be, regarding the improvement of human health?") (grading scale: 1 = minimum to 5 = maximum).

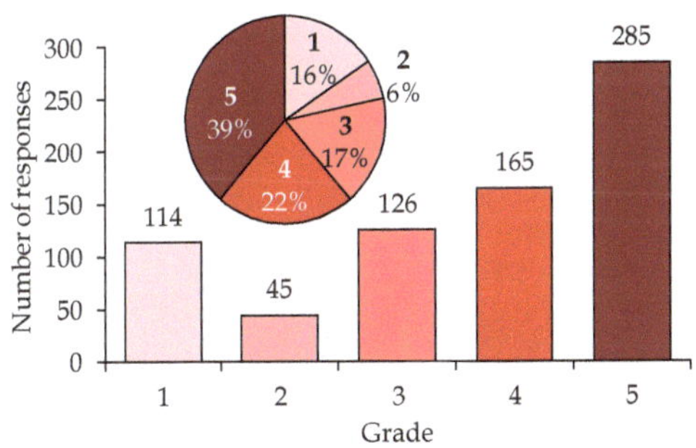

Figure 11. Responses on the degree that lockdown measures against the spread of COVID-19 altered the perception of citizens in regard to urban green areas as a means of improving public health (research question: "How much was your perception about the UGAs affected, regarding the improvement of human public health after (March 2020) the restriction measures (quarantine) for the prevention of spreading COVID-19?"), (grading scale: 1 = minimum to 5 = maximum).

In a trade-off question (Figure 12) asking them to select one of the two options in case a service was offered them gratis in the form of a public good, participants chose by great majority (82%) their access to urban green areas instead of having free access to home internet for each month of lockdown measures. This finding highlights the "use value" of urban green areas as a public good, hinting at high welfare values derived from the existence of urban green areas. It is also interesting to note that about 18% of the participants stated their preference for free internet access at home during the quarantine, against the option for accessing UGAs. About half (52%) of these responders believe that visiting or staying in UGAs is not very safe for the transmittance of diseases and they gave scores less than 3 (1 = least safe–5 = most safe) to the relevant question. In addition, only 12% of this group visits UGAs on a daily frequency, which is half the respective percentage (24%) for the other group of respondents that prefer to have access to UGAs during quarantine.

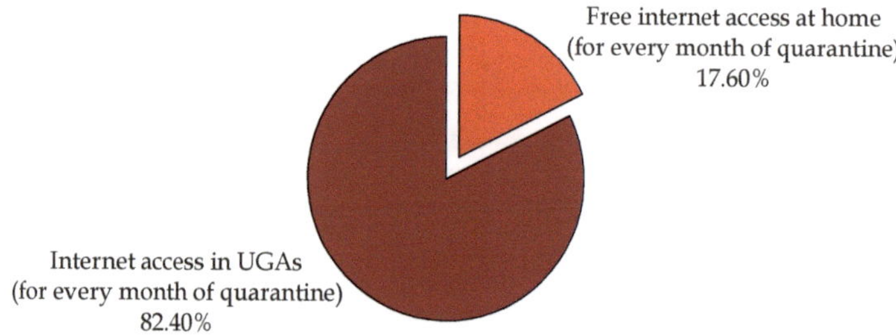

Figure 12. Responses on the selection of trade-off between two options considering lockdown measures against the spread of COVID-19. Research question: "Which of the following 2 options would you choose in case of restriction measures (quarantine) for stopping the spread of COVID-19?"

The next question (In case of restriction measures (quarantine) to help minimize the spread of COVID-19, which of the following services would you choose to be available for the general public? Figure 13) revealed that the great majority of the respondents would opt for access to public parks (41.2%) and urban forests (34.6%). This is mainly influenced by the curfew measures due to COVID-19, where residents were restricted to their homes. Therefore, the need of going to open spaces and spend some time there was preferable to other options that included visits to other spaces, such as sport centers and busy streets with shops. The high preference of respondents for the green (76%) against the blue (17%) infrastructures is also notable, and is possibly attributed to the season (late autumn and early winter) of the research, since during this period of the year, the general public indicates an increase interest for green-related activities. In all cases, however, the green and blue infrastructure facilities comprised about 93% of the respondents' preferences.

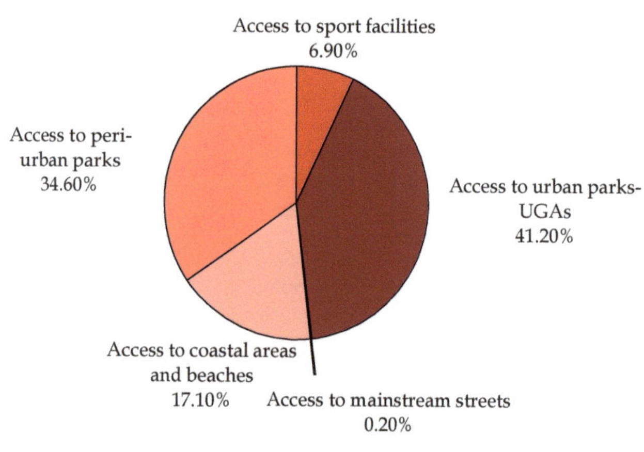

Figure 13. Responses on the preference of services that citizens prefer to be available to the wider public in case of lockdown measures against the spread of COVID-19 (research question: "In case of restriction measures (quarantine) to help minimize the spread of COVID-19, which of the following services would you choose to be available for the general public?").

Lastly, citizens appear to feel quite safe in regard to COVID-19 transmission during their presence in urban green areas (Figure 14). More specifically, 72.2% of the respondents assessed their visit to urban green areas as very safe (4) or even safest (5). It is worth noting that 54% of these citizens considered that in their municipality, many and small UGAs were safer regarding the transmission of COVID-19 against fewer and larger ones, whereas the rest (46%) supported the opposite. For the 28% of the respondents that gave a score less than 3 (1 = least safe to 5 = most safe) when evaluating how safe they feel in UGAs, the abovementioned percentages concerning the number and the magnitude of the UGAs give a clearer picture, since this citizens' group considers the many and small UGAs safer in terms of preventing COVID-19 transmission by 63%.

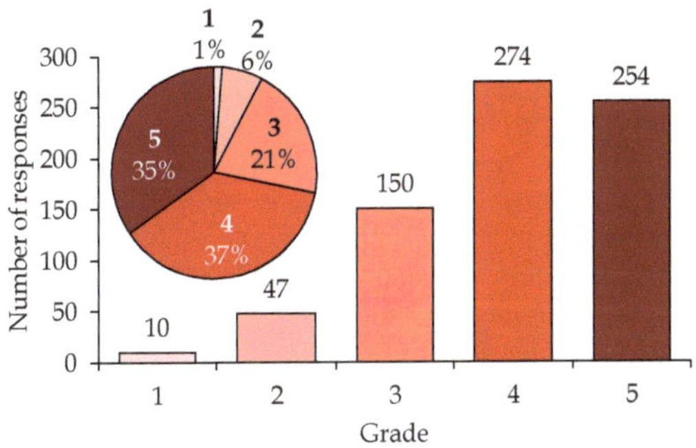

Figure 14. Responses on citizens' perception of safety of presence in urban green areas in regard to disease transmission (research question: "How safe do you consider the visiting/staying in urban green areas regarding the transmission of diseases?"), (grading scale: 1 = least safe to 5 = most safe).

5. Discussion

Among the key findings of the study, of notable importance is that citizens saw UGAs as an important means for improving public health, while citizens were willing to accept a certain increase in their municipal taxes for improving the services offered by UGAs. It is interesting to note that the main reasons behind this acceptance were basically environmentally related, such as climate change or environmental conservation. A further important finding of the survey is the dilemma between free internet access and free access to UGAs. The clear preference for the latter option, i.e., free access to UGAs, was partially biased by the COVID-19 restrictions. It should be noted that conducting this study in the period following the strictest curfews of the COVID-19 pandemic in Greece led to a significant bias towards the perception of UGAs in citizens' perception. The choice of the trade-off question was utilized as a means for controlling this bias, and indeed it was found that citizens indicated a much higher degree of preference towards the free access to UGAs. Combined with the tandem acceptance of an increase in their municipal taxes, this clear preference reveals that in situations of public health emergencies, such as the COVID-19 pandemic, UGAs emerge as a means to fulfil the citizens' physical, cognitive and emotional benefits. In that way, the role of UGAs as a public good and the "vehicle" for increasing the citizens' well-being is highlighted. Further interesting findings include the citizens' acknowledgment that UGAs are critical for public health, while the majority supported that the COVID-19 restrictions substantially affected their perception of UGAs. To that end, the vast majority of the respondents perceived UGAs as safe spaces regarding the transmission of COVID-19.

It has already been presented how UGAs have a positive influence on citizens' well-being [2,3,10,11] and how difficult its codification through a single metric can be [12]. This influence is surely multidimensional and there is further need for research [13,14]. In general, this preliminary study should take into careful account the contribution of UGAs to the urban environment along with the possible positive effects on citizens' quality of life and positive influence on citizens' well-being through the activities that could be carried out in UGAs, e.g., recreation and socialization [14]. Additionally, it has been apparent from other studies, as well as this research, that COVID-19 curfews gave a "window of opportunity" so as to highlight the importance of UGAs [37–42]. More specifically, it was shown that UGAs emerged as places where citizens' physical, cognitive and emotional need were fulfilled and satisfied. On the contrary, access restriction to UGAs or a lack of UGAs in the vicinity of citizens' residence resulted in feelings of isolation and consequently to lower well-being [34–36,42]. Furthermore, the importance of UGAs is also expressed in increased rates of WTP for the construction and maintenance of UGAs [46,47]. Consequently, it can be assumed that the COVID-19 curfew triggered a "crisis," thanks to which the multidimensional and (until then) hidden advantages of UGAs were unveiled.

6. Conclusions

This study's findings highlight the utmost importance of urban green areas for achieving the Sustainable Development Goals, specifically "Goal 3: Good Health and Well-Being" and "Goal 11: Sustainable Cities and Communities." Our results indicate that in a period of both climatic and public health crises, healthy and green urban environments can play a seminal role in alleviating and mitigating different challenges and impacts, while at the same time ensuring sustainability of urban ecosystems. In regard to policy recommendations, the current study's findings emphasize the need to further investigate the socioeconomic importance of urban green areas from an ecosystemic and public health perspective considering the emerging challenges of COVID-19. The results of this study could be utilized both by researchers and policy makers and similar studies should be mainstreamed for policy makers and experts in order to assist the design of healthier and safer municipalities according to existing initiatives at national and international levels.

It should be noted that responsible local and/or regional authorities and experts should take into account all the necessary parameters related to the realization of UGAs, i.e., the benefits, the possible spillover effects, the emergent threats, as well as the management costs. Therefore, a comprehensive study regarding the realization and maintenance of UGAs of all categories, i.e., urban parks or forests, should be carried out beforehand. This should be the starting point before the composition and implementation of any plan regarding UGA management.

It becomes evident that policy makers should incorporate climate governance in the management of green infrastructure at the local level through the establishment of integrated policy frameworks focusing on urban green areas. This implies considering open urban spaces not as isolated units, but as vital elements of the urban landscape with their own contribution to the goals of sustainability and the adaptation and mitigation of the effects of climate change.

The key limitations of this study are associated with the both the respondents' profile, mainly male, university graduates and the exact place of residence of participants, since the city typology might influence the accessibility to UGAs, citizens' perceptions of UGA importance, as well as respondents' age, which in web surveys could be linked to relatively younger respondents.

Author Contributions: Conceptualization, A.K., G.M., N.P. and A.D.S.; methodology, A.K., G.M., N.P., A.D.S., E.V.A., E.K., A.B.K. and C.G.; formal analysis, A.K., G.M., and N.P.; investigation, A.K., G.M., A.D.S., N.P., A.B.K. and C.G.; resources, A.K., G.M. and N.P.; data curation, A.K., G.M., E.P. and N.P.; writing—original draft preparation, A.K., G.M. and N.P.; writing—review and editing, A.K., A.D.S., N.P., E.V.A., E.K., G.K., A.B.K., K.K., C.G., K.L., K.T., G.M. and E.P.; visualization, A.K., K.T. and A.D.S.; supervision, A.K., A.D.S. and K.T.; funding acquisition, K.T. All authors have read and agreed to the published version of the manuscript.

Funding: This research was funded by the LIFE GrIn project "Promoting urban Integration of GReen INfra-structure to improve climate governance in cities" LIFE17 GIC/GR/000029, which is co-funded by the European Commission under the Climate Change Action-Climate Change Governance and Information component of the LIFE Programme and the Greek Green Fund.

Data Availability Statement: Not applicable.

Acknowledgments: The authors greatly acknowledge all the involved partners of the the GrIn project "Promoting urban integration of GReen INfrastructure to improve climate governance in cities" LIFE17 GIC/GR/000029.

Conflicts of Interest: The authors declare no conflict of interest.

References

1. Špes, M. Mesto kot ekosistem. *Dela* **2009**, *31*, 5–20. [CrossRef]
2. Tyrväinen, L.; Väänänen, H. The economic value of urban forest amenities: An application of the contingent valuation method. *Landsc. Urban Plan.* **1998**, *43*, 105–118. [CrossRef]
3. Langemeyer, J.; Baró, F.; Roebeling, P.; Gómez-Baggethun, E. Contrasting values of cultural ecosystem services in urban areas: The case of park Montjuïc in Barcelona. *Ecosyst. Serv.* **2015**, *12*, 178–186. [CrossRef]
4. *WHO Manifesto for a Healthy Recovery from COVID-19: Prescriptions and Actionables for a Healthy and Green Recovery*; World Health Organization: Geneva, Switzerland, 2020. Available online: https://www.who.int/news-room/feature-stories/detail/who-manifesto-for-a-healthy-recovery-from-covid-19 (accessed on 31 October 2022).
5. UNFCCC-Decision-/CP.27 Sharm el-Sheikh Implementation Plan. Sharm el-Sheikh, Egypt, 19 November 2022. Available online: https://unfccc.int/sites/default/files/resource/cop27_auv_2_cover%20decision.pdf (accessed on 23 November 2022).
6. Millennium Ecosystem Assessment. World Resources Institute. In *Ecosystems and Human Well-Being: Synthesis*; Island Press: Washington, DC, USA, 2005.
7. Plan, NYC Green Infrastructure. *NYC Green Infrastructure Plan: A Sustainable Strategy for Clean Waterways*; Plan, NYC Green Infrastructure: New York, NY, USA, 2010; Available online: https://smartnet.niua.org/sites/default/files/resources/NYCGreenInfrastructurePlan_LowRes.pdf (accessed on 10 June 2021).
8. Hartje, V. Assessment and evaluation of economic effects of green urban infrastructure. In Proceedings of the 6th Sino-German Workshop on Biodiversity Conservation, Bonn, Germany, 15–18 July 2013.
9. Haasse, D.; Frantzeskaki, N.; Elmqvist, T. Ecosystem services in urban landscapes: Practical Applications and governance implications. *Ambio* **2014**, *43*, 407–412. [CrossRef] [PubMed]
10. Harnik, P.; Crompton, J.L. Measuring the total economic value of a park system to a community. *Manag. Leis.* **2014**, *19*, 188–211. [CrossRef]
11. Brander, L.M.; Koetse, M.J. The value of urban open space: Meta-analyses of contingent valuation and hedonic pricing results. *J. Environ. Manag.* **2011**, *92*, 2763–2773. [CrossRef]
12. Kabisch, N.; Strohbach, M.; Haase, D.; Kronenberg, J. Urban green space availability in European cities. *Ecol. Indic.* **2016**, *70*, 586–596. [CrossRef]
13. Kolimenakis, A.; Solomou, A.D.; Proutsos, N.; Avramidou, E.V.; Korakaki, E.; Karetsos, G.; Maroulis, G.; Papagiannis, E.; Tsagkari, K. The Socioeconomic Welfare of Urban Green Areas and Parks; A Literature Review of Available Evidence. *Sustainability* **2021**, *13*, 7863. [CrossRef]
14. Solomou, A.D.; Topalidou, E.T.; Germani, R.; Argiri, A.; Karetsos, G. Importance, Utilization and Health of Urban Forests: A Review. *Not. Bot. Horti Agrobot.* **2019**, *47*, 10–16. [CrossRef]
15. Tsiros, I.X.; Nastos, P.; Proutsos, N.D.; Tsaousidis, A. Variability of the aridity index and related drought parameters in Greece using climatological data over the last century (1900–1997). *Atmos. Res.* **2020**, *240*, 104914. [CrossRef]
16. Proutsos, N.; Tigkas, D. Growth response of endemic black pine trees to meteorological variations and drought episodes in a Mediterranean region. *Atmosphere* **2020**, *11*, 554. [CrossRef]
17. Proutsos, N.; Liakatas, A.; Alexandris, S.; Tsiros, I. Carbon fluxes above a deciduous forest in Greece. *Atmósfera* **2017**, *30*, 311–322. [CrossRef]
18. Proutsos, N.; Korakaki, E.; Bourletsikas, A.; Solomou, A.; Avramidou, E.V.; Georgiadis, C.; Kontogianni, A.B.; Tsagari, K. Urban temperature trends in east Mediterranean: The case of Heraklion-Crete. *Eur. Water* **2020**, *69*, 3–14.
19. Proutsos, N.; Tsagari, C.; Tsaousidis, A.; Tsiros, I.X. Water availability changes for natural vegetation development in the mountainous area of Metsovo (N. Greece) for the period 1960–2000. In Proceedings of the 15th International Conference on Meteorology, Climatology and Atmospheric Physics—COMECAP 2021, Ioannina, Greece, 26–29 September 2021; pp. 564–568. Available online: https://www.conferre.gr/allevents/comecap2020/Proceedings_Final.pdf (accessed on 31 October 2022).
20. Proutsos, N.; Korakaki, E.; Bourletsikas, A.; Kaoukis, K.; Georgiadis, C. Analyzing temperature attributes for the last half century in Heraklion–Crete, Greece. In Proceedings of the 11th World Congress on Water Resources and Environment: Managing Water Resources for a Sustainable Future—EWRA 2019 Proceedings, Madrid, Spain, 25–29 June 2019; pp. 477–478. Available online: https://inis.iaea.org/search/search.aspx?orig_q=RN:52098631 (accessed on 31 October 2022).
21. Proutsos, N.D.; Tigkas, D.; Tsevreni, I.; Tsevreni, M. Drought assessment in Nestos river basin (N. Greece) for the period 1955–2018. In Proceedings of the 10th International Conference on Information and Communication Technologies in Agriculture, Food and Environment, HAICTA 2022, CEUR Workshop Proceedings, Athens, Greece, 22–25 September 2022; pp. 429–437. Available online: https://ceur-ws.org/Vol-3293/paper85.pdf (accessed on 11 October 2022).
22. Proutsos, N.D.; Solomou, A.D.; Koulelis, P.P.; Bourletsikas, A.; Chatzipavlis, N.E.; Tigkas, D. Detecting changes in annual precipitation trends during the last two climatic periods (1955–1984 and 1985–2018) in Nestos River basin, N Greece. In Proceedings of the 10th International Conference on Information and Communication Technologies in Agriculture, Food and Environment, HAICTA 2022, CEUR Workshop Proceedings, Athens, Greece, 22–25 September 2022; pp. 456–463. Available online: https://ceur-ws.org/Vol-3293/paper90.pdf (accessed on 11 October 2022).

23. Proutsos, N.D.; Solomou, A.D.; Bourletsikas, A.; Chatzipavlis, N.E.; Petropoulou, M.; Bourazani, K.; Nikolopoulos, J.N.; Georgiadis, C.; Kontogianni, A.K. Assessing drought for the period 1955–2021 in Heraklion-Crete (S. Greece) urban environment. In Proceedings of the 10th International Conference on Information and Communication Technologies in Agriculture, Food and Environment, HAICTA 2022, CEUR Workshop Proceedings, Athens, Greece, 22–25 September 2022; pp. 464–471. Available online: https://ceur-ws.org/Vol-3293/paper91.pdf (accessed on 11 October 2022).
24. Proutsos, N.D.; Solomou, A.D.; Tigkas, D. Decadal variation of aridity and water balance attributes at the urban and peri-urban environment of Attica-Greece. In Proceedings of the 10th International Conference on Information and Communication Technologies in Agriculture, Food and Environment, HAICTA 2022, CEUR Workshop Proceedings, Athens, Greece, 22–25 September 2022; pp. 472–477. Available online: https://ceur-ws.org/Vol-3293/paper92.pdf (accessed on 11 October 2022).
25. Giannopoulou, K.; Livada, I.; Santamouris, M.; Saliari, M.; Assimakopoulos, M.; Caouris, Y. On the characteristics of the summer urban heat island in Athens, Greece. *Sustain. Cities Soc.* **2011**, *1*, 16–28. [CrossRef]
26. Livada, I.; Santamouris, M.; Niachou, K.; Papanikolaou, N.; Mihalakakou, G. Determination of places in the great Athens area where the heat island effect is observed. *Theor. Appl. Climatol.* **2002**, *71*, 219–230. [CrossRef]
27. Skoulika, F.; Santamouris, M.; Kolokotsa, D.; Boemi, N. On the thermal characteristics and the mitigation potential of a medium size urban park in Athens, Greece. *Landsc. Urban Plan.* **2014**, *123*, 73–86. [CrossRef]
28. Zoulia, I.; Santamouris, M.; Dimoudi, A. Monitoring the effect of urban green areas on the heat island in Athens. *Environ. Monit. Assess.* **2009**, *156*, 275–292. [CrossRef]
29. Proutsos, N.D.; Solomou, A.D.; Petropoulou, M.; Chatzipavlis, N.E. Micrometeorological and Hydraulic Properties of an Urban Green Space on a Warm Summer Day in a Mediterranean City (Attica–Greece). *Land* **2022**, *11*, 2042. [CrossRef]
30. Melas, E.; Tsiros, I.; Thoma, E.; Proutsos, N.; Pantavou, K.; Papadopoulos, G. An assessment of microclimatic conditions inside vegetated and non-vegetated small-scale open spaces in the Athens urban environment. In Proceedings of the 15th International Conference on Meteorology, Climatology and Atmospheric Physics—COMECAP 2021, Ioannina, Greece, 26–29 September 2021; pp. 269–273. Available online: https://www.conferre.gr/allevents/comecap2020/Proceedings_Final.pdf (accessed on 31 October 2022).
31. Tsiros, I.X.; Hoffman, M.E. Thermal and comfort conditions in a semi-closed rear wooded garden and its adjacent semi-open spaces in a Mediterranean climate (Athens) during summer. *Archit. Sci. Rev.* **2014**, *57*, 63–82. [CrossRef]
32. Tsiros, I.X. Assessment and energy implications of street air temperature cooling by shade tress in Athens (Greece) under extremely hot weather conditions. *Renew. Energy* **2010**, *35*, 1866–1869. [CrossRef]
33. WHO. Urban Green Spaces and Health. A Review of Evidence. World Health Organization WHO—Regional Office for Europe Report. 2017. Available online: https://www.euro.who.int/__data/assets/pdf_file/0005/321971/Urban-green-spaces-and-health-review-evidence.pdf (accessed on 1 April 2021).
34. Vatavali, F.; Gareiou, Z.; Kehagia, F.; Zervas, E. Impact of COVID-19 on Urban Everyday Life in Greece. Perceptions, Experiences and Practices of the Active Population. *Sustainability* **2020**, *12*, 9410. [CrossRef]
35. Mouratidis, K.; Yiannakou, A. COVID-19 and urban planning: Built environment, health, and well-being in Greek cities before and during the pandemic. *Cities* **2022**, *121*, 103491. [CrossRef] [PubMed]
36. Mayen Huerta, C.; Gianluca, C. Snapshot of the Use of Urban Green Spaces in Mexico City during the COVID-19 Pandemic: A Qualitative Study. *Int. J. Environ. Res. Public Health* **2021**, *18*, 4304. [CrossRef] [PubMed]
37. Venter, Z.S.; Barton, D.N.; Gundersen, V.; Figari, H.; Nowell, M.S. Back to nature: Norwegians sustain increased recreational use of urban green space months after the COVID-19 outbreak. *Landsc. Urban Plan.* **2021**, *214*, 104175. [CrossRef]
38. Venter, Z.S.; Barton, D.N.; Gundersen, V.; Figari, H.; Nowell, M. Urban nature in a time of crisis: Recreational use of green space increases during the COVID-19 outbreak in Oslo, Norway. *Environ. Res. Lett.* **2020**, *15*, 104075. [CrossRef]
39. Derks, J.; Giessen, L.; Winkel, G. COVID-19-induced visitor boom reveals the importance of forests as critical infrastructure. *For. Policy Econ.* **2020**, *118*, 102253. [CrossRef]
40. Day, B.H. The Value of Greenspace Under Pandemic Lockdown. *Environ. Resour. Econ.* **2020**, *76*, 1161–1185. [CrossRef]
41. Noszczyk, T.; Gorzelany, J.; Kukulska-Kozieł, A.; Hernik, J. The impact of the COVID-19 pandemic on the importance of urban green spaces to the public. *Land Use Policy* **2022**, *113*, 105925. [CrossRef]
42. Ugolini, F.; Massetti, L.; Calaza-Martínez, P.; Cariñanos, P.; Dobbs, C.; Ostoić, S.K.; Marin, A.M.; Pearlmutter, D.; Saaroni, H.; Šaulienė, I.; et al. Effects of the COVID-19 pandemic on the use and perceptions of urban green space: An international exploratory study. *Urban For. Urban Green.* **2020**, *56*, 126888. [CrossRef]
43. Thornthwaite, C.W. An approach toward a rational classification of climate. *Geogr. Rev.* **1948**, *38*, 55–94. [CrossRef]
44. UNEP. *World Atlas of Desertification*; United Nations Environment Program: London, UK, 1992.
45. Proutsos, N.D.; Tsiros, I.X.; Nastos, P.; Tsaousidis, A. A note on some uncertainties associated with Thornthwaite's aridity index introduced by using different potential evapotranspiration methods. *Atmos. Res.* **2021**, *260*, 105727. [CrossRef]
46. Kalfas, D.; Chatzitheodoridis, F.; Loizou, E.; Melfou, K. Willingness to Pay for Urban and Suburban Green. *Sustainability* **2022**, *14*, 2332. [CrossRef]
47. Mandziuk, A.; Stangierska, D.; Fornal-Pieniak, B.; Gębski, J.; Żarska, B.; Kiraga, M. Preferences of Young Adults concerning the Pocket Parks with Water Reservoirs in the Aspect of Willingness to Pay (WTP) in Warsaw City, Poland. *Sustainability* **2022**, *14*, 5043. [CrossRef]

land

MDPI

Article

Old-Growth Forests in Urban Nature Reserves: Balancing Risks for Visitors and Biodiversity Protection in Warsaw, Poland

Andrzej Długoński [1], Thilo Wellmann [2,*] and Dagmar Haase [2,3]

1 Institute of Biological Sciences, Faculty of Biology and Environmental Sciences, Cardinal Stefan Wyszynski University, 01-938 Warsaw, Poland
2 Landscape Ecology Lab, Geography Department, Humboldt-Universität zu Berlin, Unter den Linden 6, 10099 Berlin, Germany
3 Department of Computational Landscape Ecology, Helmholtz Centre for Environmental Research, UFZ, 04318 Leipzig, Germany
* Correspondence: thilo.wellmann@geo.hu-berlin.de

Highlights:

- Airborne LiDAR data are used to assess defoliation in old-growth forest stands in Warsaw, Poland
- It was found that defoliated trees are often of old age and cause severe risks to visitors in the two assessed urban nature reserves.
- Both reserves were affected by a severe winter storm and closed to visitors and old trees valuable for biodiversity were cut at a high rate, respectively.
- The proposed methodology helps predict risks to visitors and can be used to develop precise visitor management, visitor information, and other targeted measures to ensure trade-offs between biodiversity potential and visitor risks are minimal.

Abstract: Urban nature reserves in Poland are precious relics of ancient nature with preserved biodiversity. They consist of valuable trees several 100 years old, are biodiverse, and are valuable recreational spaces right in and around cities. It is therefore critical to manage tradeoffs between visitor safety due to, e.g., falling dead branches and the need for old-grown trees for biodiversity conservation. This study aimed to determine whether airborne laser scanning data (LiDAR) can confirm that trees exhibiting the worst crown defoliation are the first to be damaged in storms. Our results show that during Storm Eunice in 2022, the detected defoliated trees, in fact, were damaged the most. Despite such evidence available to the city, no targeted changes to the management of the reserves were taken after the storm. One of the forests was completely closed to visitors; in the other forest, areas with damaged trees were fenced off, and then, the remaining branches and fallen trees were removed to make the forest available for recreation. Using available evidence such as LiDAR data, we propose more targeted and nuanced forms of managing biodiversity conservation in conjunction with visitor safety. This includes the establishment of priority areas, visitor information, and visitor management. This way, airborne laser scanning and Geographic Information Systems can be used to balance management needs accounting for both biodiverse old-grown forest structures while at the same time providing added safety for visitors.

Keywords: LiDAR; urban forest; biodiversity protection; urban green infrastructure; risk management; remote sensing

1. Introduction

Urban forests are valuable sources of nature and biodiversity of flora and fauna in cities, providing psychological and recreational benefits that contribute positively to the well-being of individuals [1–5]. These benefits often indicate their importance as a place of rest, such as events, tourism, and the way to get to work or spend free time [1,6].

Urban forests can generate significant ecosystem services (ES) such as ecological benefits (e.g., wildlife habitat and connectivity) [7,8], economic benefits (e.g., improved hospital recovery times and reduced crime rates) [9,10], offsetting carbon emission [1], removing air pollutants, reducing noise, regulating the local climate [11–13], health [13,14] cultural benefits (e.g., recreation), esthetic values, and amenity [5,15–17]. In this direction, linking ES and people's preferences to forest recreation can be a component of supporting and informing socio-ecological interfaces around recreation in biodiverse areas [13,18].

Forests are of outstanding importance in biodiversity protection in the urban environment, especially in highly urbanized areas where natural habitat remains scarce otherwise. Multiple red-list and endangered species can be found in urban forests [18,19]. Especially important in this regard are old trees, amplifying the value of urban forests as refuges of biodiversity. At the same time those forests are often also active places of recreation Maintaining the biodiversity that old trees can support is associated with limited interference through care works, e.g., cutting down old trees or fallen branches. This in turn can pose threats to the safety of visitors, can hinder recreation in forests, and can cause injuries. Managers often exclude these places from recreation, which makes outdoor moving (such as running, walking, and cycling) in urban forests difficult and causes public dissatisfaction [20].

This highlights that urban forests face a multitude of management challenges, ranging from implications of climate change that lead to an uncertain and complex future for urban forests [21,22] to impoverished or disturbed soils to strategic challenges such as the lack of relevant policies and inadequate operating budgets [23,24]. The combination of the issues of management and benefits of urban forests and natural protected areas with their recreational function, especially in Poland, is an increasing problem and challenges city managers [18,25].

Remote sensing (RS) has been developing intensively in recent years and provides a range of solutions that can be successfully used for inventory and monitoring of greenery in the city, mainly trees. RS is engaged in the acquisition and processing of data obtained though the registration of emitted electromagnetic radiation with the use of specialized sensors. This data can be recorded from different platforms: satellites, aircraft, air and ground unmanned vessels [26–30].

In this study, two forests in Warsaw were analyzed: the Bielański Forest and the Kabacki Forest, which are under species protection (nature reserves) and are recreational areas of Warsaw. Moreover, due to numerous storms, together with compound hot and dry events in recent years, urban forests in Poland have been damaged. Therefore, research on LiDAR (light detection and ranging) and its products, such as the High Structural Diversity Tree Crown Map for Warsaw from 2018 to 2020, is a new tool that can find new applications in the monitoring of larger urban green infrastructures and urban forests [26].

In this study, we focus on the biodiversity and recreation by inhabitants in two urban forests (nature reserves) in Warsaw in Poland. This dissimilarity needs to be addressed especially in the cases included, where the forest area has recreational functions but also has intensive tree growth, where human interference in care is limited due to legal regulations. Hence, the areas where compromises appear or may appear and the priority setting for forest health, people's safety, tree species patterns, and forest management were specified in this study.

In urban nature reserves, both the value for biodiversity and the accessibility for visitors need to be regarded in management. In cities the need for recreation and leisure in natural areas is high due to the lack of similar areas for active recreation in cities, and the only way for some urban residents to get in contact with nature which could be limited due to nature preservation of local law [4,18,20,27–34].

Safeguarding visitors is particularly important in the context of the increasing frequency of weather anomalies, the "new climate normal" [4,35]. Knowing that nature reserves are subjected to minimal care activities in Poland, also their health monitoring is rather marginal, which distinguishes these places from other green parks or street trees, which are much better monitored and examined in the context of plant health needs [4,31,32]. It is this issue that we decided to undertake the research in this article; specifically, we ask the following research questions:

1. What is the species diversity of the defoliated trees against the background of the total forest trees in the areas under consideration?
2. What is the risk pattern in both forest reserves associated with defoliated trees?
3. What is the impact of Storm Eunice (Feb. 2022), and which species were most severely affected?
4. How accurately can we predict the risks for visitors using ALS (airborne laser scanning) data, and what suggestions for management can we derive from them for the combined protection of visitors and biodiversity?

2. Literature Review: The Use of Tree Crown Mapping to Monitor Urban Nature Reserves

Biodiversity in urban forests faces increasing pressures due to human and natural influences that alter vegetation structure. Because of the inherent difficulty of measuring forested vegetation's multidimensional structure on the ground assessments are often restricted to local measurements [33]. Our study fills this gap using remote sensing in the management of biodiversity-related issues, which is underrepresented in urban RS literature [27–29,34–37]. Sensor data of places known to people can broaden discussion regarding climate change or the role of forest trees in cities [35,38,39], making monitoring technologies and the data they produce valuable [40]. Thus, forest management in urban and peri-urban areas can largely benefit from air-/space-borne methods as the effects of LiDAR can be widespread, whereas forest staff methods and tools to monitor the health of trees are often limited [40–50]. What is more, this subject seems reasonable to find a compromise between the protection of forest biodiversity and the possible ways of active recreation of visitors and their safety [13]. It can be presumed that a clear reason for considering the problems of urban ecology is that a "forest is not equal to a forest" due to threats and disruptions of forest functions understood as ecosystem services.

Airborne laser scanning is a tool for detailed local assessments and can be used for tree crown mapping [26–29,51–53]. In the RS literature, we can find papers using LiDAR as a measure of vegetation cover [54]. These data are often understood under various names, e.g., as airborne LiDAR data [55], and each time require checking the data acquisition, e.g., Sentinel-2 [45,56] or ALS [57]. In Poland, LiDAR is most often used in environmental monitoring by foresters [56–59]. Globally, it is often used to measure forest density, e.g., in Australia [54], or measurements of woodlands using GEOBIA and the program from Worldview 2 [60] and ecosystems by using high-resolution imagery through object-based image analysis [61], e.g., forests in Africa. A category of novelty in environmental monitoring is mapping the biodiversity of heterogeneous landscapes using LiDAR and ecoacoustics [62] and mapping canopy dominant trees based on some indicators, i.e., temperature [63]. It is becoming more and more common to measure street trees using hyperspectral imagery [50,64] e.g., in some cities in Central Europe like Leipzig [4]; as well as to measure selected green areas using UAV (Unmanned Aerial Vehicle) [65], e.g., SOWA, often in Poland use to control also AQI (Air Quality Index) [57].

As is presented in RS literature, the papers point out that LiDAR technology could be used to obtain spatial information on forest characteristics, but tree species identification is still challenging. However, there are a few papers indicating the biodiversity of green areas, especially large ones, e.g., on a regional scale in China [53] or on a city scale in Poland in Warsaw [26–29]. In the case of the High Structural Diversity Tree Crown Map for Warsaw, due to the large area of the city (517 km^2), ALS is divided into two stages. First, in 2018, the northern part of the city was raided, including the Bielański Forest. Secondly, in 2020, the southern part of the city, including Kabacki Forest, was raided. In total, an inventory of 9,000,000 pts. of trees was measured by ALS in Warsaw [65]. During the ALS, the following conditions had to be observed: adequate sunny weather during the vegetation period and clear skies. The map is originally called the Tree Crown Map (pl. "Mapa koron drzew"). It is also available online in basic views. In our research, its name is refined due to the parameters showing the species differentiation of trees [66]. This makes it possible to distinguish the ALS map from similar maps such as the National Tree Crown Map (pl. "Krajowa mapa koron drzew" or "mapa koron drzew"), which shows the site globally accessible through RS online monitoring but shows only the height and distribution of trees based on Sentinel-2 satellite data [67]. Besides all those developments in remote sensing the topic of biodiversity assessments in urban nature reserves with the application of LiDAR is still to be explored.

3. Case Introduction

In our study we assessed the Bielański Forest and the Kabacki Forest, which are both located in the city of Warsaw (area 52,720 ha, population: 1,860,281) in Central Poland. Both cases are relics of the former Mazovian fluff, with the oldest trees ranging in age from 300 to 400 years [68–71]. These forests are nature reserves where human interference is ought to be reduced. Care treatments in these forests are carried out sporadically, which are mostly limited to the removal of dry fallen branches on pedestrian and bike paths. The guidelines for landscape protection (pl. Plan ochrony lasu) [72] of nature reserves are based on minimizing these threats by establishing a new protection plan, which is an obligatory document for nature reserves in Poland.

In the Bielański Forest, nature protection is additionally necessary due to excessive visitor movements because of the location of a university in the center of the forest and active forms of recreation by residents. In Poland, it is common to use forests for recreational or tourist purposes due to the lack of green areas that may relieve reserves from anthropogenic pressure [73,74]. The Bielański Forest nature reserve is in the Bielany District in the northern part of Warsaw (Figure 1A). It covers an area of 130.35 ha. It is under the protection of the Nature 2000 Area (PLH 140041), which protects important European forest habitats and species. It is the remnant of the former Mazovia Primeval Forest, which has 300–400-year-old oak trees and many other tree nature monuments [71].

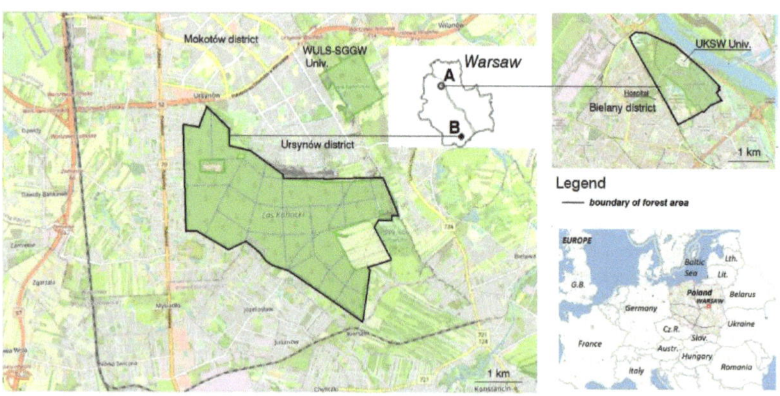

Figure 1. Location of forest areas in Warsaw, Poland: **A**—the Bielański Forest, **B**—the Kabacki Forest (author's elaboration based on: Open Street Maps [75] and Geoportal [76]).

The Kabacki Forest nature reserve is in the Ursynów District in the southern part of Warsaw (Figure 1B). It covers an area of 903.53 ha, which is seven times larger than the Bielański Forest. It is also under nature protection (Warsaw Landscape Protection Area site), with numerous natural monuments and a valuable natural landscape of Mazovia [71].

The effects that Storm Eunice had on Bielański Forest and the Kabacki Forest were analyzed in this study. Storm Eunice was an intense extratropical cyclone in February 2022 doing damage in Western and Central Europe with recorded wind speeds of over 130km/h.

In the Kabacki Forest, Storm Eunice knocked down multiple old trees. This led the local authorities to temporarily close the forest for about one month. In the Bielański Forest, the effect of the storm was weaker, and only a few ill trees were knocked down [69,70]. Hence, it was not closed, but the trees posed a real threat to the inhabitants, similar to the case of the Kabacki Forest.

Both forests are located in the Warsaw escarpment along the Vistula River. They represent two major green spaces for recreation for Warsaw residents, the Bielański Forest in the north and the Kabacki Forest in the south. These two forest areas were most strongly affected by the storm.

4. Methodology

4.1. Research Framework

The research framework of the present study is described below. It consists of, first, a literature review on biodiversity and LiDAR regarding tree crown mapping. Secondly, two similar urban forest areas (urban nature reserves) were selected, both of which are important for recreation in Warsaw. Second, spatial analysis was performed in QGIS, using the High Structural Diversity Crown Map of the selected forests, finding areas with the highest index of defoliation (presented in Tables S1 and S2). Based on the collected data, two steps were done, as follows: (1) biodiversity maps for defoliated trees (range 65–100% defoliation), and (2) the maps of the dangerous trees (defoliated trees) were created for both forests. Based on the highlighted locations, the coordinates of the weakest trees in both cases were localized in the database and mapped in QGIS (an Open-Source Geographic Information System). Using these locations, site observations were performed in tow steps, as follows: (1) field workshops with health condition analyzes of the selected trees, in which deadwood, broken branches, humus diseases, leaning trees, and threats to human health and recreation were evaluated, (2) supplementing the tables of defoliated trees of their health characteristic including height and health conditions were done. Based on GIS information and field mapping hazard maps with dangerous trees were created with safety buffer zones around them. Finally, further directions for development and guidelines for local authorities and citizens and the application of this research in the future in similar cases and scientists in the fields of urban ecology, landscape architecture, forestry, dendrology, and RS were presented (Figure 2).

4.2. Materials

In the years 2018–2020, the municipality of Warsaw city acquired three products from MGGP Aero SA company: (1) laser scanning (ALS), (2) hyperspectral imaging, and (3) multispectral imaging. The Hyspex hyperspectral scanner recorded radiation in the 400–2500 nm range using three sensors. The multispectral resolution of the data was 0.1 m. The number of spectral channels was 3 (RGB). The average scanning density of the LiDAR dataset was 14 points per 1 square meter. Tree crown information (CHM) was obtained from the point cloud [26].

Finally, the database provided by the City of Warsaw included (1) tree crown coverage maps with a table of attributes, (2) tree taxon maps, (3) tree biological condition taxon maps, including indicator maps of discoloration and defoliation, and also Normalized Difference Vegetation index (NDVI), Red Edge Normalized Difference Vegetation Index, Modified Chlorophyll Absorption Ratio Index (MCARI), Structure Intensive Pigment Index, Plant Senescence Reflectance Index (PSRI), and Moisture Stress Index (MSI) [26,77]. The spatial database for both cases (Eng. High Structural Diversity Trees Crown Map, pl. Mapa Koron Drzew m.st. Warszawy, GIS version), consisting of maps developed in the form of individual layers, was searched for the following parameters: tree crown extent maps with attribute tables, tree taxa maps, indicator tree defoliation maps (% defoliation), and indicator tree health maps (ALS) [78]. The High Structural Diversity Tree Crown Map is made available for scientific and didactic purposes for the Bielański Forest (ALS recorded in 2018) and the Kabacki Forest (ALS recorded in 2020). The database is provided as appendices (tables of attributes of defoliated trees for the Bielański Forest—Table S1, and for the Kabacki Forest—Table S2).

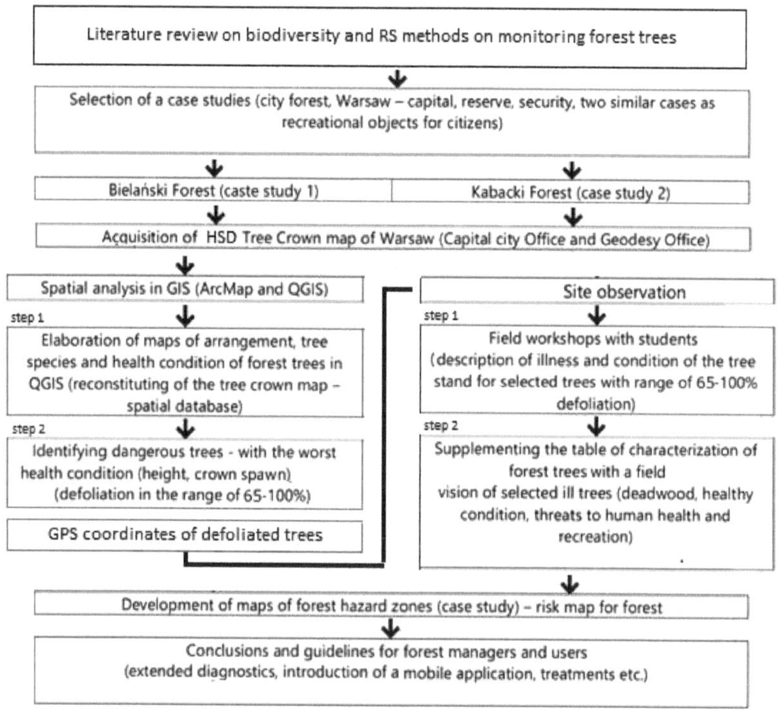

Figure 2. The research framework (author's own elaboration).

4.3. Methods

A mixed-method approach was used in the present study, which is based on the following factors:

- Mapping of trees with high defoliation (65–100%) based on the LiDAR derived information in the High Structural Diversity Tree Crown Map of Warsaw [78]. To classify defoliation levels, 10% steps and then classification using natural breaks depending on data distribution was employed. ArcMap 10.5 and QGIS 9.12. Software was used for map creation. Class breaks were calculated using the natural breaks optimization procedure described by Jenks as illustrated and described in the studies by North et al., Chen, Khamis, et al., and others [77,79,80].
- Elaboration of the database for defoliated trees and risk maps. ArcMap 10.5 and QGIS 9.12.
- Fieldwork: GPS localization of the selected defoliated trees in both forests was carried out. GPS data was recorded with a Garmin GPSMAP 60CSx GPS receiver) and converted using the MapSource software.
- Site observation of the selected defoliated trees in both forests was made according to landscape architecture and urban ecology research conducted by Haase et al. [4] Łukaszkiewicz et al. [31], and Rosłon-Szeryńska et al. [32]. This observation was based on field workshops with students (August 2022). The students verify the defoliated trees selected by LiDAR, determine the health condition characteristics of the defoliated trees (damage to trees, dry branches, tree inclination, tree static, tree diseases, etc. [20,43] as shown in Tables S1 and S2), and indicate the walking and

cycling paths on forest maps, under the supervision of scientists (interdisciplinary team of dendrologists, landscape architects, and foresters).

- The multi-mixed method used in this pilot study allowed us to characterize and assess forest trees in both urban forests. This method allowed first to select the weakest trees (defoliated trees) using the High Structural Diversity Tree Crown Map (ALS) and then to determine the GPS data of these trees. The trees were inspected, and their health condition was assessed to confirm their poor health condition indicated by LiDAR. This way, this method can be applied in forest management. Furthermore, this method is faster than the classic monitoring of trees using observation methods and the detailed inventory of trees as the subject of a separate dendrological or landscape architecture expertise.

5. Results

5.1. Characteristic of Trees in High Structural Diversity Tree Map

The Bielański Forest consists of 12,346 trees. The most frequent taxa are *Quercus, Alnus, Carpinus, Ulmus, Robinia,* and *Acer*. Less common taxa include *Pinus, Betula, Populus, Fraxinus, Aesculus, Salix,* and fruit trees, e.g., *Malus, Pyrus,* or *Prunus*. The oldest tree in the reserves is a *Quercus,* with an age of 400 years.

In the Bielański Forest, out of a total of 12,227 trees, only 83 trees (0.7%) were defoliated trees indicated on the map as ill trees with a range of 65–100% defoliation shown in the High Structural Diversity Tree Crown Map for Warsaw (Figure 3a). Other defoliated trees were *Populus, Acer, Betula, Alnus, Salix, Carpinus,* and fruit trees, e.g., *Malus, Prunus,* and *Pyrus*.

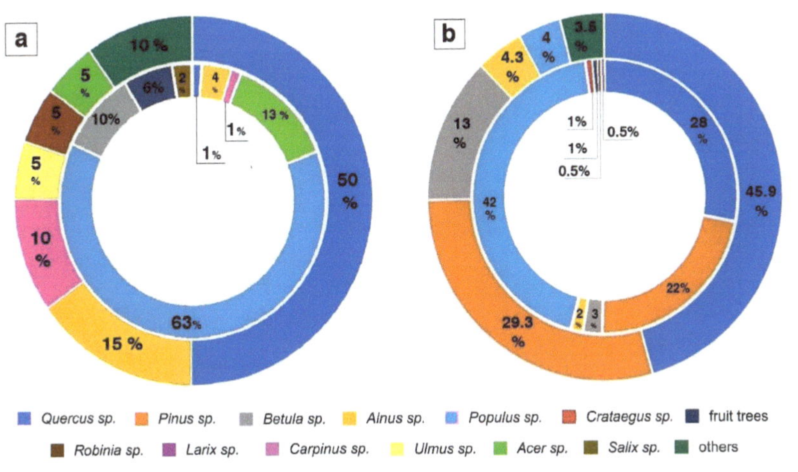

Figure 3. Percentage of defoliated trees (inside ring) on the background of all trees of the forest (outside ring), (**a**)—the Bielański Forest, (**b**)—the Kabacki Forest (author's elaboration based on Database of the High Structural Diversity Tree Crown Map for Warsaw).

The Kabacki Forest consists of 228,734 trees. The most frequent taxa are *Quercus, Pinus, Betula, Alnus,* and *Populus*. Less common trees were *Robinia, Carpinus, Larix, Fraxinus, Ulmus, Crataegus* and fruit trees, e.g., *Malus* and *Prunus*.

Out of a total of 228,734 trees in the Kabacki Forest, only 264 trees (0.2%) were indicated on the map as ill trees with a range of 65–100% defoliation shown in the High Structural Diversity Tree Crown Map for Warsaw. Some trees such as the old specimens of *Quercus* (with an age of 300 years on average) and younger specimens of trees aged about 100 years and below, were short-lived specimens with brittle wood characterization: *Pinus* sp. and *Populus* sp., which are dominant trees. The remaining species of defoliated trees were: *Betula, Alnus, Crataegus, Robinia, Larix,* and fruit trees, e.g., *Malus, Pyrus,* and *Prunus*. The compilation of trees of both forest sites is presented in Table 1 and Figure 3a,b.

Table 1. List of all trees and defoliated trees in the Bielański Forest and the Kabacki Forest.

	Bielański Forest				**Kabacki Forest**			
	All Trees		**Defoliated Trees**		**All Trees**		**Defoliated Trees**	
	picts	%	Picts	%	picts	%	picts	%
Acer	610	5	11	13				
Aesculus	121	1						
Alnus	1867	15			9761	4.3	4	2
Betula	238	2	8	10	29,581	13	6	3
Carpinus	1245	10	1	1	1920	0.8		
Crataegus					230	0.1	2	1
Fraxinus	123	1			1331	0.5		
fruit trees	61	0.5	5	6	829	0.3	2	1
Larix sp.					1420	0.5	1	0.5
Quercus	6226	50	1	1	10,4478	45.9	74	28
Pinus	370	3	3	4	66,674	29.3	59	22
Populus.	125	1.5	52	63	9052	4	115	42
Robinia	618	5			2700	1.1	1	0.5
Salix.	119	1	2	2				
Ulmus.	623	5			758	0.2		
Total	**12,346**	**100**	**83**	**100**	**228,734**	**100**	**263**	**100**

5.2. Field Observation of Defoliated Trees from ALS Data

Field analysis of a total of 345 defoliated trees in both forests selected using the ALS method (range 65–100% defoliation) indicated their health characteristics (Tables S1 and S2). A summary of the poor health condition of the defoliated trees is presented in Tables A1–A3 (Appendix A) and Figures 4 and 5. Based on these data, 16 features of poor health condition of trees were observed: These features and their description are illustrated in Table 2.

Table 2. List of features of poor health condition of trees observed in Bielański Forest and Kabacki Forest.

Feature Number	Feature Description	Most Represented Tree Species	Picts
1	broken or dry falling branches	*Acer, Populus, Quercus*	139
2	roots growing around the trunk	*Pinus, Populus*	25
3	tree inclined adjacent to the path	*Populus*	21
4	conductor broken off	*Populus, Quercus*	16
5	cracks on the trunk	*Populus, Quercus*	9
6	a decaying tree	*Populus, Quercus*	20
7	a rickety tree	*Populus*	26
8	decayed hallows	*Populus, fruit trees*	3
9	dry tree	*Populus, Quercus*	20
10	split at the base of the trunk	*Populus, Pinus, Quercus*	6
11	slightly bent tree	*Populus*	5
12	a thinning tree	*Populus*	11

Table 2. *Cont.*

Feature Number	Feature Description	Most Represented Tree Species	Picts
13	trunks intersect with each other's (with inclination sometimes greater than 45°)	*Populus*, *Quercus*	5
14	a bulky growth on the trunk	*Quercus*, *Alnus*, *Betula*	5
15	tree torn from the ground	*Populus*, *Pinus*, *Betula*	14
16	signs of disease, partially dry needles or leaves	*Populus*, *Quercus*	39

Figure 4. Defoliated trees with damages in the Bielański Forest in Warsaw, Poland: (**A**) fallen branches, (**B**) dried branches, (**C**) tree surgery (concrete guide), (**D**) old fallen branches are not removed from the reserve and are left to nature, (**E**) broken branch of old trees makes it difficult to walk the path in the forest (Photographs by A. Długoński 2022).

Figure 5. Damage to defoliated trees in the Kabacki Forest: (**A**) breakage of conductors because of Storm Eunice; (**B**) breakage of conductors prevents passage of the path; (**C**) fallen conductors and branches during storms ordered by the forest manager; (**D**) tree cuttings; (**E**) tree cancer; (**F**) inclined trees posing a threat to the safety of users (Photographs by A. Długoński 2022).

Among these 16 characteristics, feature 1 (broken or dry falling branches) was the most noticeable in both forests (Figures 4A,D,E, and 5E). These features were most evident in the following tree species: *Acer*, *Quercus*, and *Populus*. Due to the nature of forests as reserves, no attempt was made to care for them. Features 2–16 were less noticeable, mainly due to sanitary cuts made in connection with difficult recreation in more representative parts of both forests (Figures 4B,C and 5B–F). These features were most evident in the following tree species: *Populus*, *Quercus*, *Pinus*, and *Betula*. Based on our field observations, it appears that features 1 and 13 (trunk intersection) were especially dangerous. They were inclined adjacent to the forest paths, posing a threat to forest safety and recreation (Figures 4A,D,E and 5B,F). These features were most evident in the following tree species: *Acer*, *Populus*, and *Quercus*.

As presented in Tables A1–A3 (Appendix A), 59 trees (71% of all defoliated trees) were observed in the Bielański Forest, and 125 trees (48% of defoliated trees) were affected by Storm Eunice because of fractures caused by poor health due to broken branches (feature 1, Figures 4A,D,E and 5C,F), conductor broken off (feature 4, Figure 5B), cracks of the trunk (feature 5), split at the base of the trunk (feature 10, Figure 5A,B), and tree torn from the ground (feature 15, Figure 5D).

The distribution of individual tree species of defoliated trees in the Bielański Forest and the Kabacki Forest is illustrated on maps (Figures 6 and 7). In addition, trees that were near pedestrian and bicycle paths that posed a direct threat to visitors are indicated by red rings (Figures 8 and 9). A total of 25 such trees (30.1% of defoliated trees) were observed in the Bielański Forest, and 66 trees (25.2% of defoliated trees) were observed in the Kabacki Forest.

Figure 6. Location and species differentiation of defoliated trees in the Bielański Forest in Warsaw (Poland) (author's elaboration based on the High Structural Diversity Tree Crown Map database [78]).

5.3. Hazard Maps for Urban Forests

Based on the location of defoliated trees and their characteristics, threat maps were created for both forests. In this map, zones with ill trees that pose a threat to recreation and require monitoring were indicated (Figures 8 and 9) by the local authorities, and the dangerous parts of both sites were restricted to residents.

As illustrated in Figures 8 and 9, around each defoliated tree marked on the map, a hazard zone was created. The range of the hazard zone was calculated in both forests based on the average height of the trees (height average—14.7 m—multiplied by 2 from the center of a given defoliated tree shown on maps). Thus, the average distance for each zone was 29.7 m (a rounded value of 30 m was used), which is a safe distance that should be maintained when securing defoliated trees and separating danger zones.

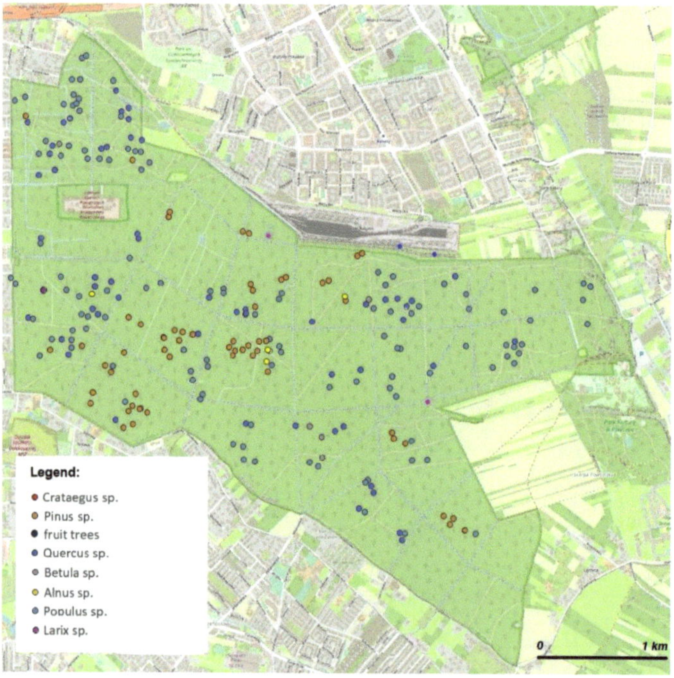

Figure 7. Location and species differentiation of ill trees in the Kabacki Forest in Warsaw (Poland) (author's elaboration based on the High Structural Diversity Tree Crown Map database [78]).

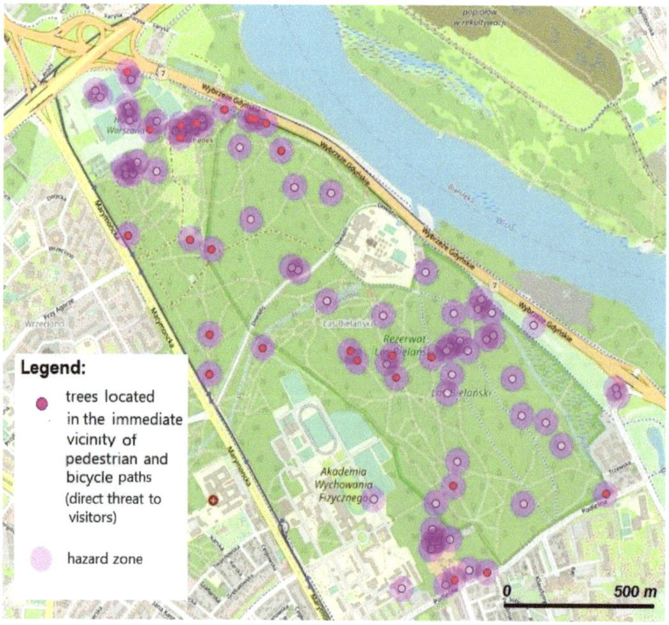

Figure 8. The risk map of the Bielański forest with hazard zones for ill trees (author's elaboration based on the High Structural Diversity Tree Crown Map database [78]).

The location of endangered trees obtained using ALS can determine the range of hazard zones of a given forest area. Examples of such dangers may be shown on maps and smartphone applications (Figure 10) [81]. It is worth noting that such practical applications of urban green infrastructure elements seem to be more popular and are already being tested by researchers, including the development of tourist routes in forests, e.g., Endomondo, MapIT, mLas, tMap or Gpies.com, and tree information systems in urban green spaces in cities, in which the residents themselves are involved [82–85]. This way, in determining the fate of trees, a map of risk zones in forests seems to be a useful instrument in informing the local community about spatial barriers and security threats in forests.

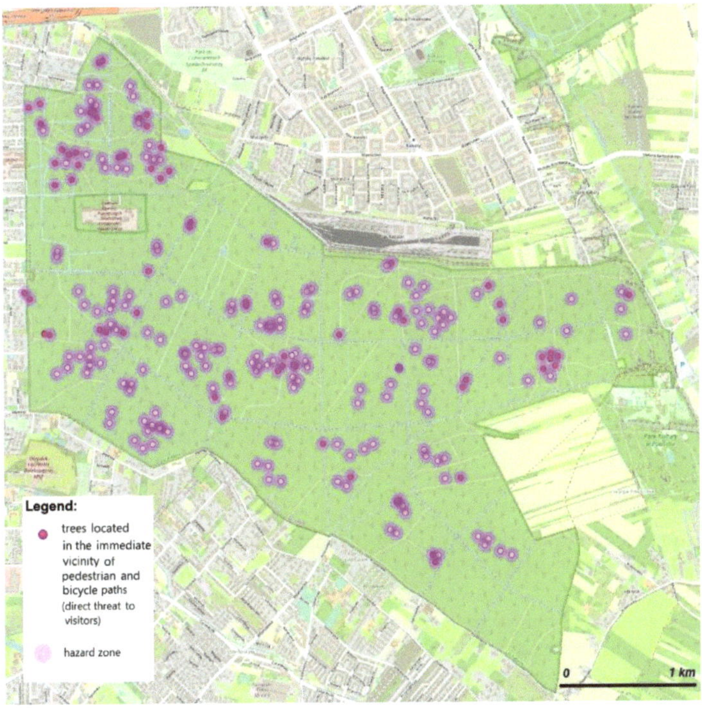

Figure 9. The risk map of the Kabacki Forest with hazard zones for ill trees (author's elaboration based on Open Street Maps [76] and High Structural Diversity Tree Crown Map database [78]).

Finally, analysis of the High Structural Diversity Tree Crown Map for both forests showed that trees in nature reserves are mostly in good condition, and only 345 trees selected as having 65–100% defoliation using ALS (0.1% of all forest trees) are ill trees in both forests. However, they pose threats to visitors, and action must be taken to secure them for recreation and to ensure safety in forests. It is especially necessary to secure leaning and broken trees nearby pedestrian and bike paths to avoid endangering visitors.

Figure 10. (**A**)—Example of prototype smartphone application with inventory and information of trees in given area in Warsaw city, (**B**)—Example of an inventory of trees of the area with numbering correlated with the mobile application (Photographs by Zarząd Zieleni Miejskiej w Warszawie) [81].

6. Discussion

6.1. Remote Sensing for Old Growth Forests

Remote sensing data and tools can be beneficial for a wide range of tasks around urban green infrastructure management [38]. As this article shows, this can also be applied in biodiverse old-growth forest stands where differing ecosystem services must be balanced. These services encompass habitat provisioning and recreational functions in particular. However, the state and diversity of Polish urban forests are mainly measured by the usage of traditional methods of land inventory [31,32]. As this is a time-consuming method, inventories through remote sensing methods like LiDAR can accelerate the gathering of information on the state of urban forests, e.g., in time-critical moments around major storms or after hot droughts [57].

Besides elaborate airborne products, in Poland, the height and location of trees can be also analyzed based on satellite data available on Tree Crown Map (pl. "Mapa koron drzew")/National Tree Crown Map (pl. "Krajowa mapa koron drzew") [67,86,87]. In the case of Warsaw, it is possible to analyze wider parameters of greenery, i.e., their distribution height, species composition, and health condition (leaf defoliation sensor) based on aerial data (ALS). The ALS map presents only the trees within the city of Warsaw [66,78,81] and currently has no wider comparative application against the background of similar cities in Poland. However, it is a valuable source of up-to-date knowledge about trees in the city and their parameters, including tree health, which can serve as guidelines for further maintenance orders and administrative decisions. At present, unfortunately, there is little research that discusses this given issue more extensively, for instance there is no study that uses the acquired ALS data for Warsaw, for environmental monitoring purposes. The local authorities of Warsaw city however began to implement RS solutions in inventorying, monitoring, and management of urban greenery throughout the administered city area [26–29]. Therefore, supportive research on the matters is important and timely.

It is worth noting that the High Structural Diversity Tree Crown Map for Warsaw may find a new application as an example of ALS in detecting the poor health condition of trees to study detailed characteristics of defoliated trees in the fields of urban ecology, forestry, dendrology, and landscape architecture [26,56,87–93]. In this direction, there is current research in large cities such as Wrocław, Kraków or Poznań within the framework of the LIFE project (LIFECOOLCITY) funded by the European Commission. The main objective is to increase the adaptive capacity of at least 10,000 EU cities by implementing two innovative RS systems to manage blue-green infrastructure [89,90].

It is worth noting that LiDAR data could classify both the overstory and dominant understory species and thus play a crucial role in identifying forest biodiversity. This approach will be useful for forestry, landscape architecture, and urban ecology to plan for sustainable urban forests, rich in biodiversity, and produce interactive maps for monitoring species of interest used by local authorities like city councils or interested citizens).

The trade-offs between visitor safety and biodiversity protection in forests can be reduced through education programs for forest managers and residents that can be based on ALS monitoring data. Of significant importance are educational activities, guidelines for managers on what actions should be taken in forest areas with intensive recreation. Further, training of staff in the application of computer techniques, and the current knowledge base of trees in GIS platforms (LiDAR and ALS, e.g., High Structural Diversity Tree Crown Map) is seen as beneficial. Applications about forests, recreation, and natural resources (providing information about spatial zoning, mosaics of management intensity, favoring more resistant species next to paths, etc.) can be helpful for residents, which are clear options for balancing safety and biodiversity [81–85,91,93].

6.2. Management Suggestions

Storm Eunice in 2022 damaged over half of the defoliated trees keeping much of the storm's damages in comparatively confined areas. Still, one of the forests was completely closed to visitors and in the other forest, areas with damaged trees were fenced off, and the remaining branches and fallen trees were removed to make the forest areas available as soon as possible for recreation, which is not necessarily the best option for biodiversity conservation. Therefore, based on our evidence, we suggest more nuanced and targeted forms of management of the reserves to mitigate trade-offs between visitor safety and biodiversity benefits.

Buffer paths. A system of protective paths for recreation with a safety distance of 50 m from the edge of the road is suggested. Such a solution will ensure a safe distance for the pedestrian from ill trees with amplified risks of falling. The distance of 50 m was an average value considering the height of trees in the forest determined using ALS (Tables S1 and S2).

Buffer defoliated trees. The use of protection zones around the defoliated trees is also suggested in this study. Trees with the worst state in health should be mechanically secured, e.g., supports, and should be undercut with dried or broken branches (or nets reinforcing the tree trunk in the case of trees with breaks). A circular zone of 30 m around these trees should be required to separate this place from the access of pedestrians and cyclists. This will ensure a safe distance between individuals and the ill tree with the risk of falling. The distance of 50 m was an average value considering the height of trees in the forest determined using a table of attributes with data on defoliated trees using ALS (Tables S1 and S2). These zones are shown in Figures 8 and 9.

Tree species that are more affected. This study indicates that it is necessary to protect tree species from defoliated trees that are particularly susceptible to breaks or dry or falling branches, such as *Betula*, *Alnus*, *Acer*, *Pinus*, *Larix*, and over 300-year-old *Quercus* sp. These trees had a shallow root system or brittle wood prone to damage and fractures, as illustrated in Figures 4 and 5. These trees should be secured with ropes to the ground or with supports that prevent the branches from falling. Furthermore, these trees can be cut down in sensitive places, e.g., in rest areas or near pedestrian walking or cycling roads, as their health is poor, or they are old and prone to breakage. Special attention should also be paid to old oak species (*Quercus*) with numerous fractures or cracks. They need to be fenced off from the visitor access space or supported by special supports and nets strengthening the trunk and root system to minimize the risk of falling heavy branches on fences or walking/driving trails. Forest managers need to observe these trees and provide their expert opinion. This way, forest tree monitoring can lead to constant monitoring, stabilization, adaptation to removal, and reducing the risk of danger to individuals visiting these sites.

Visitor Information. We propose to develop a GIS-based forest information system. First, initial information about the condition of trees and their health status and threats can be given and made accessible on the basis of QR code technology. A forest visitor with a smartphone can scan the code, which will redirect the person to information about applications and thematic maps showing threats (hazard maps) and possibilities and limitations of recreation in the forest. Moreover, those interactive maps may inform visitors about the value an old-growth forest has, and very much the fact that in case of a storm, it is not a safe place. Such information may be also distributed via GIS applications (e.g., Endomondo, MapIT, mLas, tMap, or Gpies.com), which are currently available and become increasingly more widespread among forest visitors in Poland.

Visitor management. The results of field analyzes using LiDAR technology can provide a range of information to forest managers. They can also be helpful in managing forest trees, as well as in determining core and buffer areas where dead branches and risks are allowed. Such areas can be very quickly demarcated from the recreation area, which can increase the level of safety of forest visitors.

The implementation of a green infrastructure map service in the capital city of Poland. Currently, there is a growing demand for ecological monitoring of forest in Poland, e.g., the Forest Data Bank (pl. Bank danych o lasach) and the Large Area Forest Inventory (pl. Wielkoobszarowa Inwentaryzacja Stanu Lasu). This information is collected based on continuous or periodic observations of forest land use. Reports prepared by research institutes (pl. Instytut Badwczy Leśnictwa) on behalf of the General Directorate of Forests (pl. Generalna Dyrekcja Lasów Państwoych) can be used as a comprehensive source of the health status of forests However, while increasingly more data is available it seems that appropriate forms of portraying and communicating it are not yet available. As data portals and interactive maps have potential to promote sustainable development around urban green infrastructure this is unfortunate [13]. For instance, the data presentation on the portal of the municipality of Warsaw, defoliation and species related information are shown in separate views that cannot easily be brought together [26,65,66,78]. In addition, the integration of RS specialists to analyze and interpret the available data prevents practical application of the developed High Structural Diversity Tree Crown Map in the management and protection of valuable trees. On a positive note, since March 2020, the local authorities of Warsaw city have made a map service available about Warsaw's greenery, which is the result of the LIFE project [89,90]. The goals of this project are also based on the implementation of technology on the level of green infrastructure management and social participation through iterative provision of the map service, towards the ideal of a smart city.

Remote sensing allows to obtain spatially detailed and temporally continuous information for trees and shrubs in a large area in a short time (several months) regarding the extent of crowns, occupied area, height, taxonomic differentiation, and health condition. These data can be successfully used in functional and spatial analyzes in the field of, e.g., site planning and selection of tree and shrub species for new plantings, shaping the structure of tall greenery in individual districts, valuing, and estimating the scale of ecosystem services provided by tall greenery, but also allocating land for construction projects or planning the location of infrastructure, as well as environmental monitoring [26–29,85,89,90].

6.3. *Analysis and Prediction of Hazards Using LiDAR Methods*

Biodiversity and tree crown mapping is important in the context of climate change and increasingly strong storms. It should be emphasized that the activities based on spatial LiDAR data listed in this pilot study may also have new applications, especially regarding weather anomalies in Europe, such as unexpected short storms, intense rains, and drought [34,69,70]. The access to information about urban recreation areas should be up to date and available to visitors as soon as possible [13].

Furthermore, results of this study show that both old trees (stately oaks) and younger short-lived ones (species with poor wood compactness like ash, or pines) may pose threats to visitors, which is confirmed by many years of research by Siewniak and Bobek [25] in historical parks in cities of Poland. The High Structural Diversity Tree Crown Map for Warsaw, as an example of ALS, may find new applications in the diagnosis of safety threats, as it also contains information on species composition. Similar ALS databases have already been commissioned by other cities in Poland, e.g., Wrocław. Within the framework of the LIFE project, the authors assume that within the next dozen years, a similar green inventory will be carried out using analogical parameters as in Warsaw [26–29,89,90]. Furthermore, it is possible to estimate the amount of carbon dioxide absorbed by trees, oxygen production, or urban heat island formation, which facilitates adaptation to climate change. It is envisioned that residents have the possibility to receive comprehensive and up-to-date information about the state of urban greenery. Towards this The High Structural Diversity Tree Crown Map is an important solution that can be used by other local authorities in polish cities, as well as by landscape architects, ecologists, foresters, and scientists [26–29,34–36,89,91,92,94,95].

7. Conclusions

The research methods used in this study show a new application of remote sensing-derived maps in evaluating trees in cities, not only as common elements of urban green infrastructure such as parks or tree avenues but also as nature protection areas with old trees. This method can be useful for local authorities in identifying trees that need treatment or removal, protecting neighboring areas from material damage and human health in the case of weather anomalies such as windbreaks. This way, LiDAR scanning can guide tree care treatments in urban forest reserves, where recreation pressure is high.

In both forests, site observations indicated that more than half (59.5%) of the defoliated trees identified in the High Structural Diversity Crown Map for Warsaw were destroyed or damaged in Storm Eunice in 2022. This was validated in field research organized as workshops.

ALS can be useful in the monitoring of forested areas in cities and other areas of urban green infrastructure. These results allow local authorities to find trees in the Bielański Forest and the Kabacki Forest that should be diagnosed and secured immediately in the context of forest recreation or change the visitor management or information system accordingly to steer people away from potentially dangerous locations.

This study presents guidelines and tools for detailed information on how and where such rerouting of visitors is most needed. These are hazard maps with information on endangered trees, protective paths for recreation with a safe strip and protection zones around the defoliated trees. This way forest tree monitoring can reduce the risk of danger to individuals visiting these sites.

The location of endangered trees obtained using ALS should be used to disseminate information via private and public web-GIS platforms and smartphone Apps that control leisure traffic in recreational forest areas. New applications for urban forests that will allow residents to recognize trees and download the current information about forest safety and the health condition of trees need to be developed. This way, visitors can be made aware of the need of old-grown trees and areas that are less accessible to them, helping forest managers in public relations around biodiversity conservation.

The findings of this study emphasize that it is not possible to achieve maximal ecosystem services for both biodiversity and recreation potential in an old-grown forest reserve. Much rather, a good practice would be to balance trade-offs between visitor safety and biodiversity protection mentioned above. Critical in this regard is the localized evaluation and involvement of residents as such decisions are very much context dependent that are partially beyond the eyes of a remote sensing sensor. In conclusion, specialized sensing techniques made actionable are beneficial assets in current urban governance aiming towards healthy, biodiverse, green, and smart cities.

Supplementary Materials: The following supporting information can be downloaded at: https://www.mdpi.com/article/10.3390/land12020275/s1, The article contains additional material constituting a database (High Structural Diversity Tree Crown Map). The material is presented in Tables S1 and S2.

Author Contributions: Conceptualization, A.D., T.W. and D.H.; methodology, A.D., T.W. and D.H.; software, A.D.; validation, A.D., T.W. and D.H.; formal analysis, A.D. and T.W.; investigation, A.D.; resources, A.D.; data curation, A.D.; writing—original draft preparation, A.D.; writing—review and editing, A.D., T.W. and D.H.; visualization, A.D.; supervision, D.H.; project administration, A.D.; funding acquisition, A.D. All authors have read and agreed to the published version of the manuscript.

Funding: This research was developed within the project BiNatUr: Bringing nature back—biodiversity-friendly nature-based solutions in cities and funded through the 2020–2021 Biodiversa and Water JPI joint call for research projects, under the BiodivRestore ERA-NET Cofund (GA N°101003777), with the EU and the funding organisations Academy of Finland (Finland), Bundesministerium für Bildung und Forschung (BMBF, Germany), Federal Ministry of Education and Research (Germany), National Science Centre (Poland), Research Foundation Flanders (fwo, Belgium) and Fundação para a Ciência e Tecnologia (Portugal). Dagmar Haase works in The CLEARING HOUSE project, which has received funding from the European H2020 Research and Innovation programme under the Grant Agreement no 821242. The content of this document reflects only the author's view. The European Commission is not responsible for any use that may be made of the information it contains.

Institutional Review Board Statement: Not applicable.

Informed Consent Statement: Not applicable.

Data Availability Statement: The database for the High Structural Diversity Map is synthetically discussed in the article. The data is made available to the authors of the given article by Urząd m.st. Warszawy (The city council of Warsaw city, Poland) in their letter (No. BG-IIP.6642.24.2020.LZG, OŚ-I.604.457.2022.AMA) regarding the availability of the resource in connection with the submission of this paper for publication in the special issue of MDPI Land.

Acknowledgments: We thank Zastępca Prezydenta miasta stołecznego Warszawy; Naczelnik wydziału, from Biuro Geodezji I Katastru Wydział Infrastruktury Informacji Przestrzennej Urząd m.st. Warszawy, Zastępca Dyrektora Biuro Ochrony Środowiska Urząd m.st. Warszawy, and Główny specjalista, from Biuro Ochrony Środowiska (OŚ) Wydział Strategii i Informacji o Środowisku Urząd m.st. Warszawy, for sharing the High Structural Diversity Map database. The authors would like to thank the students and interdisciplinary team of researchers for participating in the workshops on the health of endangered trees in the Bielański and Kabacki forests. The authors would like to thank the reviewers for all useful and helpful comments on our manuscript.

Conflicts of Interest: The authors declare no conflict of interest.

Appendix A

Table A1. Selected defoliated trees of poor health condition (features 1–6) in the Bielański Forest and the Kabacki Forest in Warsaw city.

	Feature 1			Feature 2			Feature 3	
	Broken or dry branches			Roots groving around the trunk			Tree inclined adjecent to the path	
	Bielański Forest	Kabacki Forest		Bielański Forest	Kabacki Forest		Bielański Forest	Kabacki Forest
Acer	10		*Acer*	3		*Acer*	2	
Alnus	2	2	*Alnus*	3	1	*Alnus*	1	
Betula	4	3	*Betula*	1	2	*Betula*		
Carpinus	2		*Carpinus*			*Carpinus*		

Table A1. *Cont.*

	Feature 1			Feature 2			Feature 3	
	Broken or dry branches			Roots groving around the trunk			Tree inclined adjecent to the path	
	Bielański Forest	Kabacki Forest		Bielański Forest	Kabacki Forest		Bielański Forest	Kabacki Forest
fruit trees	1	2	fruit trees			fruit trees	1	
Populus	24	43	*Populus*	4	1	*Populus*	8	4
Salix	2		*Salix*			*Salix*		
Quercus	1	25	*Quercus*		3	*Quercus*		3
Crataegus			*Crataegus*			*Crataegus*		
Larix			*Larix*			*Larix*		
Pinus		17	*Pinus*		7	*Pinus*		2
total	**45**	**94**	**total**	**11**	**14**	**total**	**12**	**9**
	139			**25**			**21**	
	feature 4			**feature 5**			**feature 6**	
	Conductor broken off			**Cracks on the trunk**			**A decayed tree**	
	Bielański Forest	**Kabacki Forest**		**Bielański Forest**	**Kabacki Forest**		**Bielański Forest**	**Kabacki Forest**
Acer			*Acer*			*Acer*		
Alnus			*Alnus*			*Alnus*		
Betula	2		*Betula*	1		*Betula*	1	
Carpinus			*Carpinus*			*Carpinus*		
fruit trees			fruit trees	1		fruit trees	1	
Populus	2	1	*Populus*	1	8	*Populus*	1	8
Salix			*Salix*			*Salix*		
Quercus.		4	*Quercus*	1	7	*Quercus*	1	7
Crataegus			*Crataegus*			*Crataegus*		
Larix			*Larix*		1	*Larix*		1
Pinus			*Pinus*			*Pinus*		
total	**8**	**8**	**total**	**4**	**5**	**total**	**4**	**16**
	16			**9**			**20**	

Table A2. Selected defoliated trees of poor health condition (features 7–12) in the Bielański Forest and the Kabacki Forest in Warsaw city.

	Feature 7			Feature 8			Feature 9	
	A rickety tree			Decayed hallows			dry tree	
	Bielański Forest	Kabacki Forest		Bielański Forest	Kabacki Forest		Bielański Forest	Kabacki Forest
Acer			*Acer*			*Acer*		
Alnus		1	*Alnus*			*Alnus*		
Betula	1		*Betula*			*Betula*		1
Carpinus			*Carpinus*			*Carpinus*		
fruit trees			fruit trees	1		fruit trees	1	
Populus	6	6	*Populus*	1	1	*Populus*	7	2
Salix			*Salix*			*Salix*	1	
Quercus		2	*Quercus*			*Quercus*		5
Crataegus			*Crataegus*			*Crataegus*		
Larix			*Larix*			*Larix*		
Pinus		10	*Pinus*			*Pinus*		2
total	**7**	**19**	**total**	**2**	**1**	**total**	**9**	**11**
	26			**3**			**20**	

Table A2. *Cont.*

	feature 10			feature 11			feature 12	
	Split at the base of the trunk			Slightly bent tree			A thinning tree	
	Bielański Forest	Kabacki Forest		Bielański Forest	Kabacki Forest		Bielański Forest	Kabacki Forest
Acer			*Acer*			*Acer*		
Alnus			*Alnus*			*Alnus*		
Betula			*Betula*			*Betula*		
Carpinus			*Carpinus s*			*Carpinus*		
fruit trees			fruit trees			fruit trees		
Populus	1	1	*Populus*	3	2	*Populus*		7
Salix			*Salix*			*Salix*		
Quercus.	1	1	*Quercus*			*Quercus*		3
Crataegus			*Crataegus*			*Crataegus*		
Larix			*Larix*			*Larix*		
Pinus		3	*Pinus*			*Pinus*		1
total	**2**	**4**	**total**	**3**	**2**	**total**	**0**	**11**
	6			**5**			**11**	

Table A3. Selected defoliated trees of poor health condition (features 13–16) in the Bielański Forest and the Kabacki Forest in Warsaw city.

	Feature 13			Feature 14			Feature 15	
	Trunks intersect with each others			A bulky growth on the trunk			Tree torn from the ground	
	Bielański Forest	Kabacki Forest		Bielański Forest	Kabacki Forest		Bielański Forest	Kabacki Forest
Acer			*Acer*			*Acer*		
Alnus			*Alnus*		1	*Alnus*		
Betula			*Betula*		1	*Betula*		1
Carpinus			*Carpinus*			*Carpinus*		
fruit trees			fruit trees			fruit trees		
Populus	1	2	*Populus*			*Populus*	3	6
Salix	1		*Salix*			*Salix*		
Quercus		1	*Quercus*		2	*Quercus*		2
Crataegus			*Crataegus*			*Crataegus*		
Larix			*Larix*			*Larix*		
Pinus			*Pinus*		1	*Pinus*		5
total	**2**	**3**	**total**	**0**	**5**	**total**	**0**	**14**
	5			**5**			**14**	

	feature 16	
	Signs of desease, partially dry needles or leaves	
	Bielański Forest	Kabacki Forest
Acer		
Alnus		
Betula		
Carpinus		
fruit trees		
Populus	1	22
Salix	1	
Quercus.		10
Crataegus		
Larix		
Pinus		7
total	**2**	**39**
	39	

References

1. Ajewole, O.I.; Aiyeloja, A.A. Socio-economic analysis of benefits of Ibadan Urban Forest Reserves. *J. Trop. For. Res.* **2004**, *20*, 95–105.
2. Fleming, C.M.; Manning, M.; Ambrey, C.L. Crime, greenspace and life satisfaction: An evaluation of the New Zealand experience. *Landsc. Urban Plan.* **2016**, *149*, 1–10. [CrossRef]
3. Kabisch, N.; van den Bosch, M.; Lafortezza, R. The health benefits on nature-based solutions to urbanization hallenges for children and erderly—A systematic review. *Environ. Res.* **2017**, *159*, 362–373. [CrossRef]
4. Haase, D.; Hellwig, R. Effects of heart and drought stress on the health status of six urban street tree species in Leipzig, Germany. *Trees For. People* **2022**, *8*, 100252. [CrossRef]
5. Kičić, M.; Haase, D.; Marin, A.M.; Vuletić, D.; Ostoić, S.J. Perceptions of cultural ecosystem services of tree-based green infrastructure: A focus group participatory mapping in Zagreb, Croatia. *Urban For. Urban Green.* **2021**, *78*, 127767. [CrossRef]
6. Webb, R. Learning from Urban Forestry Programmes in South East Asia. *Arboric. J.* **1999**, *23*, 39–56. [CrossRef]
7. Kowarik, I.; Hiller, A.; Planchuelo, G.; Seitz, B.; von der Lippe, M.; Buchholz, S. Emerging Urban Forests: Opportunities for Promoting the Wild Side of the Urban Green Infrastructure. *Sustainability* **2019**, *11*, 6318. [CrossRef]
8. Steenberg, J.W.N.; Duinker, P.N.; Charles, J.D. The neighbourhood approach to urban forest management: The case of Halifax, Canada. *Landsc. Urban Plan.* **2013**, *117*, 135–144. [CrossRef]
9. Troy, A.; Grove, J.M.; O'Neil-Dunne, J. The relationship between tree canopy and crime rates across an urban–rural gradient in the greater Baltimore region. *Landsc. Urban Plan.* **2012**, *106*, 262–270. [CrossRef]
10. Ulrich, R.S.; Simons, R.F.; Losito, B.D.; Fiorito, E.; Miles, M.A.; Zelson, M. Stress recovery during exposure to natural and urban environments. *J. Environ. Psychol.* **1991**, *11*, 201–230. [CrossRef]
11. Bolund, P.; Hunhammar, S. Ecosystem Services in Urban Areas. *Ecol. Econ.* **1999**, *29*, 293–301. [CrossRef]
12. Almenar, J.B.; Elliot, T.; Rugani, B.; Philippe, B.; Gutierrez, T.N.; Sonnemann, G.; Geneletti, D. Nexus between nature-based solutions, ecosystem services and urban challenges. *Land Use Policy* **2021**, *100*, 104898. [CrossRef]
13. Wellmann, T.; Andersson, E.; Knapp, S.; Lausch, A.; Palliwoda, J.; Priess, J.; Scheuer, S.; Haase, D. Reinforcing nature-based solutions through tools providing so-cial-ecological-technological integration. *Ambio* **2022**. [CrossRef]
14. Marselle, M.R.; Bowler, D.; Watzema, E.J.; Eichenberg, D.; Kirsten, T.; Bonn, A. Urban street tree biodiversity and antidepressant prescriptions. *Sci. Rep.* **2020**, *10*, 1–11.
15. Baumeister, C.F.; Gerstenberg, T.; Plieninger, T.; Schraml, U. Exploring cultural ecosystem services hotspots: Linking multiple urban forest features with public participation mapping data. *Urban For. Urban Green.* **2020**, *48*, 126561. [CrossRef]
16. Jim, C.Y.; Chen, W.Y. Ecosystem services and valuation of urban forests in China. *Cities* **2009**, *26*, 187–194. [CrossRef]
17. Schafer, C.S.; Scott, D.; Baker, J.; Winemiller, K. Recreation and Amenity Values of Urban Stream Corridors: Implications for Green Infrastructure. *J. Urban Des.* **2013**, *18*, 478–493. [CrossRef]
18. Fornal-Pieniak, B.; Żarska, B.; Zaraś-Januszkiewicz, E. Natural evaluation of landscape in urban area comprising Bielański Forest nature reserve and surroundings, Warsaw, Poland. Directions for landscape protection and planning. *Ann. Warsaw Univ. of Life Sci.-SGGW Land Reclam.* **2018**, *50*, 327–338. [CrossRef]
19. Alvey, A.A. Promoting and preserving biodiversity in the urban forest. *Urban For. Urban Green.* **2006**, *5*, 195–201. [CrossRef]
20. Długoński, A. Wykorzystanie Mapy Koron Drzew do oceny stanu zdrowotności drzewostanu rezerwatów miejskich. Przykład Lasu Bielańskiego w Warszawie [The use of the High Structural Diversity Tree Crown Map to assess the health status of forest trees in urban reserves. An example of the Bielański Forest in Warsaw]. In Proceedings of the V Edycja Konferencji Środowisko Informacji, Ministerstwo Klimatu Środowiska, Warsaw, Poland, 25 November 2021.
21. Ordónez, C.; Duinker, P.N.; Steenberg, J.W.N. Climate change mitigation and adaptation in urban forests: A framework for sustainable urban forest management. In Proceedings of the 18th Commonwealth Forestry Conference, Edinburgh, UK, 28 June–2 July 2010.
22. Haase, D.; Annegret Haase, A.; Wolff, M.; Dushkova, D. (Eds.) *Nature-Based Solutions (NBS) in Cities and Their Interaction with Urban Land, Ecosysems, Built Environment and People: Debating Societal Implications*, 1st ed.; MDPI Land: Beijing, China, 2021; pp. 1–216.
23. Konijnendijk, C.C.; Nilsson, K.; Randrup, T.B.; Schipperijn, J. Urban forestry education. In *Urban Forests and Trees*, 1st ed.; Konijnendijk, C.C., Randrup, T.B., Eds.; Springer: Berlin, Germany, 2005; Volume 1, pp. 465–478.
24. Nowak, D.J.; Noble, M.H.; Sisinni, S.M.; Dwyer, J.F. Assessing the US urban forest resource. *J. For.* **2001**, *99*, 37–42.
25. Siewniak, M.; Bobek, W. Zagrożenia ludzi i mienia w parkach, metody określania stanu statycznego drzew. [Threats to people and property in parks, methods of determining the static state of trees]. *Kur. Konserw.* **2010**, *8*, 13–17. (In Polish)
26. Webinarium Mapa Koron Drzew w Zarządzaniu Zielenią Wysoką w Mieście. Prykład Wdrożenia w Mieście. Przykład Wdrożenia w m.st. Warszawa [Webinar Tree Crown Map in the Management of Tall Greenery in the City. An Example of Implementation in the City. Example of Implementation in Warsaw], 29 October 2020, MGGP Aero. Available online: https://www.youtube.com/watch?v=S7HOlA7hDRg (accessed on 3 January 2022).
27. Wylazłowska, J.; Sławik, Ł.; Kopeć, D. Teledetekcja nowe perspektywy w zarządzaniu zielenią. *Zieleń Miejska* **2000**, *5*, 38–39.
28. Wronka, M. Warszawska Mapa Koron Drzew. *Zieleń Miejska* **2019**, *2*, 15–16.
29. Sławik, Ł.; Kopeć, D. Mapa koron drzew w zarządzaniu zielenią wysoką w mieście. *Zieleń Miejska* **2020**, *10*, 28–31.

30. Laurin, G.V.; Chan, J.C.W.; Chen, Q.; Lindsell, J.A.; Coomes, D.A.; Guerriero, L.; del Frate, F.; Miglietta, F.; Valentini, R. Biodiversity mapping in a tropical West African forest with airborne hyperspectral data. *PLoS ONE* **2014**, *9*, e97910.
31. Łukaszkiewicz, J.; Fortuna-Antoszkiewicz, B.; Borowski, J. The impact of earthworks on older trees in historical parks. *J. Environ. Eng. Landsc. Manag.* **2011**, *30*, 188–194. [CrossRef]
32. Rosłon-Szeryńska, E.; Łukaszkiewicz, J.; Fortuna-Antoszkiewicz, B. The possibility of predicting the collision of trees with construction investments. VI International Conference of Science and Technology INFRAEKO 2018 Modern Cities. *Infrastruct. Environ.* **2018**, *45*, 00076.
33. Bergen, K.M.; Goetz, S.J.; Dubayah, R.O.; Henebry, G.M.; Hunsaker, C.T.; Imhoff, M.L.; Nelson, R.F.; Parker, G.G.; Radeloff, V.C. Remote sensing of vegetation 3-D structure for biodiversity and habitat: Review and implications for lidar and radar spaceborne missions. *J. Geophys. Res. Biogeosci.* **2009**, *114*, G22009. [CrossRef]
34. Lin, B.; Ossola, A.; Ripple, W.; Alberti, M.; Andersson, E.; Bai, X.; Dobbs, C.; Elmqvist, T.; Evans, K.L.; Frantzeskaki, N.; et al. Cities and the "new climate normal": Ways forward to address the growing climate challenge. Lancet Planet. *Health* **2021**, *5*, e479–e486. Available online: http://www.thelancet.com/planetary-health (accessed on 30 July 2022).
35. Żarska, B.; Fornal-Pieniak, B.; Zaraś-Januszkiewicz, E. Areas designed for afforestation and areas excluded from afforestation-selected aspects related to the protection of the landscape in view of Poland experience. *Ecol. Quest.* **2015**, *22*, 17–22.
36. Fornal-Pieniak, B.; Długoński, A. Urban landscape assessment for tourism aspect in a big-sized city in Poland (Example of Łódź city, Central Poland— preliminary study). In Proceedings of the XXVII Conference in the Series of Garden Art and Historical Dendrology, IX International Edition: Urban Ecology and Cultural Heritage in the City, Kraków, Poland, 22–23 October 2020.
37. Wellmann, T.; Lausch, A.; Andersson, E.; Knapp, S.; Cortinovis, C.; Jache, J.; Scheuer, S.; Kremer, P.; Mascarenhas, A.; Kraemer, R.; et al. Remote sensing in urban planning: Contributions towards ecologically sound policies? *Landsc. Urban Plan.* **2020**, *204*, 103921. [CrossRef]
38. Afzalan, N.; Muller, B. The role of social media in green infrastructure planning: A case study of neighborhood participation in park siting. *J. Urban Technol.* **2014**, *21*, 67–83. [CrossRef]
39. Grêt-Regamey, A.; Switalski, M.; Fagerholm, N.; Korpilo, S.; Juhola, S.; Kyttä, M.; Käyhkö, N.; McPhearson, T.; Nollert, M.; Rinne, T.; et al. Harnessing sensing systems towards urban sustainability transformation. *Npj Urban Sustain.* **2021**, *1*, 1–9. [CrossRef]
40. Kamoske, A.G.; Dahlin, K.M.; Read, Q.D.; Record, S.; Stark, S.C.; Serbin, S.P.; Zarnetske, P.L. Towards mapping biodiversity from above: Can fusing lidar and hyperspectral remote sensing predict taxonomic, functional, and phylogenetic tree diversity in temperate forests? *Glob. Ecol. Biogeogr.* **2022**, *31*, 1239–1465. [CrossRef]
41. Nagendra, H. Using remote sensing to assess biodiversity. *Int. J. Remote Sens.* **2001**, *22*, 2377–2400. [CrossRef]
42. Anderson, C.B. Biodiversity monitoring, earth observations and the ecology of scale. *Ecol. Lett.* **2018**, *21*, 1572–1585. [CrossRef]
43. Schäfer, E.; Heiskanen, J.; Heikinheimo, V.; Pellikka, P. Mapping tree species diversity of a tropical montane forest by unsupervised clustering of airborne imaging spectroscopy data. *Ecol. Indic.* **2016**, *64*, 49–58. [CrossRef]
44. Wardhana, W.; Widyatmanti, W.; Soraya, E.; Soeprijadi, D.; Larasati, B.; Umarhadi, D.A.; Hutomo, Y.H.T.; Idris, F.; Wirabuana, P.Y.A.P. A hybrid approach of remote sensing for mapping vegetation biodiversity in a tropical rainforest. *Biodiversitas* **2020**, *21*, 3946–3953. [CrossRef]
45. Kampouri, M.; Kolokoussis, P.; Argiala, D.; Karathanassi, V. Mapping of forest tree distribution and estimation of forest biodiversity using Sentinel-2 imagery in the University Research Forest Taxiarchis in Chalkidiki, Greece. *Geocarto Int.* **2019**, *34*, 1273–1285. [CrossRef]
46. Bagaram, B.M.; Giuliarelli, D.; Chirici, G.; Giannetti, F.; Barbati, A. UAV remote sensing for biodiversity monitoring: Are forest canopy gaps good covariates? *Remote Sens.* **2018**, *10*, 1397. [CrossRef]
47. Colgan, M.S.; Baldeck, C.A.; Féret, J.-B.; Asner, G.P. Mapping savanna tree species at ecosystem scales using support vector machine classification and BRDF correction on airborne hyperspectral and LiDAR data. *Remote Sens.* **2012**, *4*, 3462–3480. [CrossRef]
48. Dees, M.; Straub, C.; Koch, B. Can biodiversity study benefit from information on the vertical structure of forests? Utility of LiDAR remote sensing. *Curr. Sci.* **2012**, *102*, 1181–1187.
49. Yadav, B.K.V.; Lucieer, A.; Baker, S.C.; Jordan, G.J. Tree crown segmentation and species classification in a wet eucalypt forest from airborne hyperspectral and LiDAR data. *Int. J. Remote Sens.* **2021**, *42*, 7952–7977. [CrossRef]
50. Agarwal, S.; Vailshery, L.S.; Jaganmohan, M.; Harini Nagendra, H. Mapping urban tree species using very high resolution satellite imagery: Comparing pixel-based and object-based approaches. *ISPRS Int. J. Geo-Inf.* **2013**, *2*, 220–236. [CrossRef]
51. Baldeck, C.A.; Asner, G.P.; Martin, R.E.; Anderson, C.B.; Knapp, D.E.; Kellner, J.R.; Wright, S.J. Operational tree species mapping in a diverse tropical forest with airborne imaging spectroscopy. *PLoS ONE* **2015**, *10*, e0118403. [CrossRef]
52. Bunting, P.; Lucas, R.M.; Jones, K.; Bean, A.R. Characterisation and mapping of forest communities by clustering individual tree crowns. *Remote Sens. Environ.* **2010**, *114*, 2536–2547. [CrossRef]
53. Feng, X.; Li, P. A tree species mapping method from UAV images over urban area using similarity in tree-crown object histograms. *Remote Sens.* **2019**, *11*, 1982. [CrossRef]
54. Lee, A.C.; Lucas, R.S. A LiDAR-derived canopy density model for tree stem and crown mapping in Australian forests. *Remote Sens. Environ.* **2007**, *111*, 493–518. [CrossRef]
55. Guo, X.; Coops, N.C.; Tompalski, P.; Nielsen, S.C.; Bater, C.W.; Stadt, J.J. Regional mapping of vegetation structure for biodiversity monitoring using airborne lidar data. *Ecol. Inform.* **2017**, *38*, 50–61. [CrossRef]

56. Bartold, M. Development of forest cover mask to monitor the health condition of forests in Poland using long-term satellite observation. *For. Res. Pap.* **2016**, *77*, 141–150. [CrossRef]
57. Długoński, A. Modern digital techniques used to record changes in the environment. In *Microbial Biotechnology in the Laboratory and Practice Theory, Exercises and Specialist Laboratories*, 1st ed.; Długoński, J., Ed.; Jagiellonian University Press: Łódź, Poland, 2021; Volume 1, pp. 124–133.
58. Górski, F.; Łaskarzewska-Średzińska, M. (Eds.) *Biocity Tom I [Biocity Volume 1]*, 1st ed.; Fundacja Wydziału Architektury Politechniki Warszawskiej NKA: Warsaw, Poland, 2015; pp. 1–56.
59. Okła, K. Możliwości wykorzystania teledetekcji i fotogrametrii w Lasach Państwowych [Possibilities of using remote sensing and photogrammetry in State Forests]. In *Geomatyka w Lasach Państwowych Część, I. Podstawy*; Centrum Informacyjne Lasów Panstwoych: Bedoń, Poland, 2010; Volume 1, pp. 428–429.
60. Karlson, M.; Reese, H.; Ostwald, M. Tree crown mapping in managed woodlands (parklands) of semi-arid West Africa using WorldView-2 imagery and geographic object based image analysis. *Sensors* **2014**, *14*, 22643–22669. [CrossRef]
61. Levick, S.L.; Rogers, H.R. Structural biodiversity monitoring in savanna ecosystems: Integrating LiDAR and high resolution imagery through object-based image analysis. In *Object-Based Image Analysis*; Springer: Berlin/Heidelberg, Germany, 2008; pp. 477–491.
62. Rappaport, D.I.; Royle, J.A.; Morton, D.C. Acoustic space occupancy: Combining ecoacoustics and lidar to model biodiversity variation and detection bias across heterogeneous landscapes. *Ecol. Indic.* **2020**, *113*, 106172. [CrossRef]
63. Hastings, J.H.; Ollinger, S.V.; Ouimette, A.P.; Sanders-DeMott, T.; Palace, M.W.; Ducey, M.J.; Sullivan, F.B.; Basler, D.; Orwig, D.A. Tree species traits determine the success of LiDAR-based crown mapping in a mixed temperate forest. *Remote Sens.* **2020**, *12*, 309. [CrossRef]
64. Alonzo, M.; Bookhagen, B.; Roberts, D.A. Urban tree species mapping using hyperspectral and lidar data fusion. *Remote Sens. Environ.* **2014**, *148*, 70–83. [CrossRef]
65. Zakochaj się w Warszawie. Mapa Koron Drzew w Wersji Numerycznej dla Obszaru m.st. Warszawy [Map of Tree Crowns in Numerical Version for the Area of the Capital City of Warsaw]. Available online: https://danemiejskie.plarszawt/warszawska-mapa-koron-drzew/ (accessed on 20 December 2022).
66. Millions of Warsaw's Trees on One Map. Available online: https://en.um.warszawa.pl/-/millions-of-warsaw-s-trees-on-one-map (accessed on 3 January 2023).
67. Krajowa Mapa Koron Drzew [National Tree Crown Map for Poland]. Available online: https://aplikacja.mapadrzew.com/ (accessed on 3 January 2023).
68. GUS Informacja o Wynikach Narodowego Spisu Powszechnego Ludności i Mieszkań 2021 na Poziomie Województw, Powiatów i Gmin [Eng. Information on the Results of the National Population and Housing Census 2021 at the Level of Voivodships, Powiats and Gminas]. Available online: https://stat.gov.pl/download/gfx/portalinformacyjny/pl/defaultaktualnosci/6494/8/1/1/informacja_o_wynikach_narodowego_spisu_powszechnego_ludnosci_i_mieszkan_20--09-2022.pdf (accessed on 16 September 2022). (In Polish)
69. 4 Dead in Poland as Storm Eunice Wrecks Havoc. Available online: https://english.news.cn/europe/20220220/083c3bb9bc5f46a0b4baaa47722065d4/c.html (accessed on 22 November 2022).
70. Eunice Hurricane In Poland Effects of the Eunice hurricane Seen in Warsaw on February 19, 2022 Warsaw Poland. Available online: https://www.imago-images.com/st/0151448614?__hstc=52317686.4b44870ec4a577029c49e44b73bd3bee.1659916800253.1659916800254.1659916800255.1&__hssc=52317686.1.1659916800256&__hsfp=2727673963 (accessed on 22 November 2022).
71. Lasy Warszawy [Eng. Warsaw' Urban Forests-Characteristic]. Available online: https://www.lasymiejskie.waw.pl/index.php/lasy/lasy-warszawy (accessed on 16 September 2022). (In Polish).
72. NATURA 2000 PLH140041 Las Bielański. Available online: https://natura2000.eea.europa.eu/Natura2000/SDF.aspx?site=PLH140041 (accessed on 25 November 2022).
73. Łukaszkiewicz, J.; Fortuna-Antoszkiewicz, B.; Długoński, A.; Wiśniewski, P. From the heap to the park-reclamation and adaptation of degraded urban areas for recreational functions in Poland. *Sci. Rev. Eng. Environ. Sci.* **2019**, *28*, 664–681.
74. Bamwesigye, D.; Fialová, J.; Kupec, P.; Łukaszkiewicz, J.; Fortuna-Antoszkiewicz, B. Forest Recreational Services in the Face of COVID-19 Pandemic Stress. *Land* **2021**, *10*, 1347. [CrossRef]
75. OpenStreetMap. 2020. Available online: https://www.openstreetmap.org/ (accessed on 23 November 2022).
76. Geoportal—Portal Krajowy [Eng. National GIS portal]. Location of Poland Based on Europe Map. Available online: https://mapy.geoportal.gov.pl/imap/Imgp_2.html?gpmap=gp0/ (accessed on 16 September 2022).
77. North, M.A. A Method for Implementing a Statistically Significant Number of Data Classes in the Jenks Algorithm, 6th International Conference on Fuzzy Systems and Knowledge Discovery. In *Proceedings of the FSKD'09: Proceedings of the 6th International Conference on Fuzzy Systems and Knowledge Discovery*, 1st ed.; Chen, Y., Zhang, D., Deng, H., Xiao, Y., Eds.; IEEE Press: Tianjin, China, 2009; Volume 7, pp. 35–38.
78. Mapa Koron Drzew m.st. Warszawy (Baza Danych). [High Structural Diversity Tree Crown Map Database-Warsaw City, Poland]. Mapa Powiatowego Zasobu Geodezyjnego. ID: BG-IIP.6642.24.2020.LZG . Available online: https://um.warszawa.pl/-/mapa-koron-drzew (accessed on 22 November 2022). (In Polish).
79. Chen, J.; Yang, S.T.; Li, H.; Zhang, B.; Lv, J.R. Research on geographical environment unit division based on the method of natural breaks (Jenks). Remote Sens. *Spat. Inf. Sci.* **2013**, *4*, 47–50.

80. Khamis, N.; Sin, T.C.; Hock, G.C. Segmentation of Residential Customer Load Profile in Peninsular Malaysia using Jenks Natural Breaks. In Proceedings of the 2018 IEEE 7th International Conference on Power and Energy (PECon), Johor Bahru, Malaysia, 3–4 December 2018; pp. 128–131.
81. Drzewa i Krzewy Pod Kontrolą-Trwa Inwentaryzacja Warszawskiej Zieleni. Zarząd Zieleni Miejskiej [Trees and Shrubs under Control-An Inventory of Warsaw's Greenery is Ongoing. City Green Board]. Available online: https://zzw.waw.pl/2021/12/03/drzewa-i-krzewy-pod-kontrola-trwa-inwentaryzacja-warszawskiej-zieleni/ (accessed on 3 January 2023).
82. Wietecha, M.; Sławik, Ł.; Kopeć, D.; Gadawska, A. Jak powstała Krajowa Mapa Koron Drzew, od ambitnego pomysłu dorealizacji [How the National Tree Crown Map was created, from an ambitious idea to implementation]. *Mag. Geoinformacyjny* **2022**, *1*, 28–29.
83. Ciesielski, M.; Stereńczak, K.; Bałazy, R. Wykorzystanie danych społecznościowej informacji geograficznej do monitorowania ruchu w przestrzeni leśnej [The use of social geographic information data to monitor traffic in forest space]. *Sylwan* **2019**, *163*, 80–88.
84. Mapa Drzew Łodzi (Eng. Map of Łódź City Trees). Available online: https://mapadrzewlodzi.pl/?page=page_8 (accessed on 20 November 2022).
85. Łaszkiewicz, E.; Czembrowski, P.; Kronenberg, J. Creating a Map of the Social Functions of Urban Green Spaces in a City with Poor Availability of Spatial Data: A Sociotope for Lodz. *Land* **2020**, *9*, 183. [CrossRef]
86. Powstała Mapa Koron Drzew Dla Polski [Elaboration of Tree Crown Map for Poland]. Available online: https://www.gov.pl/web/edukacja-i-nauka/powstala-mapa-koron-drzew-dla-polski (accessed on 3 January 2023).
87. Tree Crown Map for Poland. Available online: https://www.uni.lodz.pl/en/news/details/tree-crown-map-for-poland (accessed on 10 January 2023).
88. Tagliabue, G.; Cinzia Panigada, C.; Colombo, R.; Francesco, F.; Chiara, C.; Frédéric, B.; Kristin, V.; Koen, M.; Micol, R. Forest species mapping using airborne hyperspectral APEX data. *Misc. Geogr. Reg. Stud. Dev.* **2016**, *20*, 28–33. [CrossRef]
89. Teledetekcja Pomoże w Adaptacji Miast do Zmian Klimatu. Rusza Projekt za Blisko 5 mln euro [Remote Sensing Will Help Cities Adapt to Climate Change. The Project Starts for Nearly 5 Million Euros]. Available online: https://gisplay.pl/gis/10055-teledetekcja-pomoze-w-adaptacji-miast-do-zmian-klimatu.html (accessed on 3 January 2022).
90. Od Ochrony Torfowisk po Teledetekcję Zielono-Niebieskiej Infrastruktury Miejskiej: 8 Projektów Otrzyma w Programie LIFE ok. 43,7 mln zł. To Rekordowy Wynik w Dotychczasowej Historii Naborów [From Peatland Protection to Remote Sensing of Green and Blue Urban Infrastructure: 8 Projects Will Receive Approx. PLN 43.7 Million under the LIFE Programme. This Is a Record Result in the Recruitment History so Far]. Available online: https://www.wfos.krakow.pl/od-ochrony-torfowisk-po-teledetekcje-zielono-niebieskiej-infrastruktury-miejskiej-8-projektow-otrzyma-w-programie-life-ok-437-mln-zl-to-rekordowy-wynik-w-dotychczasowej-historii-naborow/ (accessed on 3 January 2022).
91. Biernacka, M.; Kronenberg, J.; Łaszkiewicz, E. An integrated system of monitoring the availability, accessibility and attractiveness of urban parks and green squares. *Appl. Geogr.* **2020**, *116*, 102152.
92. Mapa Koron Drzew-portal m.st. Warszawy. [High Structural Diversity Tree Crown Map-Warsaw City Website]. Available online: http://mapa.um.warszawa.pl/mapaApp1/mapa?service=zielen-51 (accessed on 3 January 2023).
93. mLAS Mini Instrukcja Użytkowania [mLAS Mini User Manual]. Available online: https://przedborow.poznan.lasy.gov.pl/documents/688725/21047984/mlas_mini_instrukcja_uzytkownika_mlasmini2.pdf/43 (accessed on 3 January 2022).
94. MGGP Aero-Oferty Pracy [MGGP Aero-Work Offers]. Available online: https://www.mggpaero.com/oferty-pracy-specjalista-dendrologii.html (accessed on 3 January 2023).
95. Długoński, A.; Szumański, M. *Atlas Ekourbanistyczny Zielonej Infrastruktury Miasta Łodzi, Teka 1, Tom 1a [Eco-Urban Atlas of Green Infrastructure of Lodz City, Vol. 1/1a]*, 1st ed.; Łódzkie Towarzystwo Naukowe: Łódź, Poland, 2016; pp. 1–330.

MDPI AG
Grosspeteranlage 5
4052 Basel
Switzerland
Tel.: +41 61 683 77 34

Land Editorial Office
E-mail: land@mdpi.com
www.mdpi.com/journal/land

www.ingramcontent.com/pod-product-compliance
Lightning Source LLC
Chambersburg PA
CBHW041354210926
51850CB00013B/3073
* 9 7 8 3 7 2 5 8 4 2 4 9 0 *